PRELIMINARY EDITION

Calculus

VOLUME 2

THE PRINDLE, WEBER & SCHMIDT
SERIES IN MATHEMATICS

Althoen and Bumcrot, *Introduction to Discrete Mathematics*
Bean, Sharp, and Sharp, *Precalculus*
Boye, Kavanaugh, and Williams, *Elementary Algebra*
Boye, Kavanaugh, and Williams, *Intermediate Algebra*
Burden and Faires, *Numerical Analysis, Fifth Edition*
Cass and O'Connor, *Fundamentals with Elements of Algebra*
Cass and O'Connor, *Basic Mathematics: Prealgebra and Algebra*
Cullen, *Linear Algebra with Differential Equations, Second Edition*
Dick and Patton, *Calculus, Volume 1*
Dick and Patton, *Calculus, Volume 2*
Dick and Patton, *Technology in Calculus: A Sourcebook of Activities*
Eves, *In Mathematical Circles*
Eves, *Mathematical Circles Squared*
Eves, *Return to Mathematical Circles*
Faires and Burden, *Numerical Methods: With Software*
Fletcher, Hoyle, and Patty, *Foundations of Discrete Mathematics*
Fletcher and Patty, *Foundations of Higher Mathematics, Second Edition*
Fraser, *Intermediate Algebra: An Early Functions Approach*
Gantner and Gantner, *Trigonometry*
Geltner and Peterson, *Geometry for College Students, Second Edition*
Gilbert and Gilbert, *Elements of Modern Algebra, Third Edition*
Gobran, *Beginning Algebra, Fifth Edition*
Gobran, *Intermediate Algebra, Fourth Edition*
Gordon, *Calculus and the Computer*
Hall, *Beginning Algebra*
Hall, *Intermediate Algebra*
Hall, *Algebra for College Students, Second Edition*
Hall, *College Algebra with Applications, Third Edition*
Hartfiel and Hobbs, *Elementary Linear Algebra*
Huff and Peterson, *College Algebra Activities for the TI-81 Graphics Calculator*
Humi and Miller, *Boundary-Value Problems and Partial Differential Equations*
Kaufmann, *Elementary Algebra for College Students, Fourth Edition*
Kaufmann, *Intermediate Algebra for College Students, Fourth Edition*
Kaufmann, *Algebra for College Students, Fourth Edition*
Kaufmann, *Algebra with Trigonometry for College Students, Third Edition*
Kaufmann, *College Algebra, Second Edition*
Kaufmann, *Trigonometry*
Kaufmann, *College Algebra and Trigonometry, Second Edition*
Kaufmann, *Precalculus, Second Edition*
Kennedy and Green, *Prealgebra for College Students*
Laufer, *Discrete Mathematics and Applied Modern Algebra*
Lavoie, *Liberal Arts Mathematics*
Nicholson, *Elementary Linear Algebra with Applications, Second Edition*
Nicholson, *Introduction to Abstract Algebra*
Pence, *Calculus Activities for Graphic Calculators*
Pence, *Calculus Activities for the TI-81 Graphic Calculator*
Pence, *Calculus Activities for the HP-48X and the TI-85 Calculators, Second Edition*
Plybon, *An Introduction to Applied Numerical Analysis*
Powers, *Elementary Differential Equations*
Powers, *Elementary Differential Equations with Boundary-Value Problems*
Proga, *Arithmetic and Algebra, Third Edition*
Proga, *Basic Mathematics, Third Edition*
Rice and Strange, *Plane Trigonometry, Sixth Edition*
Rogers, Haney, and Laird, *Fundamentals of Business Mathematics*
Schelin and Bange, *Mathematical Analysis for Business and Economics, Second Edition*
Sgroi and Sgroi, *Mathematics for Elementary School Teachers*
Swokowski and Cole, *Fundamentals of College Algebra, Eighth Edition*

Swokowski and Cole, *Fundamentals of Algebra and Trigonometry, Eighth Edition*
Swokowski and Cole, *Fundamentals of Trigonometry, Eighth Edition*
Swokowski and Cole, *Algebra and Trigonometry with Analytic Geometry, Eighth Edition*
Swokowski, *Precalculus: Functions and Graphs, Sixth Edition*
Swokowski, *Calculus, Fifth Edition*
Swokowski, *Calculus, Fifth Edition, Late Trigonometry Version*
Swokowski, *Calculus of a Single Variable*
Tan, *Applied Finite Mathematics, Third Edition*
Tan, *Calculus for the Managerial, Life and Social Sciences, Second Edition*
Tan, *Applied Calculus, Second Edition*
Tan, *College Mathematics, Second Edition*
Trim, *Applied Partial Differential Equations*
Venit and Bishop, *Elementary Linear Algebra, Alternate Second Edition*
Venit and Bishop, *Elementary Linear Algebra, Third Edition*
Wiggins, *Problem Solver for Finite Mathematics and Calculus*
Willard, *Calculus and Its Applications, Second Edition*
Wood and Capell, *Arithmetic*
Wood and Capell, *Intermediate Algebra*
Wood, Capell, and Hall, *Developmental Mathematics, Fourth Edition*
Zill, *Calculus, Third Edition*
Zill, *A First Course in Differential Equations with Applications, Fifth Edition*
Zill and Cullen, *Elementary Differential Equations with Boundary-Value Problems, Third Edition*
Zill and Cullen, *Advanced Engineering Mathematics*

THE PRINDLE, WEBER & SCHMIDT
SERIES IN ADVANCED MATHEMATICS

Brabenec, *Introduction to Real Analysis*
Ehrlich, *Fundamental Concepts of Abstract Algebra*
Eves, *Foundations and Fundamental Concepts of Mathematics, Third Edition*
Keisler, *Elementary Calculus: An Infinitesimal Approach, Second Edition*
Kirkwood, *An Introduction to Real Analysis*
Patty, *Topology*
Ruckle, *Modern Analysis: Measure Theory and Functional Analysis with Applications*
Sieradski, *An Introduction to Topology and Homotopy*

THE OREGON STATE UNIVERSITY
CALCULUS CURRICULUM PROJECT

PRELIMINARY EDITION

Calculus

VOLUME 2

Thomas P. Dick

OREGON STATE UNIVERSITY

Charles M. Patton

HEWLETT-PACKARD COMPANY

PWS Publishing Company

Boston

PWS PUBLISHING COMPANY
20 Park Plaza, Boston, MA 02116-4324

PWS Publishing Company is a division of Wadsworth, Inc.

Sponsoring Editor: Steve Quigley
Production Editor: Helen Walden
Manufacturing Coordinator: Ellen Glisker
Printer/Binder: Courier/Westford

ISBN 0-534-92400-X

Printed in the United States of America.
93 94 95 96 97 -- 10 9 8 7 6 5 4 3

International Thomson Publishing
The trademark ITP is used under license

Preface

Calculus occupies a particularly critical position in mathematics education as the gateway to advanced training in most scientific and technical fields. It is fitting that calculus should receive particular attention as we prepare for the needs of the twenty-first century. The Sloan Conference (Tulane, 1986) and the Calculus for a New Century Conference (Washington, 1987) sounded the call for reform in the calculus curriculum. Now the entire introductory course in calculus is being reexamined under the closest scrutiny that it has received in several years. Through a special funding initiative, the National Science Foundation has made resources available for a variety of calculus curriculum revision efforts to be tried and implemented. The Oregon State University Calculus Curriculum Project is one of these NSF-funded efforts, and this book is one of the major results of the project.

ABOUT THE MATERIAL IN VOLUME II

Calculus, Volume II covers material appropriate for a semester or two quarters of multivariable and vector calculus. A brief glance at the Table of Contents might suggest that the text does not differ radically from a traditional calculus text in terms of major topics. This is as it should be—calculus reform will not change the importance and vitality of the major ideas of calculus, and any wholesale departure from those ideas should be viewed with great skepticism. What *is possible* is a fresh approach to these important ideas in light of the availability of modern technology. In particular, the technology can invite us to change or adopt new emphases in instruction.

Chapter 12 is a basic introduction to the language, notation, and properties of vectors and matrices. Chapter 13 follows with the calculus of vector-valued functions of a single variable, including a discussion of ways a graphing calculator or software can be used to visualize parametrized curves in space. Functions of several variables are introduced in Chapter 14, with special attention to tools for the visualization of surfaces. Linear and quadratic functions of several variables receive special attention. Chapter 15 covers the differential calculus of scalar-valued functions of several variables, and the properties and applications of the gradient are the central focus. Chapter 16 turns to multiple integration, including numerical techniques for approximating multiple integrals, such as the Monte Carlo method. Finally, Chapter 17 provides an introduction to the analysis of vector-valued functions, including divergence and curl of vector fields, line and surface integrals, and the fundamental theorems of Stokes, Green, and Gauss.

Making intelligent use of technology has been one of the major themes of the Oregon State Project. Computer algebra systems, spreadsheets, and graphing calculators are just a few of the readily available technological tools providing students with new windows of understanding and new opportunities for applying calculus. However, technology should not be viewed as a panacea for calculus instruction. While being "technology-aware," the text itself does not assume the availability of any particular machine or software. To do so would invite immediate obsolescence and ignore how quickly technology advances. (While a graphing calculator is adequate for some of the activities, more sophisticated three-dimensional graphing capabilities are helpful for other problems in the book.)

ACKNOWLEDGMENTS

The Oregon State University Calculus Curriculum Project has been made possible with the support of the National Science Foundation, Oregon State University and the Lasells Stewart Foundation, the Hewlett-Packard Corporation, and PWS-KENT Publishing Company.

This book was typeset on Macintosh computers using Donald Knuth's TeX with Textures 1.2 (Blue Sky Research) and the AmS-TeX - Version 2.0 macro package (American Mathematical Society). The illustrations were produced using MacPaint and MacDraw II (Claris), Illustrator 88 (Adobe), and Grapher 881 Version 1.2 (thanks to Steve Scarborough).

Special thanks are owed to: Steve Quigley, Barbara Lovenvirth, Christian Gal, and Helen Walden of PWS-KENT Publishers, for their support and expertise in helping bring the work of the project to publication; Gregory D. Foley (Sam Houston State University), Thomas Tucker (Colgate University), Robert Moore (University of Washington), William Wickes (Hewlett-Packard), John Kenelly and Don LaTorre (Clemson University), for serving on the National Advisory Panel to the project; Clain Anderson and Ron Brooks of Hewlett-Packard, for their efforts in making technology more easily available to teachers and students in both high schools and colleges; Dianne Hart, Howard L. Wilson, Michelle Jones and Linda Simonsen, of Oregon State University, for their exceptional instructional and in-service efforts with the project; Dianne Hart for preparation of answers to the exercises; Joan McCarter, Laura Moore-Mueller, and Richard Metzler, who piloted the first draft of the materials; Marilyn Wallace, Donna Kent, George and Colleen Dick, for technical typing and preparation of some of the illustrations.

Finally, to Leslie and Colleen, Daniel, Jean, Connor, and Eamon, we dedicate this book.

Contents of Volume II

From the Publisher

It has been said that the only constant in life is change. We see it all around us. From the transformation of seasons ... to the new-found freedoms in Eastern Europe ... to individual growth and development.

The field of mathematics is no exception. Even the notion of *change* itself forms the very foundation of the calculus.

At PWS-KENT Publishing Company we are convinced that the call for reform in calculus instruction is here to stay. It may not alter much in content. It may not usher in a myriad of published product. And it may not appeal to every mathematician. It has received recognition, however, from such venerable institutions as the National Science Foundation, the American Mathematical Society and the Mathematical Association of America. And it has piqued the interest of many dedicated instructors who feel that something new and innovative is needed to revitalize the subject matter.

Issues and developments in this reform movement have been the focus of several Calculus Reform Workshops which have been jointly sponsored by PWS-KENT, the Oregon State University Calculus Curriculum Project, and the Hewlett-Packard Company (in conjunction with a grant from the NSF). The workshops, whose primary funtion has been to provide the "grass roots" of the mathematics community with an opportunity to contribute their ideas and concerns on the curriculum, also have contributed greatly to this publication by Thomas Dick and Charles Patton.

As PWS-KENT enters its twenty-seventh year of being "Partners in Education," we hope that the published works of the Oregon State University Calculus Curriculum Project will be recognized as our contribution to an evolving calculus marketplace which has for more than sixteen years been very generous to us.

With this product we invite inquiry, scrutiny, and discovery. For, together, we can seek the security of change.

The Editors

12

Fundamentals of Vectors

There are many quantities arising in applications that simply cannot be described adequately by a single number. Take, for example, the following description of a traffic accident: "I was driving my car, going about 20 miles per hour, when suddenly this other car comes out of nowhere and hits me. The other car must have been going at least 30 miles per hour."

What do you think was the extent of the damage to the first car? Almost certainly you're thinking that you really need more information, namely the *directions* the two cars were headed at the time of the collision. Did the other car come from the side, the rear, or head-on? And at what angle did the other car hit the first car? Certainly a rear-end collision at the speeds described would be much less damaging than a head-on collision.

In technical terms, we need to know the *velocities* of the two cars. In everyday language, *velocity* and *speed* are often used interchangeably. But, in mathematics and physics, there is a very important distinction to be made: *velocity* can be thought of as "speed with a direction." *Acceleration* and *force* are two other physical quantities whose descriptions require both a numerical *magnitude* and a *direction*. In mathematics, such a quantity is called a **vector quantity**, or simply a **vector**. Velocity, acceleration, and force are all examples of vector quantities. To distinguish from those quantities whose descriptions require only a numerical magnitude, such as *mass*, *distance*, and *time*, the latter are sometimes called **scalar quantities**, or simply **scalars**.

In this chapter, we introduce how vectors can be represented both geometrically and analytically. Consequently, arithmetic operations with vectors have both geometric and analytic interpretations. Operations and functions involving vectors can sometimes be described conveniently by the use of rectangular arrays of numbers called *matrices*, and we'll discuss how matrices are combined arithmetically. Finally, we'll look at some applications of vectors to space geometry.

12.1 REPRESENTATIONS OF VECTORS

First, let's discuss notation for vectors. In this book, we denote vectors in boldface to distinguish them from scalars. Thus,

$$\mathbf{v} \quad \text{denotes a vector quantity,}$$

while

$$v \quad \text{denotes a scalar quantity.}$$

Other notations for vectors you may see use some kind of "arrow" over the letter, or perhaps underlining. Alternative notations for **v** are

$$\vec{v} \quad \text{or} \quad \overrightarrow{v} \quad \text{or} \quad \underline{v}.$$

You will quite likely want to use one of these notations or a similar one, since boldface is not practical for handwritten work.

Geometric representations of vectors—directed line segments

Geometrically, a **vector** can be represented by a *directed line segment* (a line segment with an arrowhead). The *magnitude* of the vector is indicated by the length of the segment. The *direction* of the vector is indicated by the direction of the arrow. See Figure 12.1 below for an illustration.

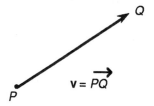

Figure 12.1 Illustration of a vector $\mathbf{v} = \overrightarrow{PQ}$.

In this illustration, P denotes the **initial point** or "tail" of the vector **v**, and Q denotes the **terminal point** or "head" of **v**. If we refer to the vector by the initial and terminal points, then we use the arrow notation:

$$\mathbf{v} = \overrightarrow{PQ}.$$

Note that $\overrightarrow{QP}$ is a *different* vector that points in the opposite direction of $\overrightarrow{PQ}$.

You can see that this geometric representation of a vector $\mathbf{v} = \overrightarrow{PQ}$ could possibly be confused with the common pictorial representation and notation for a *ray* in geometry. Since rays have infinite "length," the length of the arrow used to represent them is unimportant. In contrast, the length of the arrow used to represent a vector is crucially important. Context generally makes it clear whether arrows are being used to represent rays or vectors.

Two directed line segments having the same length and direction represent the same vector quantity. This means any single vector has infinitely many representatives as directed line segments, all of which are equivalent in the sense of equal length and same direction. For this reason, a vector is sometimes referred to as an *equivalence class* of directed line segments. What is important to keep in mind is that two arrows in different positions can represent the same vector, as long as they have the *same length* and *same direction* (see the illustrations in Figure 12.2).

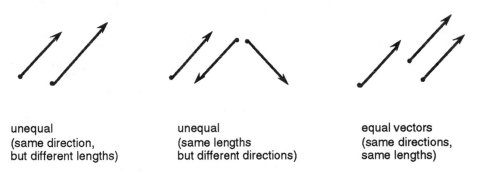

unequal
(same direction,
but different lengths)

unequal
(same lengths
but different directions)

equal vectors
(same directions,
same lengths)

Figure 12.2 Equal vectors have both the same length and same direction.

One special vector is the vector of length 0 (direction can be arbitrary on the zero vector). We denote this vector as **0**, using boldface to distinguish it from the scalar 0. The initial and terminal points of the zero vector *coincide*. So, if P is any point, then

$$\mathbf{0} = \overrightarrow{PP}.$$

Analytic representation of vectors—components

Another very useful way to represent vectors, particularly for analysis, is by *components*.

If our vectors all lie in a plane, we can set up a rectangular (Cartesian) coordinate system. Now, suppose we have a vector $\mathbf{v}$ represented by the directed line segment $\overrightarrow{PQ}$, where

$$P = (x_1, y_1) \quad \text{and} \quad Q = (x_2, y_2).$$

To determine the components of the vector **v** with respect to this coordinate system, we simply subtract the coordinates of its tail from the coordinates of its head. Hence,

the x-component of **v** is $x_2 - x_1$,

and

the y-component of **v** is $y_2 - y_1$.

Now, $\overline{PQ}$ is only one representative of this vector. We could imagine picking up this directed line segment and (without changing the direction) placing its initial point (tail) at the origin $(0,0)$. Now, we can read off the components of **v** directly as the coordinates of its terminal point (head), as shown in Figure 12.3.

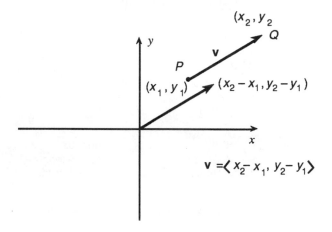

Figure 12.3 Determining the components of a vector **v**.

When we specify a vector by its components, we use the notation

$$\mathbf{v} = \langle v_1, v_2 \rangle$$

to distinguish between the *vector* $\mathbf{v} = \langle v_1, v_2 \rangle$ and the *point* (v_1, v_2).

Two vectors represented by components $\mathbf{v} = \langle v_1, v_2 \rangle$ and $\mathbf{w} = \langle w_1, w_2 \rangle$ are *equal* if and only if their components match exactly. In other words,

$$\mathbf{v} = \mathbf{w} \qquad \text{if and only if} \qquad v_1 = w_1 \text{ and } v_2 = w_2.$$

In terms of components, $\mathbf{0} = \langle 0, 0 \rangle$.

EXAMPLE 1 Find the components of $\mathbf{v} = \overrightarrow{PQ}$ if $P = (3, -4)$ and $Q = (-1.5, 2.3)$.

Solution $\mathbf{v} = \langle [(-1.5) - 3], [2.3 - (-4)] \rangle = \langle -4.5, 6.3 \rangle.$ ∎

EXAMPLE 2 Suppose this same vector $\mathbf{v}$ is represented by the directed line segment $\overrightarrow{RS}$ with the initial point $R = (-1, -2.2)$. Find the coordinates of the terminal point S.

Solution Since $\mathbf{v} = \langle -4.5, 6.3 \rangle$, we must have

$$S = (-1 + (-4.5), -2.2 + 6.3) = (-5.5, 4.1).$$

∎

Representing vectors in 3-dimensional space

There is no particular difficulty in representing vectors in 3-dimensional space, until we run into the pictorial difficulties of illustrating three dimensions on (two-dimensional) paper or blackboard. Geometrically, we can represent vectors as directed line segments in space just as we do for vectors lying in a single plane. To discuss such vectors analytically by components, we can use a rectangular coordinate system in three-dimensional space.

A rectangular coordinate system for space has three mutually perpendicular coordinate axes, with x, y, and z used to represent the first, second, and third coordinates, respectively. Generally, such a system is illustrated on paper with the positive y-axis extending to the right, the positive z-axis extending straight up, and the positive x-axis drawn at an obtuse angle to both of the other axes to suggest that it extends straight *out* from the page. If you think of the origin as a corner point on the floor of a rectangular room, then these positive axes represent the two floor edges (x and y) and the wall corner (z) that all meet at that point. Figure 12.4 illustrates a 3-dimensional rectangular coordinate system.

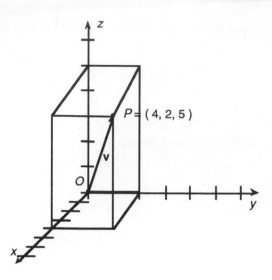

Figure 12.4 A 3-dimensional rectangular coordinate system.

Pictured are the positive x-, y-, and z-axes, along with the point $P = (4, 2, 5)$ and the origin $O = (0, 0, 0)$. The vector $\mathbf{v} = \overrightarrow{OP}$ in the picture has three components. Since its initial point is at O and its terminal point is at P, we have $\mathbf{v} = \langle 4, 2, 5 \rangle$. Notice that the point $Q = (0, 1, 3)$ would appear to be at the same place in our picture as the point P. This points out both the difficulty with representing 3-dimensional space with a 2-dimensional picture as well as the need for care in *labelling* points in our diagrams to avoid confusion.

Note that 3-space is naturally divided into eight octants (as opposed to four quadrants in 2-space). The octant where all coordinates are positive is the first octant (for example, P is in the first octant); the other seven octants have no standard numbering. For reference we note other areas of 3-space and their descriptions.

Coordinate axes	Description
x-axis	$\{(x, 0, 0) \; : \; x \in \mathbb{R}\}$
y-axis	$\{(0, y, 0) \; : \; y \in \mathbb{R}\}$
z-axis	$\{(0, 0, z) \; : \; z \in \mathbb{R}\}$

Coordinate planes	Description
xy-plane	$\{(x, y, 0) \; : \; x, y \in \mathbb{R}\}$
xz-plane	$\{(x, 0, z) \; : \; x, z \in \mathbb{R}\}$
yz-plane	$\{(0, y, z) \; : \; y, z \in \mathbb{R}\}$

Visually, the xy-plane is horizontal and perpendicular to the page. The xz-plane is vertical and perpendicular to the page, and the page itself lies in the yz-plane.

Representing vectors in n-dimensional Space

There really is no restriction in talking about vectors with n components, even if $n > 3$. For example,

$$\mathbf{v} = \langle -1, 3, 5, -4 \rangle$$

is a 4-dimensional vector.

Certainly, a geometric picture of an n-dimensional vector becomes impossible if $n > 3$. Nevertheless, the idea of a directed line segment can be very useful in guiding our intuition, even in these higher dimensions. Our analytic view of vectors allows for any number of components. Indeed, in higher mathematics vectors with *infinitely* many components may be studied.

One might ask what use there might be for such higher dimensional vectors. An example that is particularly important in physics is the notion of space-time. Space-time can be thought of as having four dimensions– the usual three dimensions of space and the fourth dimension of time.

A "point" in space-time describes a spatial location at a specific time. Additional components might be necessary if we want to specify additional "states" existing at that location and time. Some of the newer theories in physics suggest that as many as *eleven* dimensions may be needed to model all the forces at work in nature!

In some contexts, we may want to consider scalars as 1-dimensional vectors. That is, a scalar is a vector having only one component. The real number line can serve as the coordinate system ($a = \langle a \rangle$ in this notation), and we refer to this system as 1-space.

The Cartesian plane can be thought of as the Cartesian product of the set of real numbers with itself. If we write

$$\mathbb{R}^2 = \mathbb{R} \times \mathbb{R}$$

as a suggestive shorthand, we can generalize a concise notation for the various dimensional spaces of vectors with real number components:

$$\mathbb{R}^1 = 1\text{-space} = \{\langle x \rangle \ : \ x \in \mathbb{R}\}$$

$$\mathbb{R}^2 = 2\text{-space} = \{\langle x, y \rangle \ : \ x, y \in \mathbb{R}\}$$

$$\mathbb{R}^3 = 3\text{-space} = \{\langle x, y, z \rangle \ : \ x, y, z \in \mathbb{R}\}$$

$$\vdots$$

$$\mathbb{R}^n = n\text{-space} = \{\langle x_1, x_2, \ldots, x_n \rangle \ : \ x_1, x_2, \ldots, x_n \in \mathbb{R}\}.$$

EXERCISES

Suppose that in the picture below, each of the directed line segments **p, q, u,** *and* **v** *has a length of three units, while* **n** *has a length of two units. Furthermore, suppose that the initial point of* **m** *is located at the point* $(-1, 1)$ *and the terminal point of* **m** *is located at the point* $(-4, -1)$. *Use this information to answer exercises 1-10 below.*

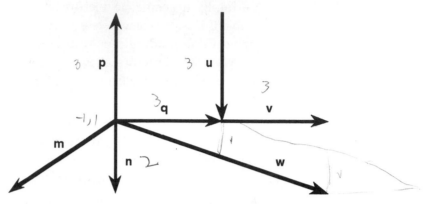

1.   Find the coordinates of the initial points and terminal points of each of the vectors **n, p, q, u, v,** and **w**.

2.   Find the components of each vector. For example, $\mathbf{p} = \langle 0, 3 \rangle$.

3.   Find the magnitudes (lengths) of the vectors **m** and **w**.

4.   Which of the vectors are equal?

5.   Which pairs of vectors have the same length, but different directions?

6.   Which pairs of vectors have the same direction, but different lengths?

7.   What is the area of the parallelogram having **q** and **w** as sides?

8.   Which vectors have perpendicular directions?

9.   Suppose you started at the origin and then proceeded to walk in the directions indicated by these vectors (in alphabetical order). At what point would you end up?

10.   Suppose you started at the origin and then proceeded to walk in the directions indicated by these vectors in reverse alphabetical order. At what point would you end up?

For exercises 11-15, suppose a vector in $\mathbb{R}^3$ *has the given initial point* P *and terminal point* Q. *Write the vector in component notation* $\langle a, b, c \rangle$.

11. $P = (1, 2, 3);$ $Q = (-2, -3, 4)$

12. $P = (-5, 1, 0);$ $Q = (5, -1, 0)$

13. $P = (2.5, -3.4, -4.1);$ $Q = (0, 0, 0)$

14. $P = (\pi, \sqrt{2}, \ln 3);$ $Q = (\pi, \sqrt{2}, \ln 3)$

15. $P = (-2, -3, 4); \quad Q = (1, 2, 3)$

---

16. Suppose $R = (-1, 2, 3)$. For each of exercises 11-15, find the point S such that $\vec{PQ} = \vec{RS}$.

17. Suppose $U = (2.5, -3.4, -4.1)$. For each of exercises 11-15, find the point T such that $\vec{PQ} = \vec{TU}$.

Vectors in a plane are sometimes represented using a distance (magnitude) and a compass direction. For example, wind velocity vectors may be indicated in this way. The convention is to assign 0° to due North, and measure positive angles *clockwise* from due North (in contrast to the counterclockwise convention for polar coordinates). Due East is therefore 90°, due South is 180°, and due West is 270° (or −90°).

Express each of the vectors in exercises 18-22 in component notation $\langle a, b \rangle$.

18. 3 units at 135°.

19. 4 units at −60°.

20. 13.57 units at 213°.

21. 0 units at 84°.

22. 84 units at 0°.

---

23. Find the compass direction of each of the vectors in the illustration for exercises 1-10.

---

24. Here's a classic riddle: You start out and walk one mile due South, then one mile due East, then one mile due North, where you run into a bear standing at your starting point. What color is the bear?

25. Here's a real challenge. Find all the points on the surface of the earth for which the directions of the previous exercise would lead you to your starting point. (Hint: there is more than one point satisfying the conditions.)

12.2 ALGEBRA OF VECTORS

Vectors can be combined algebraically, but there are some important distinctions to be made between algebraic operations on vectors and algebraic operations on real numbers. In this section we discuss the most important operations that can be performed with vectors, and both their geometric *and* analytic interpretations.

Addition

If **v** and **w** are two vectors, then we obtain **v** + **w** geometrically by placing the tail of **w** on the head of **v** and noting **v** + **w** as the **resultant vector** with the same tail as **v** and same head as **w**. You can see from Figure 12.5 why vector addition is said to follow a *Triangle Law.*

Figure 12.5 Addition of vectors.

This *resultant vector sum* also has a physical interpretation. If you imagine pushing an object the distance and direction indicated by **v**, and then followed this by pushing the distance and direction indicated by **w**, then the end position of the object is exactly the same as that obtained by the single push indicated by **v** + **w**.

In terms of components, addition of vectors is simply performed *componentwise*. For example, in $\mathbb{R}^2$, if

$$\mathbf{v} = \langle v_1, v_2 \rangle \qquad \text{and} \qquad \mathbf{w} = \langle w_1, w_2 \rangle,$$

then

$$\mathbf{v} + \mathbf{w} = \langle v_1 + w_1, v_2 + w_2 \rangle.$$

Additive inverses

If **v** is a vector, then geometrically −**v** is the vector with the same length but opposite direction (see Figure 12.6).

Figure 12.6 Additive inverse of a vector.

Analytically, the additive inverse of a vector is obtained by taking the additive inverse of each component. For example, in $\mathbb{R}^3$,

$$\text{if} \quad \mathbf{v} = \langle v_1, v_2, v_3 \rangle, \qquad \text{then} \qquad -\mathbf{v} = \langle -v_1, -v_2, -v_3 \rangle.$$

The additive inverse has the property that for any vector **v**, we have

$$\mathbf{v} + (-\mathbf{v}) = \mathbf{0}.$$

Note that $-\mathbf{0} = \mathbf{0}$.

Subtraction

One way of thinking of subtraction of real numbers is as "adding the opposite." This is one way to think of subtraction of vectors. If **v** and **w** are two vectors, then

$$\mathbf{v} - \mathbf{w} = \mathbf{v} + (-\mathbf{w}).$$

Figure 12.7 illustrates the geometric interpretation of vector subtraction.

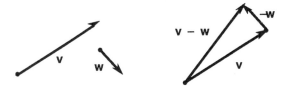

Figure 12.7 Subtraction of vectors—"adding the opposite."

Here is an equivalent method for determining the difference of two vectors **v** − **w** geometrically: If you place the tail of **w** on the tail of **v**, then **v** − **w** has its tail at the head of **w** and its head on the head of **v**.

This method corresponds to the "missing addend" approach to subtraction. That is, we can think of **v** − **w** as the missing vector to add to **w** to obtain **v**. Figure 12.8 illustrates this approach.

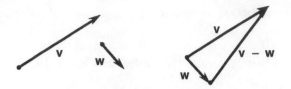

Figure 12.8 Subtraction of vectors—"missing addend" approach.

Analytically, subtraction is performed componentwise. For example, in $\mathbb{R}^2$, if

$$\mathbf{v} = \langle v_1, v_2 \rangle \qquad \text{and} \qquad \mathbf{w} = \langle w_1, w_2 \rangle,$$

then

$$\mathbf{v} - \mathbf{w} = \langle v_1 - w_1, v_2 - w_2 \rangle.$$

Scalar multiplication and division

Vectors can be multiplied by scalars. Suppose $\mathbf{v}$ is a vector and a is a scalar (in other words, a is simply a real number).

Geometrically, $a\mathbf{v}$ is a vector whose length is $|a|$ times the original length of $\mathbf{v}$. As for the direction:

if $a > 0$, then $a\mathbf{v}$ has the same direction as $\mathbf{v}$;

if $a < 0$, then $a\mathbf{v}$ has the opposite direction as $\mathbf{v}$;

if $a = 0$, then $a\mathbf{v} = 0$.

Note that this means $(-1)\mathbf{v} = -\mathbf{v}$. Figure 12.9 illustrates several scalar multiples of a particular vector $\mathbf{v}$.

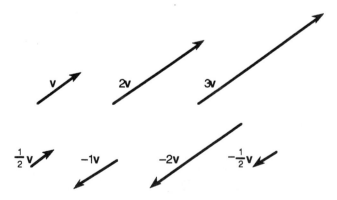

Figure 12.9 Scalar multiples of a vector.

Analytically, scalar multiplication is accomplished by distributing the scalar over each component. For example, in $\mathbb{R}^3$,

$$\text{if} \quad \mathbf{v} = \langle v_1, v_2, v_3 \rangle, \qquad \text{then} \qquad a\mathbf{v} = \langle av_1, av_2, av_3 \rangle.$$

EXAMPLE 3   Suppose $\mathbf{v} = \langle 2, 3 \rangle$ and $\mathbf{w} = \langle -1, 2 \rangle$. Represent $\mathbf{v}$ and $\mathbf{w}$ geometrically. Find $2\mathbf{v}$, $-\mathbf{w}$, $\mathbf{v}+\mathbf{w}$, and $\mathbf{v}-\mathbf{w}$ analytically by components. Illustrate each result geometrically.

Solution   Figure 12.10 illustrates the solution both analytically and geometrically.

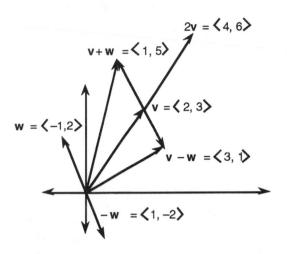

Figure 12.10 Algebraic operations on vectors.

■

To divide a vector $\mathbf{v}$ by a scalar $a \neq 0$ is equivalent to multiplying by the scalar $1/a$.

EXAMPLE 4   In $\mathbb{R}^5$, suppose

$$\mathbf{v} = \langle 5, 2, -1, 0, -5 \rangle \qquad \text{and} \qquad \mathbf{w} = \langle 0, 1, 2, -7, 4 \rangle.$$

Then,

$$\mathbf{v} + \mathbf{w} = \langle 5, 3, 1, -7, -1 \rangle$$

$$\mathbf{v} - \mathbf{w} = \langle 5, 1, -3, 7, -9 \rangle$$

$$2\mathbf{v} = \langle 10, 4, -2, 0, -10 \rangle$$

$$\frac{\mathbf{w}}{-5} = -\frac{1}{5}\langle 0, 1, 2, -7, 4 \rangle$$

$$= \langle 0, -0.2, -0.4, 3.5, -0.8 \rangle.$$

■

Algebraic properties of vectors

We can see that there are two ways of thinking of a vector:

1) geometrically, using a directed line segment of specific length and direction, or

2) analytically, using components.

Which is better? The particular situation under investigation has a lot to do with whether we use a geometric representation or an analytic representation of a vector (or both). You should strive to feel comfortable with both representations, as each has its advantages.

As an illustration, we will list some of the algebraic properties of vectors and show how these properties can be understood in either the geometric sense or the analytic sense.

Theorem 12.1

For any vectors $\mathbf{u}$, $\mathbf{v}$, and $\mathbf{w}$ (all of the same dimension) and for any scalars a and b, the following properties hold:

1) $\mathbf{u} + \mathbf{v} = \mathbf{v} + \mathbf{u}$ commutative law for addition
2) $(\mathbf{u} + \mathbf{v}) + \mathbf{w} = \mathbf{u} + (\mathbf{v} + \mathbf{w})$ associative law for addition
3) $\mathbf{u} + \mathbf{0} = \mathbf{0} + \mathbf{u} = \mathbf{u}$ additive identity law for $\mathbf{0}$
4) $\mathbf{u} + (-\mathbf{u}) = \mathbf{0}$ additive inverse law
5) $a(b\mathbf{u}) = (ab)\mathbf{u}$ associative law for scalars
6) $a(\mathbf{u} + \mathbf{v}) = a\mathbf{u} + a\mathbf{v}$ first distributive law for scalars
7) $(a + b)\mathbf{u} = a\mathbf{u} + b\mathbf{u}$ second distributive law for scalars
8) $1\mathbf{u} = \mathbf{u}$ scalar multiplication identity law for 1
9) $0\mathbf{u} = \mathbf{0}$ scalar multiplication property of 0
10) $-1\mathbf{u} = -\mathbf{u}$ scalar multiplication property of -1

Reasoning   These properties can be illustrated or verified using either geometric or analytic reasoning. We will discuss 1) and 2) here for 2-dimensional vectors. In the exercises, you are asked to verify 3) through 10) for yourself.

1) (using components) We will verify the commutative law analytically for 2-dimensional vectors. Similar reasoning can be used for vectors of higher dimensions.

Let $\mathbf{u} = \langle u_1, u_2 \rangle$ and $\mathbf{v} = \langle v_1, v_2 \rangle$ be any two vectors in $\mathbb{R}^2$.

Then,

$$\mathbf{u} + \mathbf{v} = \langle u_1, u_2 \rangle + \langle v_1, v_2 \rangle$$

$$= \langle u_1 + v_1, u_2 + v_2 \rangle$$

$$= \langle v_1 + u_1, v_2 + u_2 \rangle$$

$$= \langle v_1, v_2 \rangle + \langle u_1, u_2 \rangle$$

$$= \mathbf{v} + \mathbf{u}.$$

(Note that we used the commutativity of real number addition in the third line.)

1) (geometric illustration) If we represent $\mathbf{u}$ and $\mathbf{v}$ with directed line segments, then $\mathbf{u} + \mathbf{v}$ and $\mathbf{v} + \mathbf{u}$ represent the same (directed) diagonal of a parallelogram. Figure 12.11 illustrates this **parallelogram law** of vector addition.

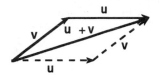

Figure 12.11 Parallelogram law for vector addition.

2) (using components) Let $\mathbf{u} = \langle u_1, u_2 \rangle$, $\mathbf{v} = \langle v_1, v_2 \rangle$, and $\mathbf{w} = \langle w_1, w_2 \rangle$.

Then,

$$(\mathbf{u} + \mathbf{v}) = (\langle u_1, u_2 \rangle + \langle v_1, v_2 \rangle) + \langle w_1, w_2 \rangle$$

$$= \langle u_1 + v_1, u_2 + v_2 \rangle + \langle w_1, w_2 \rangle$$

$$= \langle (u_1 + v_1) + w_1, (u_2 + v_2) + w_2 \rangle$$

$$= \langle u_1 + (v_1 + w_1), u_2 + (v_2 + w_2) \rangle$$

$$= \langle u_1, u_2 \rangle + \langle v_1 + w_1, v_2 + w_2 \rangle$$

$$= \mathbf{u} + (\mathbf{v} + \mathbf{w}).$$

(Notice that we used the associativity of real numbers in the fourth line.)

2) (geometric illustration) If we represent all three vectors with directed line segments, then the associative law can be seen by noting that we obtain the same resultant vector for the sum, regardless of whether we compute $\mathbf{u} + \mathbf{v}$ or $\mathbf{v} + \mathbf{w}$ first (see Figure 12.12).

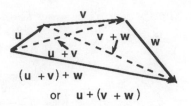

Figure 12.12 Associative law for vector addition.

Since it doesn't matter whether we write $(\mathbf{u} + \mathbf{v}) + \mathbf{w}$ or $\mathbf{u} + (\mathbf{v} + \mathbf{w})$, we can write $\mathbf{u} + \mathbf{v} + \mathbf{w}$ without ambiguity.   □

EXERCISES

Suppose vectors **m, n, p, q, u, v,** *and* **w** *can be represented as lying in the same plane as illustrated in the picture below. Use this picture to answer exercises 1-6.*

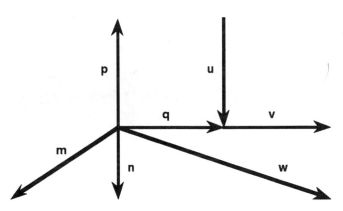

1.   What vector represents $\mathbf{p} + \mathbf{q} + \mathbf{u}$?

2.   What vector represents $\mathbf{n} + \mathbf{q} + \mathbf{v}$?

3.   What vector represents $\mathbf{q} - \mathbf{v}$?

4.   What vector represents $\mathbf{n} - \mathbf{q}$?

5.   What vector represents $\mathbf{w} - \mathbf{n}$?

6.   What vector represents $\mathbf{m} - \mathbf{n} + \mathbf{p} + \mathbf{q} + \mathbf{u} - 2\mathbf{v} + \mathbf{w}$?

Suppose six vectors of equal length are arranged to form a regular hexagon as shown below, with two of the consecutive sides labelled **v** *and* **w**. *Use this picture to answer exercises 7 and 8.*

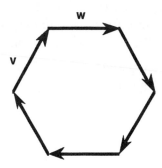

7.Express each of the other four vector sides in terms of **v** and **w**.

8.What is the vector sum of all six sides?

Suppose **v** $= \langle -4, -3 \rangle$ *and* **w** $= \langle 2.72, 3.14 \rangle$. *In exercises 9-16, find the indicated scalars or vectors.*

9.2**v**

10.-3**w**

11.**v** $+$ **w**

12.**v** $-$ **w**

13.**w** $-$ **v**

14.2**v** $-$ 3**w**

15.**w**$/5 - 4$**v**

16.**v** $+ (0.5)(\mathbf{i} - \mathbf{j})$

Suppose **v** $= \langle 2, -3, 1 \rangle$ *and* **w** $= \langle \frac{1}{2} 1, -7 \rangle$. *In exercises 17-24, find the indicated scalars or vectors.*

17.2**v**

18.-3**w**

19.**v** $+$ **w**

20.**v** $-$ **w**

21.**w** $-$ **v**

22.2**v** $-$ 3**w**

23.**w**$/5 - 4$**v**

24.**v** $+ (0.5)\langle 1, 0, -1 \rangle$

---

25. Verify properties 3-10 in Theorem 12.1 both geometrically and analytically for vectors in $\mathbb{R}^2$.

12.3 DOT PRODUCTS OF VECTORS

While we have discussed the product of a scalar and a vector, we haven't yet mentioned the *product of two vectors*. Since we have defined the sum of two vectors *componentwise*, one might guess that the same should hold true for the product of two vectors.

For example, suppose we did use a component-wise definition of vector multiplication. Then we have

$$\langle 2, 0, -1 \rangle \langle 0, 3, 0 \rangle = \langle 0, 0, 0 \rangle = \mathbf{0}.$$

Here we have two *nonzero* vectors with a product that is the *zero vector*. That is certainly different than real number multiplication! For our purposes, this component-wise idea of multiplication is not very useful.

The **dot product** is a type of product that does prove to be useful, and is sometimes referred to as an *inner product* or a *scalar product*. As we did for the other arithmetic operations, we will give both a geometric and an analytic explanation of the dot product.

Definition 1

> The **dot product** of two vectors $\mathbf{v}$ and $\mathbf{w}$ (of the same dimension) is the sum of the component-wise products, and is denoted by $\mathbf{v} \cdot \mathbf{w}$.

In $\mathbb{R}^2$: $\langle v_1, v_2 \rangle \cdot \langle w_1, w_2 \rangle = v_1 w_1 + v_2 w_2$

In $\mathbb{R}^3$: $\langle v_1, v_2, v_3 \rangle \cdot \langle w_1, w_2, w_3 \rangle = v_1 w_1 + v_2 w_2 + v_3 w_3$

In $\mathbb{R}^n$: $\langle v_1, v_2, \ldots v_n \rangle \cdot \langle w_1, w_2, \ldots w_n \rangle = v_1 w_1 + v_2 w_2 + \cdots + v_n w_n$

(In $\mathbb{R}^1$, the dot product is just the usual product of two numbers.)

 The name *scalar product* **makes sense since the result of a dot product is always a scalar, and** *not* **a vector.**

EXAMPLE 5   Here are several dot product calculations for vectors of various dimensions:

$$\langle 2, 3 \rangle \cdot \langle 5, -1 \rangle = 2 \cdot 5 + 3(-1) = 10 - 3 = 7$$

$$\langle 1, 2, 3 \rangle \cdot \langle -5, 3, -7 \rangle = -5 + 6 - 21 = -20$$

$$\langle -1, 2, 4 \rangle \cdot \langle 10, 3, 1 \rangle = -10 + 6 + 4 = 0$$

$$\langle 0, 0, 0 \rangle \cdot \langle 1000, 10000, 1000000 \rangle = 0$$

$$\langle 1, 2, 3, 4, 5 \rangle \cdot \langle 5, 4, 3, 2, 1 \rangle = 5 + 8 + 9 + 8 + 5 = 35$$

■

Some algebraic properties of the dot product are easy to verify using the definition.

Theorem 12.2

For any vectors $\mathbf{u}$, $\mathbf{v}$ and $\mathbf{w}$ (all of the same dimension) and for any scalar a, the following properties hold:

$$\mathbf{u} \cdot \mathbf{v} = \mathbf{v} \cdot \mathbf{u}$$

$$\mathbf{u} \cdot (\mathbf{v} + \mathbf{w}) = \mathbf{u} \cdot \mathbf{v} + \mathbf{u} \cdot \mathbf{w}$$

$$a(\mathbf{u} \cdot \mathbf{v}) = (a\mathbf{u}) \cdot \mathbf{v} = \mathbf{u} \cdot (a\mathbf{v}).$$

Reasoning Let's verify these for vectors in $\mathbb{R}^2$. (The verifications for higher dimensions are very similar.)

Suppose $\mathbf{u} = \langle u_1, u_2 \rangle$, $\mathbf{v} = \langle v_1, v_2 \rangle$, and $\mathbf{w} = \langle w_1, w_2 \rangle$. The first property tells us that the dot product is *commutative*. Note that

$$\mathbf{u} \cdot \mathbf{v} = u_1 v_1 + u_2 v_2 = v_1 u_1 + v_2 u_2 = \mathbf{v} \cdot \mathbf{u}.$$

The second property tells us that the dot product *distributes over vector sums*:

$$\mathbf{u} \cdot (\mathbf{v} + \mathbf{w}) = \langle u_1, u_2 \rangle \cdot \langle v_1 + w_1, v_2 + w_2 \rangle$$

$$= \langle u_1(v_1 + w_1), u_2(v_2 + w_2) \rangle$$

$$= \langle u_1 v_1 + u_1 w_1, u_2 v_2 + u_2 w_2 \rangle$$

$$= \langle u_1 v_1, u_2 v_2 \rangle + \langle u_1 w_1, u_2 w_2 \rangle$$

$$= \mathbf{u} \cdot \mathbf{v} + \mathbf{u} \cdot \mathbf{w}.$$

The last property describes how scalar multiplication and dot products behave together. Just compute each of the three quantities indicated, and you'll see that all three have the same value:

$$a(\mathbf{u} \cdot \mathbf{v}) = (a\mathbf{u}) \cdot \mathbf{v} = \mathbf{u} \cdot (a\mathbf{v}) = a u_1 v_1 + a u_2 v_2. \qquad \square$$

The dot product is an extremely useful computational tool whose worth will become more evident once we provide a geometric description.

Length of a vector—norm

The **norm** of a vector in two or three dimensions is simply its *length*. We denote the norm of a vector **v** as $||\mathbf{v}||$. If such a vector's components are given, then the norm of the vector $||\mathbf{v}||$ can be computed using the Pythagorean Theorem, as illustrated in Figure 12.13.

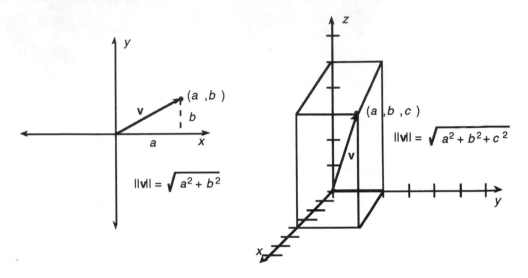

Figure 12.13 The norm of a vector is its length.

While the geometric notion of length might not seem to make sense for higher dimensional vectors, we can still generalize the Pythagorean Theorem to provide an analytic definition of norm in $\mathbb{R}^n$:

$$\text{If}\quad \mathbf{v} = \langle v_1, v_2, \ldots, v_n \rangle, \quad \text{then}\quad ||\mathbf{v}|| = \sqrt{v_1^2 + v_2^2 + \cdots + v_n^2}.$$

In the special case of $\mathbb{R}^1$, the norm of a 1-dimensional vector is simply its absolute value: $||\langle a \rangle|| = |a|$.

EXAMPLE 6   Here are several norm calculations for vectors of various dimensions:

$$||\langle 2, 3 \rangle|| = \sqrt{2^2 + 3^2} = \sqrt{4 + 9} = \sqrt{13} \approx 3.61$$

$$||\langle -1, 2, 4 \rangle|| = \sqrt{(-1)^2 + 2^2 + 4^2} = \sqrt{21} \approx 4.58$$

$$||\langle 0, 0, 0 \rangle|| = \sqrt{0^2 + 0^2 + 0^2} = 0$$

$$||\langle 1, -0.2, 3.4, 4, -1.8 \rangle|| = \sqrt{1^2 + (-0.2)^2 + (3.4)^2 + 4^2 + (-1.8)^2} \approx 5.6427$$

☞ **It is important to realize that $\|v\|$ is always a (nonnegative) scalar.**

EXAMPLE 7   How does scalar multiplication affect the norm of a vector?

Solution   If $\mathbf{v} = \langle v_1, v_2, \ldots, v_n \rangle$ is a vector, and a is a scalar, then

$$a\mathbf{v} = \langle av_1, av_2, \ldots, av_n \rangle$$

and

$$\|a\mathbf{v}\| = \sqrt{a^2 v_1^2 + a^2 v_2^2 + \cdots + a^2 v_n^2} = \sqrt{a^2}\sqrt{v_1^2 + v_2^2 + \cdots + v_n^2} = |a|\,\|\mathbf{v}\|.$$

Geometrically, $a\mathbf{v}$ points in the same or opposite direction as $\mathbf{v}$, depending on whether a is positive or negative. The length of $a\mathbf{v}$ is $|a|$ times the length of the original vector $\mathbf{v}$.   ■

Unit vectors and the standard basis vectors

A **unit vector** is a vector of norm 1.

EXAMPLE 8   In $\mathbb{R}^2$, $\mathbf{v} = \langle \frac{3}{5}, \frac{4}{5} \rangle$ is a unit vector, since

$$\|\mathbf{v}\| = \sqrt{\frac{9}{25} + \frac{16}{25}} = 1.$$

In $\mathbb{R}^3$, $\mathbf{w} = \langle -\frac{2}{3}, -\frac{1}{3}, \frac{2}{3} \rangle$ is a unit vector, since

$$\|\mathbf{w}\| = \sqrt{\frac{4}{9} + \frac{1}{9} + \frac{4}{9}} = 1.$$

■

Some special unit vectors in $\mathbb{R}^2$ are:

$$\mathbf{i} = \langle 1, 0 \rangle \qquad \text{and} \qquad \mathbf{j} = \langle 0, 1 \rangle.$$

These are called the standard basis vectors for $\mathbb{R}^2$. Note that any vector $\langle a, b \rangle$ can be written as a **linear combination** (meaning a sum of scalar multiples) of $\mathbf{i}$ and $\mathbf{j}$:

$$\langle a, b \rangle = a\mathbf{i} + b\mathbf{j}.$$

To see this, just carry out the indicated scalar multiplication and vector addition:

$$a\mathbf{i} + b\mathbf{j} = a\langle 1,0 \rangle + b\langle 0,1 \rangle = \langle a,0 \rangle + \langle 0,b \rangle = \langle a,b \rangle.$$

The standard basis vectors for $\mathbb{R}^3$ are the three vectors

$$\mathbf{i} = \langle 1,0,0 \rangle, \qquad \mathbf{j} = \langle 0,1,0 \rangle, \qquad \mathbf{k} = \langle 0,0,1 \rangle,$$

and $\langle a,b,c \rangle = a\mathbf{i} + b\mathbf{j} + c\mathbf{k}$ for any vector in $\mathbb{R}^3$.

EXAMPLE 9   Write $\langle -2, 3.2, -9 \rangle$ in terms of the standard basis vectors.

Solution $\langle -2, 3.2, -9 \rangle = -2\mathbf{i} + 3.2\mathbf{j} - 9\mathbf{k}.$ ■

Unit vectors are often used when only a particular *direction* needs to be indicated for some purpose. Since the magnitude of the vector we use in this case is immaterial, we often agree to choose a unit vector.

Given a nonzero vector $\mathbf{v}$, we can always find a unit vector having the same direction by simply dividing $\mathbf{v}$ by its own length (a positive scalar). In other words, $\mathbf{v}/\|\mathbf{v}\|$ is a unit vector having the same direction as $\mathbf{v}$, provided $\mathbf{v} \neq \mathbf{0}$.

EXAMPLE 10   Find a unit vector having the same direction as $\mathbf{v} = \langle -2, 3, -6 \rangle$.

Solution $\dfrac{\mathbf{v}}{\|\mathbf{v}\|} = \dfrac{\langle -2,3,-6 \rangle}{\sqrt{(-2)^2 + 3^2 + (-6)^2}} = \dfrac{\langle -2,3,-6 \rangle}{\sqrt{49}} = \frac{1}{7}\langle -2,3,-6 \rangle = \langle -\frac{2}{7}, \frac{3}{7}, -\frac{6}{7} \rangle.$

Since this vector is a positive scalar multiple ($\frac{1}{7}$) of our original vector, we know that it points in the same direction. We can also check that it has unit length:

$$\|\langle -\tfrac{2}{7}, \tfrac{3}{7}, -\tfrac{6}{7} \rangle\| = \sqrt{(-\tfrac{2}{7})^2 + (\tfrac{3}{7})^2 + (-\tfrac{6}{7})^2} = \sqrt{\tfrac{4}{49} + \tfrac{9}{49} + \tfrac{36}{49}} = \sqrt{\tfrac{49}{49}} = \sqrt{1} = 1.$$

Hence, the vector $\langle -\frac{2}{7}, \frac{3}{7}, -\frac{6}{7} \rangle$ satisfies both requirements.   ■

Geometric description of dot product

Since

$$||\mathbf{v}|| = \sqrt{v_1^2 + v_2^2 + \cdots + v_n^2}$$

and

$$\mathbf{v} \cdot \mathbf{v} = v_1^2 + v_2^2 + \cdots + v_n^2,$$

we can write

$$||\mathbf{v}|| = \sqrt{\mathbf{v} \cdot \mathbf{v}},$$

or

$$\mathbf{v} \cdot \mathbf{v} = ||\mathbf{v}||^2.$$

 In other words, the dot product of any vector with itself is the square of the norm.

We are now in a position to give a geometric description of the dot product of two vectors. In $\mathbb{R}^2$ or $\mathbb{R}^3$, the dot product of vectors $\mathbf{u}$ and $\mathbf{v}$ is

$$\mathbf{u} \cdot \mathbf{v} = ||\mathbf{u}|| \; ||\mathbf{v}|| \cos \theta$$

where θ is the smallest positive angle between $\mathbf{u}$ and $\mathbf{v}$ (so $0 \le \theta \le \pi$), provided both $\mathbf{u}$ and $\mathbf{v}$ are nonzero. Figure 12.14 illustrates the angle θ in several instances.

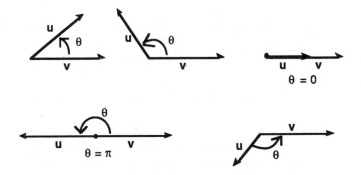

Figure 12.14 Smallest positive angle θ between two vectors.

(If $\mathbf{u} = 0$ or $\mathbf{v} = 0$, then $\mathbf{u} \cdot \mathbf{v} = 0$.)

This description of the dot product looks quite different from our original definition. Let's compute the dot product using both the definition and the geometric description in a specific example.

EXAMPLE 11   Find the dot product $\mathbf{v} \cdot \mathbf{w}$ where $\mathbf{v} = \langle 2, 2 \rangle$ and $\mathbf{w} = \langle 0, 3 \rangle$, using both the analytic definition and the geometric description.

Solution   Figure 12.15 shows the two vectors, and in this case we can see that the angle between the vectors is $45° = \pi/4$.

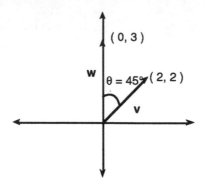

Figure 12.15 Vectors $\mathbf{v} = \langle 2, 2 \rangle$ and $\mathbf{w} = \langle 0, 3 \rangle$.

First, we compute the lengths of the two vectors:

$$\|\mathbf{v}\| = \sqrt{2^2 + 2^2} = \sqrt{8}$$

$$\|\mathbf{w}\| = \sqrt{0^2 + 3^2} = 3$$

Now, using the geometric description of dot product, we have

$$\mathbf{v} \cdot \mathbf{w} = \|\mathbf{v}\| \ \|\mathbf{w}\| \cos 45° = (\sqrt{8})(3)(\sqrt{2}/2) = 6.$$

Using the analytic definition of dot product, we have

$$\mathbf{v} \cdot \mathbf{w} = \langle 2, 2 \rangle \cdot \langle 0, 3 \rangle = 2(0) + 2(3) = 6.$$

Hence, we obtain the same result using either method.   ■

Let's see why the geometric description always provides the same result as the analytic definition, at least in $\mathbb{R}^2$.

Reasoning   Suppose $\mathbf{v} = \langle v_1, v_2 \rangle$ and $\mathbf{w} = \langle w_1, w_2 \rangle$ are any two *nonzero* vectors. Figure 12.16 illustrates two such vectors, along with their difference $\mathbf{v} - \mathbf{w}$.

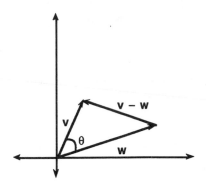

Figure 12.16 Two vectors and their difference.

Now, let's examine the side lengths of the triangle formed by the three vectors $\mathbf{v}$, $\mathbf{w}$, and $\mathbf{v} - \mathbf{w}$ (see Figure 12.17).

Figure 12.17 Triangle formed by the three vectors.

These lengths are vector norms. If we label them

$$||\mathbf{v}|| = a, \quad ||\mathbf{w}|| = b, \quad \text{and} \quad ||\mathbf{v} - \mathbf{w}|| = c,$$

then the Law of Cosines states that

$$c^2 = a^2 + b^2 - 2ab \cos \theta,$$

where θ is the angle between the sides of lengths a and b.

Now we can calculate $||\mathbf{v} - \mathbf{w}||^2$ in two different ways. First, we can use the Law of Cosines to write

$$||\mathbf{v} - \mathbf{w}||^2 = ||\mathbf{v}||^2 + ||\mathbf{w}||^2 - 2||\mathbf{v}||||\mathbf{w}|| \cos \theta.$$

If $\mathbf{v} = \langle v_1, v_2 \rangle$ and $\mathbf{w} = \langle w_1, w_2 \rangle$, then

$$||\mathbf{v}||^2 = v_1^2 + v_2^2 \quad \text{and} \quad ||\mathbf{w}||^2 = w_1^2 + w_2^2.$$

Hence, we can write

$$||\mathbf{v} - \mathbf{w}||^2 = v_1^2 + v_2^2 + w_1^2 + w_2^2 - 2||\mathbf{v}|| \, ||\mathbf{w}|| \cos \theta.$$

Of course, we can also simply compute the square of the length of $\mathbf{v} - \mathbf{w}$.

$$\|\mathbf{v} - \mathbf{w}\|^2 = (v_1 - w_1)^2 + (v_2 - w_2)^2$$
$$= v_1^2 - 2v_1w_1 + w_1^2 + v_2^2 - 2v_2w_2 + w_2^2.$$

When we equate these two ways of computing $\|\mathbf{v} - \mathbf{w}\|^2$, we obtain

$$v_1^2 - 2v_1w_1 + w_1^2 + v_2^2 - 2v_2w_2 + w_2^2 = v_1^2 + v_2^2 + w_1^2 + w_2^2 - 2\|\mathbf{v}\|\,\|\mathbf{w}\|\cos\theta.$$

After subtracting $v_1^2 + v_2^2 + w_1^2 + w_2^2$ from both sides, and dividing both sides by -2, we have

$$\underbrace{v_1w_1 + v_2w_2}_{\text{analytic definition of } \mathbf{v}\cdot\mathbf{w}} \qquad = \qquad \underbrace{\|\mathbf{v}\|\,\|\mathbf{w}\|\cos\theta}_{\text{geometric description of } \mathbf{v}\cdot\mathbf{w}}$$

so we can see that the analytic definition and the geometric description provide the same result.

Computing the angle between two vectors—orthogonal vectors

We now have a means of computing the angle between two nonzero vectors. Using

$$\mathbf{v}\cdot\mathbf{w} = \|\mathbf{v}\|\,\|\mathbf{w}\|\cos\theta,$$

we have

$$\cos\theta = \frac{\mathbf{v}\cdot\mathbf{w}}{\|\mathbf{v}\|\,\|\mathbf{w}\|}.$$

Solving for θ gives us

$$\theta = \arccos\left(\frac{\mathbf{v}\cdot\mathbf{w}}{\|\mathbf{v}\|\,\|\mathbf{w}\|}\right).$$

EXAMPLE 12   Under what conditions on $\mathbf{v}$ and $\mathbf{w}$ is the dot product $\mathbf{v}\cdot\mathbf{w} = 0$?

Solution   Certainly, if either $\mathbf{v} = 0$ or $\mathbf{w} = 0$, then $\mathbf{v}\cdot\mathbf{w} = 0$. If both $\mathbf{v}$ and $\mathbf{w}$ are nonzero, then their lengths are nonzero, and we can conclude that $\cos\theta = 0$, since

$$0 = \mathbf{v}\cdot\mathbf{w} = \|\mathbf{v}\|\,\|\mathbf{w}\|\cos\theta.$$

But $\cos\theta = 0$ means that $\theta = 90°$ or $\pi/2$ radians.   ■

☞ **The dot product of two nonzero vectors is *zero* precisely when the vectors are *perpendicular* to each other.**

Definition 2 | Two vectors **v** and **w** are said to be **orthogonal** if and only if $\mathbf{v} \cdot \mathbf{w} = 0$.

☞ **In other words, we say that two vectors are orthogonal if and only if they are perpendicular to each other, or at least one of them is the zero vector.**

Since $\cos\theta < 0$ for $\pi/2 < \theta \leq \pi$, and $\cos\theta > 0$ for $0 \leq \theta < \pi/2$, we can see how the sign of the dot product can be used to determine the type of angle θ between two nonzero vectors **v** and **w**:

$$\theta \text{ is acute } (< 90°) \qquad \text{if and only if} \qquad \mathbf{v} \cdot \mathbf{w} > 0$$
$$\theta \text{ is obtuse } (> 90°) \qquad \text{if and only if} \qquad \mathbf{v} \cdot \mathbf{w} < 0$$
$$\theta = \pi/2 (= 90°) \qquad \text{if and only if} \qquad \mathbf{v} \cdot \mathbf{w} = 0$$

EXAMPLE 13   Given the vectors $\mathbf{u} = \mathbf{i} + \mathbf{j}$, $\mathbf{v} = 2\mathbf{j} - 3\mathbf{k}$, and $\mathbf{w} = 5\mathbf{k}$ in $\mathbb{R}^3$, find the angle between each pair of vectors. Which (if any) of the vectors are orthogonal to each other?

Solution   First, we compute the norm of each vector and the dot products of each pair of vectors:

$$\|\mathbf{u}\| = \sqrt{2} \qquad \|\mathbf{v}\| = \sqrt{13} \qquad \|\mathbf{w}\| = 5$$

$$\mathbf{u} \cdot \mathbf{v} = 2 \qquad \mathbf{v} \cdot \mathbf{w} = -15 \qquad \mathbf{u} \cdot \mathbf{w} = 0$$

This tells us immediately that **u** and **w** are orthogonal to each other. (Think about the position of these vectors in space. **u** can be represented geometrically as lying in the xy-plane, while **w** can be represented as lying along the z-axis.) Hence, the angle between **u** and **w** is $\pi/2$ radians or 90°.

To find the angles between the other two pairs of vectors, we compute

$$\frac{\mathbf{u} \cdot \mathbf{v}}{\|\mathbf{u}\| \, \|\mathbf{v}\|} = \frac{2}{\sqrt{26}} \approx 0.27735 \qquad \text{and} \qquad \frac{\mathbf{v} \cdot \mathbf{w}}{\|\mathbf{v}\| \, \|\mathbf{w}\|} = \frac{-15}{5\sqrt{13}} \approx -0.83205.$$

The angle between **u** and **v** is approximately

$$\arccos(0.27735) \approx 1.29 \text{ radians or } 73.9°,$$

while the angle between **v** and **w** is approximately

$$\arccos(-0.83205) \approx 2.55 \text{ radians or } 144.6°.$$

■

Component of a vector relative to a direction

We have seen that one way to specify a vector is in terms of its standard components (relative to the standard basis vectors). For example,

$$\mathbf{v} = \langle a, b, c \rangle = a\mathbf{i} + b\mathbf{j} + c\mathbf{k}$$

has standard components $a\mathbf{i}$, $b\mathbf{j}$, and $c\mathbf{k}$. We can also refer to these three vectors as the **vector components** of $\mathbf{v}$ in the directions $\mathbf{i}$, $\mathbf{j}$, and $\mathbf{k}$, respectively. The scalars a, b, and c are called the **scalar components** of $\mathbf{v}$ in the directions $\mathbf{i}$, $\mathbf{j}$, and $\mathbf{k}$, respectively. Sometimes it is useful to describe a vector relative to a direction other than those given by the standard basis vectors. Specifically, suppose a direction is specified by some given *unit* vector $\mathbf{u}$. Then for any other vector $\mathbf{v}$, we can find its **vector component** in the direction $\mathbf{u}$.

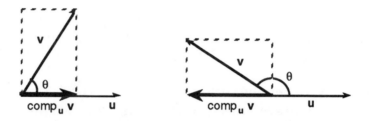

Figure 12.18 Vector component of $\mathbf{v}$ in direction $\mathbf{u}$.

We can visualize this component geometrically by placing the tail of $\mathbf{v}$ on the tail of $\mathbf{u}$ and "projecting" $\mathbf{v}$ down on the line determined by $\mathbf{u}$ (see Figure 12.18). (The line segment formed in this way is called an *orthogonal projection* because of the right triangle formed in this picture.) This vector is called the **vector component of $\mathbf{v}$ in direction** $\mathbf{u}$.

In Figure 12.18, we have indicated this vector component as $\mathrm{comp}_{\mathbf{u}}\mathbf{v}$. Note that the vector component may point in either the same or opposite direction as $\mathbf{u}$. If $\mathbf{v}$ and $\mathbf{u}$ are orthogonal, then the vector component of $\mathbf{v}$ in direction $\mathbf{u}$ is simply the zero vector $\mathbf{0}$. In any case, we always have

$$\mathrm{comp}_{\mathbf{u}}\mathbf{v} = a\mathbf{u}$$

for some scalar a. This scalar a is called the **scalar component of $\mathbf{v}$ in direction** $\mathbf{u}$. Since $\mathbf{u}$ is a unit vector,

$$|a| = \|\mathrm{comp}_{\mathbf{u}}\mathbf{v}\|.$$

The scalar component is negative if the vector component points in the opposite direction of $\mathbf{u}$, and is positive if the vector component points in the same direction as $\mathbf{u}$.

Now we'll look at a very convenient way of computing vector and scalar components. Notice that the angle θ in Figure 12.18 is the angle between **v** and **u**. From triangle trigonometry, we can see that

$$\cos \theta = \frac{\text{adjacent}}{\text{hypotenuse}} = \frac{\|\text{comp}_\mathbf{u}\mathbf{v}\|}{\|\mathbf{v}\|} = \frac{a}{\|\mathbf{v}\|}.$$

On the other hand, we know from the geometric description of the dot product that

$$\cos \theta = \frac{\mathbf{v} \cdot \mathbf{u}}{\|\mathbf{v}\| \, \|\mathbf{u}\|}.$$

Equating these two expressions for $\cos\theta$, and using the fact that $\|\mathbf{u}\| = 1$ (remember, **u** is a *unit* vector), we can find a formula for a:

$$a = \|\text{comp}_\mathbf{u}\mathbf{v}\| = \|\mathbf{v}\| \cos \theta = \|\mathbf{v}\|\frac{\mathbf{v} \cdot \mathbf{u}}{\|\mathbf{v}\|} = \mathbf{v} \cdot \mathbf{u}.$$

That's neat! To find the scalar component of **v** in the direction of unit vector **u**, we need only take the dot product of the two vectors. To find the vector component, we simply multiply this scalar by the unit vector **u**:

$$\text{comp}_\mathbf{u}\mathbf{v} = a\mathbf{u} = (\mathbf{v} \cdot \mathbf{u})\mathbf{u}.$$

EXAMPLE 14 Find the scalar and vector components of $\mathbf{v} = \langle 1, -2, 3 \rangle$ in the direction of the unit vector $\mathbf{u} = \langle -\frac{2}{3}, \frac{1}{3}, \frac{2}{3} \rangle$.

Solution First, note that **u** really is a unit vector, since

$$\|\mathbf{u}\| = \sqrt{(-\frac{2}{3})^2 + (\frac{1}{3})^2 + (\frac{2}{3})^2} = \sqrt{\frac{4}{9} + \frac{1}{9} + \frac{4}{9}} = 1.$$

The scalar component of **v** in the direction **u** is

$$\mathbf{v} \cdot \mathbf{u} = \langle 1, -2, 3 \rangle \cdot \langle -\frac{2}{3}, \frac{1}{3}, \frac{2}{3} \rangle = -\frac{2}{3} - \frac{2}{3} + \frac{6}{3} = \frac{2}{3}.$$

The vector component is

$$(\mathbf{v} \cdot \mathbf{u})\mathbf{u} = \frac{2}{3}\langle -\frac{2}{3}, \frac{1}{3}, \frac{2}{3} \rangle = \langle -\frac{4}{9}, \frac{2}{9}, \frac{4}{9} \rangle.$$

∎

If the direction is specified by a vector **w** that does not have unit length, then we can simply use $\mathbf{u} = \mathbf{w}/\|\mathbf{w}\|$ for the purposes of determining scalar and vector components in that direction.

EXAMPLE 15   Find the scalar and vector components of $\mathbf{v} = \langle 3.5, -2.4, \rangle$ in the direction of the vector $\mathbf{w} = \langle -3, 4 \rangle$.

Solution   First, we find a unit vector $\mathbf{u}$ in the same direction as $\mathbf{w}$:

$$\mathbf{u} = \frac{\mathbf{w}}{\|\mathbf{w}\|} = \frac{\langle -3, 4 \rangle}{\sqrt{(-3)^2 + 4^2}} = \frac{\langle -3, 4 \rangle}{5} = \langle -0.6, 0.8 \rangle.$$

Now, the scalar component of $\mathbf{v}$ in the direction $\mathbf{w}$ is

$$\mathbf{v} \cdot \mathbf{u} = \langle 3.5, -2.4, \rangle \cdot \langle -0.6, 0.8 \rangle = (3.5)(-0.6) + (-2.4)(0.8) = -4.02,$$

and the vector component is

$$(\mathbf{v} \cdot \mathbf{u})\mathbf{u} = (-4.02)\langle -0.6, 0.8 \rangle = \langle 2.412, -3.216 \rangle.$$

■

Once we have found the vector component of $\mathbf{v}$ in the direction of unit $\mathbf{u}$, we can find another vector component in the direction orthogonal or perpendicular to $\mathbf{u}$ (and lying in the same plane as both $\mathbf{u}$ and $\mathbf{v}$). These two vector components sum up to the original vector $\mathbf{v}$, so that

component of $\mathbf{v}$ orthogonal to $\mathbf{u}$ = $\mathbf{v} - \text{comp}_{\mathbf{u}}\mathbf{v}$.

The illustrations in Figure 12.19 indicate this orthogonal vector component.

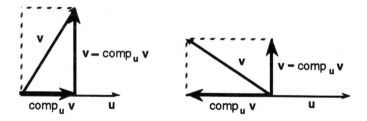

Figure 12.19 $\mathbf{v} - \text{comp}_{\mathbf{u}}\mathbf{v}$ is the orthogonal vector component of $\mathbf{v}$.

■■■■■ **EXERCISES**

*Suppose vectors **m, n, p, q, u, v,** and **w** can be represented as lying in the same plane as illustrated in the picture below. Use this picture to answer exercises 1-6, assuming that **p, q, u, v** have a length of three units, and **n** has a length of two units.*

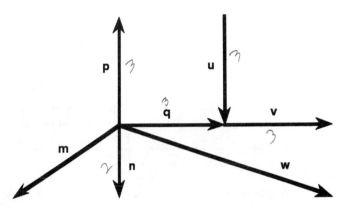

1.   Find $p \cdot n$.

2.   Find $p \cdot q$.

3.   Find $u \cdot p$.

4.   Find $w \cdot n$.

5.   Find the vector component of **m** in the direction of **u**.

6.   Which pairs of vectors in the picture have a positive dot product? negative dot product? zero dot product?

_____

*Suppose six vectors of unit length are arranged to form a regular hexagon as shown below, with two of the consecutive sides labelled **v** and **w**. Use this picture to answer exercises 7 and 8.*

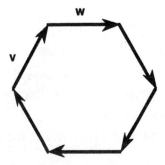

7.   Find $v \cdot w$.

8.   Find the dot product of **v** with each of the other four sides.

_____

Suppose $\mathbf{v} = \langle -4, -3 \rangle$ *and* $\mathbf{w} = \langle 2.72, 3.14 \rangle$. *In exercises 9-22, find the indicated scalars or vectors.*

9. $\mathbf{v} \cdot \mathbf{w}$

10. $\mathbf{w} \cdot (0.5)(\mathbf{i} - \mathbf{j})$

11. $\|\mathbf{v}\|$

12. $\|\mathbf{w}\|$

13. $\|\mathbf{v} + \mathbf{w}\|$

14. $\|\mathbf{v}\| + \|\mathbf{w}\|$

15. $\|\mathbf{v} - \mathbf{w}\|$

16. $\|\mathbf{v}\| - \|\mathbf{w}\|$

17. A unit vector having the same direction as $\mathbf{v}$.

18. A unit vector orthogonal to $\mathbf{w}$.

19. The angle θ between $\mathbf{v}$ and $\mathbf{w}$.

20. The scalar component of $\mathbf{v}$ in the direction $\mathbf{w}$.

21. The vector component of $\mathbf{v}$ in the direction $\mathbf{w}$.

22. The area of the triangle determined by the origin and the terminal points of $\mathbf{v}$ and $\mathbf{w}$ (if each has its initial point at the origin).

Suppose $\mathbf{v} = 2\mathbf{i} - 3\mathbf{j} + \mathbf{k}$ *and* $\mathbf{w} = \dfrac{\mathbf{i}}{2} + \mathbf{j} - 7\mathbf{k}$. *In exercises 23-36, find the indicated scalars or vectors.*

23. $\mathbf{v} \cdot \mathbf{w}$

24. $\mathbf{w} \cdot (0.5)(\mathbf{i} - \mathbf{k})$

25. $\|\mathbf{v}\|$

26. $\|\mathbf{w}\|$

27. $\|\mathbf{v} + \mathbf{w}\|$

28. $\|\mathbf{v}\| + \|\mathbf{w}\|$

29. $\|\mathbf{v} - \mathbf{w}\|$

30. $\|\mathbf{v}\| - \|\mathbf{w}\|$

31. A unit vector having the same direction as $\mathbf{v}$.

32. A unit vector orthogonal to both $\mathbf{v}$ and $\mathbf{w}$.

33. The angle θ between $\mathbf{v}$ and $\mathbf{w}$.

34. The scalar component of $\mathbf{v}$ in the direction $\mathbf{w}$.

35. The vector component of $\mathbf{v}$ in the direction $\mathbf{w}$.

36. The area of the triangle determined by the origin and the terminal points of $\mathbf{v}$ and $\mathbf{w}$ (if each has its initial point at the origin).

---

37. If $\mathbf{v} \cdot \mathbf{w} = 0$, must $\mathbf{v} = 0$ or $\mathbf{w} = 0$?

38. If $\|\mathbf{v} - \mathbf{w}\| = 0$, must $\mathbf{v} = \mathbf{w}$?

39. If $\mathbf{u}$ is a unit vector and $\mathbf{v}$ is any other vector, verify that $\mathbf{v} - \mathrm{comp}_{\mathbf{u}}\mathbf{v}$ is orthogonal to $\mathbf{u}$ by using the dot product.

40. Show that the scalar components of the vector $\langle a, b, c \rangle$ in the directions of the positive coordinate axes are simply the standard components a, b, and c.

41. Suppose $\mathbf{v} = \langle x, y \rangle$. If $\mathbf{v}$ has its tail at the origin, then its head is at the point (x, y). Express the polar coordinates (r, θ) of this point in terms of **v**. (Recall, r is the distance to the origin, and θ is the angle made with the positive x-axis.)

The **direction cosines** of a vector **u** are the cosines of the angles **u** makes with the positive coordinate axes. Since the standard basis vectors are unit vectors that lie in these directions, we can use them to compute the direction cosines as follows (where either $\mathbf{u} = \langle u_1, u_2 \rangle$ in $\mathbb{R}^2$ or $\mathbf{u} = \langle u_1, u_2, u_3 \rangle$ in $\mathbb{R}^3$):

$$\cos \alpha = \frac{\mathbf{u} \cdot \mathbf{i}}{\|\mathbf{u}\| \, \|\mathbf{i}\|} = \frac{u_1}{\|\mathbf{u}\|},$$

$$\cos \beta = \frac{\mathbf{u} \cdot \mathbf{j}}{\|\mathbf{u}\| \, \|\mathbf{j}\|} = \frac{u_2}{\|\mathbf{u}\|},$$

$$\cos \gamma = \frac{\mathbf{u} \cdot \mathbf{k}}{\|\mathbf{u}\| \, \|\mathbf{k}\|} = \frac{u_3}{\|\mathbf{u}\|}.$$

The **direction angles** are α, β, and γ, respectively.

Use the formulas for the direction cosines given above to answer exercises 42-48.

42. Find the direction angles α and β for the vectors **v** and **w** given in the instructions for exercises 9-22.

43. Compute $\cos^2 \alpha + \cos^2 \beta$ for the vectors **v** and **w** given in the instructions for exercises 9-22.

44. Verify that the result of the previous exercise is true for any nonzero vector **u** in $\mathbb{R}^2$.

45. Find the direction angles α, β, and γ for the vectors **v** and **w** given in the instructions for exercises 23-36.

46. Compute $\cos^2 \alpha + \cos^2 \beta + \cos^2 \gamma$ for the vectors **v** and **w** given in the instructions for exercises 23-36.

47. Verify that the result of the previous exercise is true for any nonzero vector **u** in $\mathbb{R}^3$.

48. Is it possible for a nonzero vector in $\mathbb{R}^2$ or $\mathbb{R}^3$ to be perpendicular to *all* the coordinate axes at once? If so, give an example. If not, explain why it is not possible.

12.4 MATRICES

A **matrix** is a rectangular array of numbers. If the matrix has n rows and m columns, we call it an $n \times m$ matrix ("n by m" matrix). We call the element in the ith row and j^{th} column the ij-entry of the matrix. We will be concerned with *real* matrices, whose entries are all real numbers.

EXAMPLE 16   Here are three matrices:

$$A = \begin{bmatrix} 2 & 1 & -3 \\ -2 & 4 & 1 \end{bmatrix}, \quad B = \begin{bmatrix} -3 & 2 & 7 \\ 1 & 4 & -2 \end{bmatrix} \quad \text{and} \quad C = \begin{bmatrix} 1 & 2 & 3 \\ -2 & -1 & 0 \\ 0 & 4 & -3 \end{bmatrix}$$

A and B are 2×3 matrices; C is a 3×3 matrix.   ■

If a matrix has the same number of rows as columns, we call it a **square matrix**. In the example above, C is a 3×3 square matrix.

Two matrices are considered equal if and only if they are exactly the same size and they match entry-by-entry throughout.

If a matrix has 0's in all its entries, then we call it a **zero matrix**.

EXAMPLE 17 $O = \begin{bmatrix} 0 & 0 \\ 0 & 0 \\ 0 & 0 \end{bmatrix}$ is a 3×2 zero matrix.   ■

If a *square* matrix has 1's along the *main diagonal* (from upper left to lower right), we call it an **identity matrix**.

EXAMPLE 18 $I = \begin{bmatrix} 1 & 0 & 0 \\ 0 & 1 & 0 \\ 0 & 0 & 1 \end{bmatrix}$ is a 3×3 identity matrix.   ■

We'll reserve the following notation for zero and identity matrices:

I_n represents an $n \times n$ identity matrix.

O_n represents an $n \times n$ zero matrix.

EXAMPLE 19 $I_3 = \begin{bmatrix} 1 & 0 & 0 \\ 0 & 1 & 0 \\ 0 & 0 & 1 \end{bmatrix}$ and $O_2 = \begin{bmatrix} 0 & 0 \\ 0 & 0 \end{bmatrix}$.   ■

Algebra of matrices

We can perform arithmetic operations on matrices in a manner very similar to the arithmetic operations we perform on vectors. We summarize these operations below, and illustrate each with some examples.

Scalar multiplication: We can multiply a matrix by a scalar simply by multiplying each entry by the scalar.

EXAMPLE 20   Using the matrices A and B from the previous example, we have

$$2A = 2 \begin{bmatrix} 2 & 1 & -3 \\ -2 & 4 & 1 \end{bmatrix} = \begin{bmatrix} 4 & 2 & -6 \\ -4 & 8 & 2 \end{bmatrix}$$

$$(-1)B = (-1) \begin{bmatrix} -3 & 2 & 7 \\ 1 & 4 & -2 \end{bmatrix} = \begin{bmatrix} 3 & -2 & -7 \\ -1 & -4 & 2 \end{bmatrix}.$$

Matrix addition and subtraction: Two matrices can be added or subtracted only if they are the same size (same number of rows, same number of columns). We add and subtract matrices "entry-wise."

EXAMPLE 21   The matrices A and C *cannot* be added because they are not the same size. Similarly, B and C *cannot* be added because they are not the same size. We can, however, add A and B:

$$A + B = \begin{bmatrix} 2 & 1 & -3 \\ -2 & 4 & 1 \end{bmatrix} + \begin{bmatrix} -3 & 2 & 7 \\ 1 & 4 & -2 \end{bmatrix} = \begin{bmatrix} -1 & 3 & 4 \\ -1 & 8 & -1 \end{bmatrix}.$$

EXAMPLE 22   We call $(-1)B = -B$ the additive inverse of B, since

$$B + (-1)B = \begin{bmatrix} 0 & 0 & 0 \\ 0 & 0 & 0 \end{bmatrix},$$

a zero matrix.

Matrix multiplication: Two matrices can be multiplied if and only if *the number of columns of the first matrix is the same as the number of rows of the second matrix.* To get the ij entry of the matrix product, we take the *dot product* of the ith *row* of the first matrix with the jth *column* of the second (as if they were vectors).

EXAMPLE 23   The matrix product AB does *not* make sense. (A has 3 columns, B has only 2 rows.) On the other hand, the matrix product AC does make sense. (A has 3 columns and C has 3 rows.)

The first entry of AC (first row, first column) is the dot product of the first row of A and the first column of C:

$$\langle 2, 1, -3 \rangle \cdot \langle 1, -2, 0 \rangle = 2(1) + 1(-2) + -3(0) = 0.$$

The middle entry of the first row is the dot product of the first row of A and the second column of C:

$$\langle 2, 1, -3 \rangle \cdot \langle 2, -1, 4 \rangle = 2(2) + 1(-1) + (-3)(4) = -9.$$

Continuing in this manner, we can complete the computation of the matrix product:

$$AC = \begin{bmatrix} 2 & 1 & -3 \\ -2 & 4 & 1 \end{bmatrix} \begin{bmatrix} 1 & 2 & 3 \\ -2 & -1 & 0 \\ 0 & 4 & -3 \end{bmatrix} = \begin{bmatrix} 0 & -9 & 15 \\ -10 & -4 & -9 \end{bmatrix}$$

☞   **If we multiply an $(n \times m)$ matrix times an $(m \times p)$ matrix, the result is an $(n \times p)$ matrix.**

Properties of matrix algebra

Here we summarize some of the important properties of matrix algebra. Below, S, T, and R are matrices and a and b are scalars.

Matrix addition is associative:

$$S + (T + R) = (S + T) + R$$

Matrix addition is commutative:

$$S + T = T + S$$

Scalar multiplication is distributive over matrix addition:

$$a(S + T) = aS + aT$$

In all three of the properties above, we assume that S, T, and R are the same size (same number of rows in each, same number of columns in each).

Scalar multiplication is associative with matrix multiplication:

$$a(ST) = (aS)T = S(aT).$$

Matrix multiplication is associative:

$$S(TR) = S(TR)$$

Matrix multiplication is distributive over matrix addition:

$$S(T + R) = ST + SR \quad \text{and} \quad (S + T)R = SR + TR$$

In all three of these properties, we assume that the products make sense (the numbers of rows and columns in each matrix must be appropriate).

Notice that we have said nothing about matrix multiplication being *commutative*. Is it? (Try finding the products BC and CB using the matrices B and C from the previous example.)

If S is a square $n \times n$ matrix, then

$$S + O_n = O_n + S = S \quad \text{and} \quad SI_n = I_n S = S,$$

so the zero and identity matrices behave in a similar way to the real numbers 0 and 1, respectively.

Vectors as matrices

When multiplying two matrices together, we essentially treat the rows and columns as if they are vectors and use the dot product to compute the entries of the product matrix. In general, a vector can be considered either as a matrix with a single row or a matrix with a single column.

EXAMPLE 24 Suppose $\mathbf{v} = \langle 2, 1, -5 \rangle = 2\mathbf{i} + \mathbf{j} - 5\mathbf{k}$. If it is convenient, we could consider $\mathbf{v}$ as a matrix with one row

$$\mathbf{v} = \begin{bmatrix} 2 & 1 & -5 \end{bmatrix}$$

or as a matrix with one column

$$\mathbf{v} = \begin{bmatrix} 2 \\ 1 \\ -5 \end{bmatrix}$$

To distinguish these two ways of writing the vector, we say that in the first case, $\mathbf{v}$ is a **row vector**, and in the second case, $\mathbf{v}$ is a **column vector**. ∎

Even a 1-dimensional vector $\langle a \rangle$ can be considered as a 1×1 matrix $[a]$.

When written as a row or column vector, we can multiply the vector by a matrix (provided the matrix has the appropriate size).

EXAMPLE 25 Using $A = \begin{bmatrix} 2 & 1 & -3 \\ -2 & 4 & 1 \end{bmatrix}$ and $\mathbf{v} = \langle 2, 1, -5 \rangle$ as a column vector, find $A\mathbf{v}$.

Solution

$$A\mathbf{v} = \begin{bmatrix} 2 & 1 & -3 \\ -2 & 4 & 1 \end{bmatrix} \begin{bmatrix} 2 \\ 1 \\ -5 \end{bmatrix} = \begin{bmatrix} 2(2) + 1(1) + (-3)(-5) \\ -2(2) + 4(1) + 1(-5) \end{bmatrix} = \begin{bmatrix} -10 \\ -5 \end{bmatrix}.$$

Note that since A is a 2×3 matrix, and $\mathbf{v}$ is a 3×1 matrix (3 rows and 1 column), the product $A\mathbf{v}$ is a 2×1 matrix. We could interpret this result as a 2-dimensional column vector. ■

EXAMPLE 26 Using $A = \begin{bmatrix} 2 & 1 & -3 \\ -2 & 4 & 1 \end{bmatrix}$ and $\mathbf{w} = \langle -1, 4 \rangle$ as a row vector, find $\mathbf{w}A$.

Solution

$$\mathbf{w}A = \begin{bmatrix} -1 & 4 \end{bmatrix} \begin{bmatrix} 2 & 1 & -3 \\ -2 & 4 & 1 \end{bmatrix} = \begin{bmatrix} (-1)(2) + 4(-2) \\ (-1)(1) + 4(4) \\ (-1)(-3) + 4(1) \end{bmatrix} = \begin{bmatrix} -10 & 15 & 7 \end{bmatrix}.$$

The result can be considered a 3-dimensional row vector. ■

EXAMPLE 27 Using $\mathbf{v}$ and $\mathbf{w}$ as in the previous two examples, compute $\mathbf{v}A$ and $A\mathbf{w}$.

Solution Neither product makes sense, regardless of whether we write the vectors in row or column format. (The sizes are inappropriate for matrix products).■

Determinants

The **determinant** of a square matrix is a single number that gives us some very important information about the matrix. If A is a square matrix, we denote its determinant by

$$\det A = |A|.$$

The determinant of a 1×1 matrix is simply the value of its single entry:

$$\det [a] = a.$$

The determinant of a 2×2 matrix is computed by taking the difference of the products of the diagonal entries:

$$\det \begin{bmatrix} a & b \\ c & d \end{bmatrix} = \begin{vmatrix} a & b \\ c & d \end{vmatrix} = ad - bc.$$

EXAMPLE 28   Find the determinant of $\begin{bmatrix} -2 & 3 \\ -1 & -4 \end{bmatrix}$

Solution

$$\det \begin{bmatrix} -2 & 3 \\ -1 & -4 \end{bmatrix} = \begin{vmatrix} -2 & 3 \\ -1 & -4 \end{vmatrix} = (-2)(-4) - 3(-1) = 11.$$

■

The determinant of a 3×3 matrix has a more complicated formula:

$$\det \begin{bmatrix} a_1 & a_2 & a_3 \\ b_1 & b_2 & b_3 \\ c_1 & c_2 & c_3 \end{bmatrix} = \begin{vmatrix} a_1 & a_2 & a_3 \\ b_1 & b_2 & b_3 \\ c_1 & c_2 & c_3 \end{vmatrix}$$

$$= a_1(b_2 c_3 - b_3 c_2) - a_2(b_1 c_3 - b_3 c_1) + a_3(b_1 c_2 - b_2 c_1).$$

EXAMPLE 29   Find the determinant of $C = \begin{bmatrix} 1 & 2 & 3 \\ -2 & -1 & 0 \\ 0 & 4 & -3 \end{bmatrix}$

Solution

$$\det C = \begin{vmatrix} 1 & 2 & 3 \\ -2 & -1 & 0 \\ 0 & 4 & -3 \end{vmatrix}$$

$$= 1(-1(-3) - 0(4)) - 2(-2(-3) - 0(0)) + 3(-2(4) - (-1)(0))$$

$$= -33.$$

■

There is a handy method of remembering this determinant formula for 3×3 matrices, called **expansion by minors** or **cofactors**. Here's how it works:

If we pick any entry in the 3×3 matrix and cross out all the entries in the same row and column, we are left with a 2×2 matrix. The *determinant* of this smaller matrix is called the **minor** of that entry.

EXAMPLE 30   Find the minor of b_2 in the 3×3 matrix above.

Solution   The minor of b_2 is $a_1 c_3 - a_3 c_1$.   ■

Now, we multiply the minor by $+1$ or -1 to obtain the **cofactor** of the entry. To determine which sign to use, just take the row number i and the column number j of the entry and compute $(-1)^{i+j}$.

EXAMPLE 31   Find the cofactor of b_2 in the 3×3 matrix above.

Solution   The cofactor of b_2 is $(-1)^{2+2}(a_1 c_3 - a_3 c_1) = a_1 c_3 - a_3 c_1$, since b_2 is in the second row and second column.

Some people refer to a "checkerboard" of "$+$" and "$-$" signs to determine the choice $+1$ or -1. If you make a 3×3 matrix with a "$+$" in the upper left and alternate signs in both directions, you obtain the pattern

$$\begin{bmatrix} + & - & + \\ - & + & - \\ + & - & + \end{bmatrix}.$$

Notice that the sign corresponding to b_2 (second row and second column) tells us that $+1$ is the correct choice.   ■

Finally, the determinant of the 3×3 matrix is computed by

1) choosing any single row or any single column,

2) multiplying each entry in that row or column by its cofactor, and

3) adding the results.

If you examine the formula for the determinant of the 3×3 matrix above, you'll see that it is written using the first row and expansion by minors (see Figure 12.20).

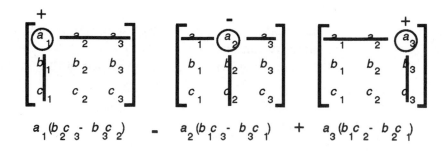

Figure 12.20 Calculating the determinant using expansion by minors.

For 4×4, 5×5, 6×6, and higher dimensions, the same expansion by minors technique is used. Notice that for a 4×4 matrix, the minors are determinants of 3×3 matrices, each of which may be calculated using expansion by minors! Hence, you can see that the computation of determinants quickly gets more complicated and tedious as the size of the matrix increases.

Fortunately, certain calculators and computer software are equipped to calculate determinants of matrices.

☞ **Beware, however, that the numerical limitations of calculators and computers can be sorely tested by the calculation of determinants. This is because the calculation requires finding several differences of products, whose values have been rounded off. This leads to cancellation errors much like what we can experience when computing difference quotients with a machine.**

EXERCISES

Exercises 1-15 refer to the following matrices:

$$S = \begin{bmatrix} 2 & -1 \\ -3 & 4 \\ 0 & 1 \end{bmatrix}, \quad T = \begin{bmatrix} 3 & -2 \\ 7 & 0 \\ 1 & -3 \end{bmatrix}, \quad R = \begin{bmatrix} 1 & 4 \\ 2 & 2 \\ -6 & -8 \end{bmatrix}$$

as well as the previous matrix examples

$$A = \begin{bmatrix} 2 & 1 & -3 \\ -2 & 4 & 1 \end{bmatrix}, \quad B = \begin{bmatrix} -3 & 2 & 7 \\ 1 & 4 & -2 \end{bmatrix} \quad and \quad C = \begin{bmatrix} 1 & 2 & 3 \\ -2 & -1 & 0 \\ 0 & 4 & -3 \end{bmatrix}.$$

1. Find $U = S + T$.

2. Find $V = U + R$.

3. Find $X = T + R$.

4. Find $Y = S + X$.

5. Compare matrix V and matrix Y. What matrix algebra property does this illustrate?

6. Illustrate that matrix addition is commutative by verifying that $S+T = T + S$.

7. Find $W = -2S - 2T$.

8. Find $Z = -2U$.

9. Compare matrix W and matrix Z. What matrix algebra property does this illustrate?

10. Find $D = AS$ and $E = SA$. Is matrix multiplication commutative?

11. Find $F = AT$.

12. Find ET and SF. What matrix algebra property does this illustrate?

13. Find I_3C and CI_3.

14. Find $\det(D)$.

15. Find $\det(E)$.

---

16. Find $\det(I_2)$.

17. Find $\det(I_3)$.

18. Find $\det(I_4)$. What is det (I_n) for any n?

19. What is $\det(O_n)$ for any n?

20. Let $M = \begin{bmatrix} a & b \\ c & d \end{bmatrix}$ and $N = \begin{bmatrix} d & e \\ f & g \end{bmatrix}$. Find $|M|$, $|N|$, MN, and $|MN|$, and verify that $\det(MN) = \det(M)\det(N)$.

Let

$$P = \begin{bmatrix} 1 & 2 & 3 \\ -2 & 4 & -1 \\ ? & ? & ? \end{bmatrix}.$$

Exercises 21-30 ask you to consider this matrix for different choices of entries in its third row.

21. If the third row is replaced by 1 1 1, what is $\det(P)$?

22. If the third row is replaced by 2 2 2, what is $\det(P)$?

23. If the third row is replaced by $-3 \ -3 \ -3$, what is $\det(P)$?

24. If the third row is replaced by 0 0 0, what is $\det(P)$?

25. If the third row is replaced by $a\ a\ a$, what is $\det(P)$?

26. If the third row is replaced by 1 1 1, but the first two rows are interchanged, what is $\det(P)$?

27. If the third row is the same as the first row, what is $\det(P)$?

28. If the third row is the same as the second row, what is $\det(P)$?

29. Replace the third row by the difference of twice the first row and three times the second row. What is $\det(P)$?

30. Replace the third row by the sum of a times the first row and b times the second row. What is $\det(P)$?

Can you generalize your results and make conjectures about the behavior of determinants of 3×3 matrices?

12.5 GEOMETRIC APPLICATIONS OF VECTORS IN $\mathbb{R}^3$

In this section we'll examine how vectors can be used to solve geometric problems in space, and the various ways we can represent lines and planes in space. First, let's introduce some special tools used just for vectors in $\mathbb{R}^3$.

Cross product of 3-dimensional vectors

The *dot product* is defined for vectors in $\mathbb{R}^n$ for any dimension n.

In $\mathbb{R}^3$, if

$$\mathbf{v} = \langle v_1, v_2, v_3 \rangle \quad \text{and} \quad \mathbf{w} = \langle w_1, w_2, w_3 \rangle,$$

then

$$\mathbf{v} \cdot \mathbf{w} = v_1 w_1 + v_2 w_2 + v_3 w_3.$$

If $\mathbf{v}$ and $\mathbf{w}$ are nonzero, then we also have a nice geometric description of the dot product:

$$\mathbf{v} \cdot \mathbf{w} = ||\mathbf{v}|| \, ||\mathbf{w}|| \cos\theta$$

where θ is the smallest angle between $\mathbf{v}$ and $\mathbf{w}$. This allows us to use the dot product to find the angle between two vectors. In particular, we can use the dot product to test the orthogonality of two vectors. (Recall that two vectors are orthogonal if and only if their dot product is zero.)

The **cross product** is a product defined *only* for vectors in $\mathbb{R}^3$. Unlike the dot product, which produces a scalar result, the cross product produces a *vector* result. The notation and analytic formula for the cross product of two 3-dimensional vectors $\mathbf{v}$ and $\mathbf{w}$ is

$$\mathbf{v} \times \mathbf{w} = (v_2 w_3 - v_3 w_2)\mathbf{i} - (v_1 w_3 - v_3 w_1)\mathbf{j} + (v_1 w_2 - v_2 w_1)\mathbf{k}$$

$$= \langle v_2 w_3 - v_3 w_2, \; v_3 w_1 - v_1 w_3, \; v_1 w_2 - v_2 w_1 \rangle.$$

This somewhat cumbersome formula is most easily remembered in the form of a "determinant." If we create a 3 by 3 matrix having the standard basis vectors $\mathbf{i}$, $\mathbf{j}$, and $\mathbf{k}$ in the first row, the components of $\mathbf{v}$ in the second row, and the components of $\mathbf{w}$ in the third row, then we can compute $\mathbf{v} \times \mathbf{w}$ as if it were the determinant:

$$\mathbf{v} \times \mathbf{w} = \begin{vmatrix} \mathbf{i} & \mathbf{j} & \mathbf{k} \\ v_1 & v_2 & v_3 \\ w_1 & w_2 & w_3 \end{vmatrix}.$$

EXAMPLE 32   Find $\mathbf{v} \times \mathbf{w}$ if $\mathbf{v} = \langle 1, 2, 3 \rangle$ and $\mathbf{w} = \langle -3, 1, 5 \rangle$.

Solution

$$\mathbf{v} \times \mathbf{w} = \langle 1, 2, 3 \rangle \times \langle -3, 1, 5 \rangle$$

$$= \begin{vmatrix} \mathbf{i} & \mathbf{j} & \mathbf{k} \\ 1 & 2 & 3 \\ -3 & 1 & 5 \end{vmatrix}$$

$$= (2(5) - 3(1))\mathbf{i} - (1(5) - 3(-3))\mathbf{j} + (1(1) - 2(-3))\mathbf{k}$$

$$= 7\mathbf{i} - 14\mathbf{j} + 7\mathbf{k} = 7 \langle 1, -2, 1 \rangle.$$

■

Theorem 12.3

For any vectors $\mathbf{u}$, $\mathbf{v}$, and $\mathbf{w}$ in $\mathbb{R}^3$, and for any scalar a, the following properties hold:

$$\mathbf{v} \times \mathbf{w} = -(\mathbf{w} \times \mathbf{v})$$

$$(a\mathbf{v}) \times \mathbf{w} = \mathbf{v} \times (a\mathbf{w}) = a(\mathbf{v} \times \mathbf{w})$$

$$\mathbf{u} \times (\mathbf{v} + \mathbf{w}) = \mathbf{u} \times \mathbf{v} + \mathbf{u} \times \mathbf{w}$$

$$\mathbf{v} \times \mathbf{v} = \mathbf{0}.$$

Reasoning   If we let $\mathbf{u} = \langle u_1, u_2, u_3 \rangle$, $\mathbf{v} = \langle v_1, v_2, v_3 \rangle$, and $\mathbf{w} = \langle w_1, w_2, w_3 \rangle$, then all of these properties can be verified by simply computing the indicated vector quantities on both sides of the "=" sign, and checking that the same result is obtained.   □

Some of these properties deserve some special mention. Note that the first property tells us that the cross product is definitely *not* commutative. Indeed, the cross product is **anti-commutative**, meaning that a reversal of the order of the factors changes the sign of the result. The third property tells us that the cross product does distribute over vector sums. The last property is actually a consequence of the first ($\mathbf{u} \times \mathbf{u} = -(\mathbf{u} \times \mathbf{u})$ implies that $\mathbf{u}$ must be the zero vector). Together with the second property, this means

$$(a\mathbf{u}) \times \mathbf{u} = \mathbf{u} \times (a\mathbf{u}) = a(\mathbf{u} \times \mathbf{u}) = \mathbf{0}.$$

In other words, the cross product of a vector with any scalar multiple of itself is the zero vector.

Just as the dot product can be used to check that two vectors are *perpendicular*, the cross product can be used to check that two vectors in $\mathbb{R}^3$ are *parallel*.

 The cross product $v \times w = 0$ if v and w have the same or opposite directions.

Geometric description of the cross product

Just as the dot product has a nice geometric interpretation, so does the cross product.

For two nonzero vectors v and w, the magnitude (length) of $v \times w$ is

$$\|\mathbf{v} \times \mathbf{w}\| = \|\mathbf{v}\| \, \|\mathbf{w}\| \sin\theta,$$

where, as before, θ is the smallest angle between v and w. Notice that $\sin\theta = 0$ when $\theta = 0$ or $\theta = 180° = \pi$ radians. This is consistent with our earlier observation that the cross product of parallel vectors is the zero vector.

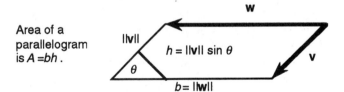

Figure 12.21 The area of the parallelogram is $|\mathbf{v} \times \mathbf{w}| = \|\mathbf{v}\| \, \|\mathbf{w}\| \sin\theta$.

Using trigonometry, Figure 12.21 shows that we can interpret this magnitude as the *area of the parallelogram* determined by v and w.

The *direction* of the cross product $v \times w$ is orthogonal to both v and w. If v and w are not parallel, there are exactly two possible directions (both perpendicular to the plane containing both v and w). The correct choice is determined by the **right-hand rule**: If you hold your right hand so that your fingers point in the direction of v and can curl toward w, then your extended thumb will point in the direction of $v \times w$.

Figure 12.22 illustrates the right-hand rule.

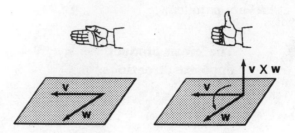

Figure 12.22 The right-hand rule determines the direction of the cross product.

You can check that using the right-hand rule for $\mathbf{w} \times \mathbf{v}$ forces your thumb to point in the opposite direction.

EXAMPLE 33   Compute all the possible dot and cross products of the standard basis vectors $\mathbf{i}$, $\mathbf{j}$, and $\mathbf{k}$.

Solution

$$\mathbf{i} \cdot \mathbf{j} = \mathbf{j} \cdot \mathbf{i} = 0 \qquad \mathbf{i} \cdot \mathbf{i} = 1$$

$$\mathbf{i} \cdot \mathbf{k} = \mathbf{k} \cdot \mathbf{i} = 0 \qquad \mathbf{j} \cdot \mathbf{j} = 1$$

$$\mathbf{j} \cdot \mathbf{k} = \mathbf{k} \cdot \mathbf{j} = 0 \qquad \mathbf{k} \cdot \mathbf{k} = 1$$

$$\mathbf{i} \times \mathbf{j} = \mathbf{k} \qquad \mathbf{j} \times \mathbf{k} = \mathbf{i} \qquad \mathbf{k} \times \mathbf{i} = \mathbf{j}$$

$$\mathbf{i} \times \mathbf{i} = \mathbf{j} \times \mathbf{j} = \mathbf{k} \times \mathbf{k} = 0$$

$$\mathbf{j} \times \mathbf{i} = -\mathbf{k} \qquad \mathbf{k} \times \mathbf{j} = -\mathbf{i} \qquad \mathbf{i} \times \mathbf{k} = -\mathbf{j}$$

Some people like the following diagram for remembering cross products of the standard basis vectors:

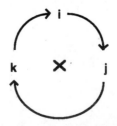

The cross product of any two standard basis vectors is either $+1$ (if the order goes clockwise) or -1 (if the order goes counterclockwise) times the remaining basis vector. ∎

EXAMPLE 34   Find the area of the parallelogram determined by $\mathbf{u} = \mathbf{j} + 2\mathbf{k}$ and $\mathbf{v} = \mathbf{i} - 2\mathbf{j}$.

Solution $\mathbf{u} \times \mathbf{v} = \begin{vmatrix} \mathbf{i} & \mathbf{j} & \mathbf{k} \\ 0 & 1 & 2 \\ 1 & -2 & 0 \end{vmatrix} = 4\mathbf{i} + 2\mathbf{j} - \mathbf{k}$. The area of the parallelogram is

$$\|\mathbf{u} \times \mathbf{v}\| = \sqrt{4^2 + 2^2 + (-1)^2} = \sqrt{21} \approx 4.58.$$

∎

EXAMPLE 35   Find all unit vectors that are perpendicular to both $\mathbf{u} = \langle 1, 2, 3 \rangle$ and $\mathbf{v} = \langle -2, 0, -4 \rangle$.

Solution   We can find a vector perpendicular to both $\mathbf{u}$ and $\mathbf{v}$ by means of their cross product:

$$\mathbf{u} \times \mathbf{v} = \begin{vmatrix} \mathbf{i} & \mathbf{j} & \mathbf{k} \\ 1 & 2 & 3 \\ -2 & 0 & -4 \end{vmatrix} = -8\mathbf{i} - 2\mathbf{j} + 4\mathbf{k}.$$

Now, we can scale this vector to unit length. Since

$$\|\mathbf{u} \times \mathbf{v}\| = \sqrt{64 + 4 + 16} = \sqrt{84} = 2\sqrt{21},$$

$\langle \frac{-8}{2\sqrt{21}}, \frac{-2}{2\sqrt{21}}, \frac{4}{2\sqrt{21}} \rangle = \langle -\frac{4}{\sqrt{21}}, -\frac{1}{\sqrt{21}}, \frac{2}{\sqrt{21}} \rangle$ is a unit vector.

We check that this vector is orthogonal to both $\mathbf{u}$ and $\mathbf{v}$ by means of the dot product:

$$\langle 1, 2, 3 \rangle \cdot \left\langle -\frac{4}{\sqrt{21}}, -\frac{1}{\sqrt{21}}, \frac{2}{\sqrt{21}} \right\rangle = -\frac{4}{\sqrt{21}} - \frac{2}{\sqrt{21}} + \frac{6}{\sqrt{21}} = 0,$$

and

$$\langle -2, 0, -4 \rangle \left\langle -\frac{4}{\sqrt{21}}, -\frac{1}{\sqrt{21}}, \frac{2}{\sqrt{21}} \right\rangle = \frac{8}{\sqrt{21}} + 0 - \frac{8}{\sqrt{21}} = 0.$$

The additive inverse of this vector is $\langle \frac{4}{\sqrt{21}}, \frac{1}{\sqrt{21}}, -\frac{2}{\sqrt{21}} \rangle$, and this is the only other unit vector perpendicular to both $\mathbf{u}$ and $\mathbf{v}$. ∎

The right-hand rule for cross products of vectors is dependent on our choice of a *right-handed coordinate system* for $\mathbb{R}^3$. A different labelling for the coordinate axes can result in a left-handed coordinate system, in the sense that we must use a left-hand rule for cross products. Figure 12.23 illustrates a left-handed coordinate system.

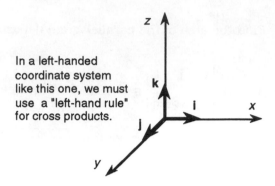

In a left-handed coordinate system like this one, we must use a "left-hand rule" for cross products.

Figure 12.23 A left-handed coordinate system.

Note that the right-hand rule applied to this coordinate system would result in $\mathbf{i} \times \mathbf{j} = -\mathbf{k}$, while the left-hand rule furnishes the usual result $\mathbf{i} \times \mathbf{j} = \mathbf{k}$.

Triple scalar product

Given three vectors $\mathbf{u}$, $\mathbf{v}$, and $\mathbf{w}$ in $\mathbb{R}^3$, we can combine the dot and cross products to obtain what is known as the **triple scalar product** $\mathbf{u} \cdot (\mathbf{v} \times \mathbf{w})$. Computationally, the triple scalar product is simply the *determinant* of the 3×3 matrix obtained by using the components of $\mathbf{u}$, $\mathbf{v}$, and $\mathbf{w}$ as the rows of the matrix:

$$\mathbf{u} \cdot (\mathbf{v} \times \mathbf{w}) = \begin{vmatrix} u_1 & u_2 & u_3 \\ v_1 & v_2 & v_3 \\ w_1 & w_2 & w_3 \end{vmatrix}.$$

Note that the final result of this computation is a *scalar*.

Geometrically, the absolute value of the triple scalar product gives us the volume of the parallelepiped (slanted box) determined by the three vectors $\mathbf{u}$, $\mathbf{v}$, and $\mathbf{w}$ (see Figure 12.24).

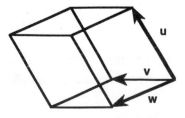

Figure 12.24 The volume of "box" is $|\mathbf{u} \cdot (\mathbf{v} \times \mathbf{w})|$.

The triple product is 0 exactly when the three vectors all lie in the same plane (resulting in a "flat" box). Can you see why?

Parametric equations of lines

You are already familiar with the **slope-intercept equation**

$$y = mx + b$$

and the **general equation** for a line in the Cartesian plane:

$$ax + by + c = 0.$$

A line can also be represented by parametric equations expressing both x and y in terms of a parameter t.

EXAMPLE 36 Graph and find both the slope-intercept and the general equations for the line represented by the parametric equations

$$x = 2t + 1,$$

$$y = 3t - 2.$$

Solution We can find two points on this line very quickly by substituting $t = 0$ and $t = 1$. The corresponding points are $(1, -2)$ and $(3, 1)$. Figure 12.25 shows the graph of the line.

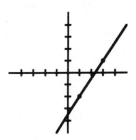

Figure 12.25 Graph of line represented by $x = 2t + 1$ and $y = 3t - 2$.

We can compute the slope of the line as $m = \frac{1 - (-2)}{3 - 1} = \frac{3}{2}$ and see that the y-intercept is $b = -\frac{7}{2}$. The slope-intercept form is

$$y = \frac{3}{2}x - \frac{7}{2},$$

and the general equation is

$$3x - 2y - 7 = 0.$$

You know that a line is determined by its slope and one point on the line. In terms of vectors, we can think of a line as being determined by a *direction vector* **v** and an initial *position vector* **u**. The points of the line are determined by the terminal points (heads) of all vectors of the form

$$\mathbf{u} + t\mathbf{v}$$

where t is any real number. This is called the **vector parametric form** of the line.

EXAMPLE 37 Find a position vector **u** and a direction vector **v** for the line of the previous example, and express it in vector parametric form.

Solution The initial position vector can be obtained by substituting $t = 0$:

$$\mathbf{u} = \langle 1, -2 \rangle,$$

and a direction vector is given by the coefficients of t in the equations for x and y:

$$\mathbf{v} = \langle 2, 3 \rangle.$$

The vector parametric form of the line is

$$\mathbf{u} + t\mathbf{v} = \langle 1, -2 \rangle + t \langle 2, 3 \rangle,$$

where t is any real number. Notice that the components of the vector parametric form are simply the original parametric equations:

$$\langle x(t), y(t) \rangle = \langle 1 + 2t, -2 + 3t \rangle.$$

∎

EXAMPLE 38 Find a vector parametric form for the line passing through the point $(-3, 5)$ and with direction vector $\langle 2, -1 \rangle$.

Solution $\langle x(t), y(t) \rangle = \langle -3 + 2t, 5 - t \rangle.$ ∎

EXAMPLE 39 Find a vector parametric form for the line passing through the points $P = (2, 2)$ and $Q = (-2, 3)$.

Solution We can determine a direction vector by taking

$$\overrightarrow{PQ} = \langle -2 - 2, \ 3 - 2 \rangle = \langle -4, 1 \rangle.$$

Either point can serve as the initial position vector. If we use $P = (2, 2)$, we have

$$\langle x(t), y(t) \rangle = \langle 2 - 4t, 2 + t \rangle.$$

We can check that $\langle x(1), y(1) \rangle = \langle -2, 3 \rangle$ corresponds to point Q. ■

Vector parametric equations for lines in 3-dimensional space can be found in exactly the same way. All we need is an initial position vector and a direction vector.

EXAMPLE 40 Find a vector parametric equation for the line that passes through the point $(-1, -3, 0)$ and with direction vector $\langle 5, -1, 2 \rangle$.

Solution $\langle x(t), y(t), z(t) \rangle = \langle -1 + 5t, -3 - t, 2t \rangle.$ ■

EXAMPLE 41 Find a vector parametric form for the line passing through the two points $P = (1, 1, 1)$ and $Q = (-1, -2, -3)$.

Solution Using $\overrightarrow{PQ} = \langle -2, -3, -4 \rangle$ as the direction vector and $\langle 1, 1, 1 \rangle$ as the initial position vector, we have

$$\langle x(t), y(t), z(t) \rangle = \langle 1 - 2t, 1 - 3t, 1 - 4t \rangle.$$

Note that $\langle x(1), y(1), z(1) \rangle = \langle -1, -2, -3 \rangle$ corresponds to Q. ■

Equations for a plane

A line in space is 1-dimensional in the sense that all three of its components can be expressed in terms of a single parameter t. A plane in space is 2-dimensional, so we will need two parameters s and t to adequately describe it.

Just as a line is determined by an initial position vector and a direction vector, a plane is determined by an initial position vector $\mathbf{u}$ and *two* nonparallel direction vectors $\mathbf{v}$ and $\mathbf{w}$ that lie in the plane. The vector parametric form of the plane is

$$\mathbf{u} + s\mathbf{v} + t\mathbf{w}.$$

As the values of s and t range over the set of real numbers, this form specifies the points in the plane.

EXAMPLE 42 Find the vector parametric form for the plane containing the point $(1, 2, 3)$ and with direction vectors $\mathbf{v} = \mathbf{i} + \mathbf{j}$ and $\mathbf{w} = \mathbf{j} - \mathbf{k}$.

Solution $\mathbf{u} = \langle 1, 2, 3 \rangle$ is the initial position vector, $\mathbf{v} = \langle 1, 1, 0 \rangle$ and $\mathbf{w} = \langle 0, 1, -1 \rangle$ are the direction vectors. We can write the vector parametric form of the plane as

$$\langle x(s,t), y(s,t), z(s,t) \rangle = \langle 1 + s, 2 + s + t, 3 - t \rangle.$$

We can locate three noncollinear points in the plane by substituting $(0, 0)$, $(1, 0)$, and $(0, 1)$ for (s, t) in the vector parametric form:

$$\langle x(0,0), y(0,0), z(0,0) \rangle = \langle 1, 2, 3 \rangle$$

$$\langle x(1,0), y(1,0), z(1,0) \rangle = \langle 2, 3, 3 \rangle$$

$$\langle x(0,1), y(0,1), z(0,1) \rangle = \langle 1, 3, 2 \rangle.$$

∎

A plane in space also has a **general equation** of the form

$$ax + by + cz + d = 0.$$

The points (x, y, z) in the plane are precisely those that satisfy this equation.

EXAMPLE 43 $x + 2y - 3z - 6 = 0$ is the general equation of a plane. Three points that satisfy the equation are $(6,0,0)$, $(0,3,0)$, and $(0,0,-2)$. ∎

Remarkably, the vector $\langle a,b,c \rangle$ must be orthogonal to every vector that lies in this plane! (In the previous example, this means that the vector $\langle 1,2,-3 \rangle$ is orthogonal to any vector lying in the plane.) The general reasoning goes like this: If we pick any two points in the plane $ax + by + cz + d = 0$, say $P = (x_1, y_1, z_1)$ and $Q = (x_2, y_2, z_2)$, then we know that

$$ax_1 + by_1 + cz_1 + d = 0$$

$$ax_2 + by_2 + cz_2 + d = 0.$$

If we subtract the first equation from the second, we obtain

$$a(x_2 - x_1) + b(y_2 - y_1) + c(z_2 - z_1) = 0.$$

Notice that the left-hand side of this equation is the dot product of $\langle a,b,c \rangle$ and $\overrightarrow{PQ} = \langle (x_2 - x_1), (y_2 - y_1), (z_2 - z_1) \rangle$. Since this dot product is zero, the two vectors must be orthogonal.

The vector $\langle a,b,c \rangle$ is said to be a **normal vector** to the plane. This observation gives a way of finding the general equation of a plane, provided we can determine a normal vector and a point in the plane.

EXAMPLE 44   Find the general equation of the plane containing the point $(-2,3,-5)$ and with normal vector $\langle 1,-3,2 \rangle$.

Solution   From the normal vector, we can see that the general equation must be of the form

$$x - 3y + 2z + d = 0.$$

We can determine d by substituting the point and solving

$$(-2) - 3(3) + 2(-5) + d = 0.$$

This tells us that $d = 21$, and the general equation of the plane is

$$x - 3y + 2z + 21 = 0.$$

∎

EXERCISES

For exercises 1-15, refer to the following vectors:

$$\mathbf{a} = 3\mathbf{i} + \mathbf{j} - \mathbf{k}$$

$$\mathbf{b} = 2\mathbf{i} - \mathbf{j} + \mathbf{k}$$

$$\mathbf{c} = \mathbf{i} + 2\mathbf{j} + 3\mathbf{k}$$

1. Find $\mathbf{b} \times \mathbf{a}$.

2. Find $\mathbf{c} \times \mathbf{a}$.

3. Find $(\mathbf{b} - \mathbf{a}) \times (\mathbf{c} - \mathbf{a})$.

4. Find a unit vector perpendicular to both $\mathbf{a}$ and $\mathbf{c}$.

5. Find the vector parametric form of the line containing the origin and with direction vector $\mathbf{a}$.

6. Find a vector parametric form for the line determined by the two points $(1, 4, 7)$ and $(-2, 1, -6)$.

7. If $\mathbf{b}$ and $\mathbf{c}$ both have their initial points at the origin, find the vector parametric form of the line passing through their terminal points.

8. If $\mathbf{a}$, $\mathbf{b}$, and $\mathbf{c}$ all have their initial points at the origin, find both the vector parametric form and the general equation of the plane passing through their terminal points.

9. Find the area of the parallelogram determined by $\mathbf{b}$ and $\mathbf{c}$.

10. Find the volume of the parallelepiped determined by all three vectors.

11. Show that these three vectors do not lie in the same plane. (Hint: consider the triple scalar product.)

12. Find a vector parametric form for the plane determined by the three points $(1, 4, 7)$, $(-2, 1, -6)$ and $(3, 0, -5)$.

13. Find a general equation for the plane determined by the three points $(1, 4, 7)$, $(-2, 1, -6)$ and $(3, 0, -5)$.

14. Determine the angle between $\mathbf{a}$ and $\mathbf{c}$ using the dot product.

15. Determine the sine of the angle between $\mathbf{b}$ and $\mathbf{c}$ using the cross product. What are the two possible angles (between 0 and π) that have this sine value. Now determine which is the actual angle between $\mathbf{b}$ and $\mathbf{c}$.

For exercises 16-26, answer "TRUE" or "FALSE."

16. For any two vectors **a**, **b** in $\mathbb{R}^3$, $\mathbf{a} \cdot \mathbf{b} = \mathbf{b} \cdot \mathbf{a}$.

17. For any two vectors **a**, **b** in $\mathbb{R}^3$, $\mathbf{a} \times \mathbf{b} = \mathbf{b} \times \mathbf{a}$.

18. For any vector **a** in $\mathbb{R}^3$, $\mathbf{a} \times \mathbf{a} = 0$.

19. If $\mathbf{a} \cdot \mathbf{b} = 0$, then either $\mathbf{a} = 0$ or $\mathbf{b} = 0$.

20. For any two vectors **a**, **b** in $\mathbb{R}^3$, $\mathbf{a} \times \mathbf{b} = \|\mathbf{a}\| \, \|\mathbf{b}\| \sin \theta$.

21. If $\mathbf{a} \times \mathbf{b} = 0$, then either $\mathbf{a} = 0$ or $\mathbf{b} = 0$.

22. For any three vectors in $\mathbb{R}^3$, **a**, **b**, and **c** we have $\mathbf{a} \cdot (\mathbf{b} \times \mathbf{c}) = (\mathbf{a} \cdot \mathbf{b}) \times (\mathbf{a} \cdot \mathbf{c})$.

23. $\mathbf{i} \cdot (\mathbf{j} \times \mathbf{k}) = 1$.

24. For any two vectors in $\mathbb{R}^3$, **a** and **b**, $|\mathbf{a} \cdot \mathbf{b}| \leq \|\mathbf{a}\| \, \|\mathbf{b}\|$.

25. If **a** and **b** are orthogonal, then $\mathbf{a} \times \mathbf{b} = 0$.

26. If **a** and **b** lie in the same plane, then $\mathbf{a} \cdot \mathbf{b} = 0$.

13

Calculus of Curves

A **vector-valued function** produces a vector as an output. In the last chapter, we discussed a very special type of vector function called a *linear transformation*. Recall that a linear transformation takes one vector as an input and produces another vector as an output.

In this chapter, we discuss vector-valued functions that take a *single real number as an input* and produce a *vector for an output*. We'll often denote the independent variable as t (to suggest *time*), and the output vector $\mathbf{r}(t)$. Note the use of boldface:

$$t \quad \text{represents a } scalar \text{ input}$$

and

$$\mathbf{r} \quad \text{is a function producing a } vector \text{ output.}$$

You can think of these vector-valued functions as describing some vector quantity that changes with time. We've already used calculus to study real-valued functions, and now we'll extend the use of derivatives and integrals to study these new vector-valued functions.

13.1 POSITION FUNCTIONS AND GRAPHICAL REPRESENTATIONS

If $\mathbf{r}$ is a function that takes a real number as input from a domain D, and whose outputs are 2-dimensional vectors, we write

$$\mathbf{r} : D \longrightarrow \mathbb{R}^2.$$

The assignment process defining this function has two components corresponding to the two **coordinate functions** x and y, each of which is a real-valued function:

$$\mathbf{r} : t \longmapsto \mathbf{r}(t) = \langle x(t), y(t) \rangle = x(t)\mathbf{i} + y(t)\mathbf{j}.$$

Here, the single independent variable is t, while $x(t)$ and $y(t)$ represent the two real number components of $\mathbf{r}(t)$.

Similarly, if the outputs of a function $\mathbf{r}$ are 3-dimensional, we could write

$$\mathbf{r} : D \longrightarrow \mathbb{R}^3,$$

and the assignment process defining $\mathbf{r}$ has three components:

$$\mathbf{r} : t \longmapsto \mathbf{r}(t) = \langle x(t), y(t), z(t) \rangle = x(t)\mathbf{i} + y(t)\mathbf{j} + z(t)\mathbf{k}.$$

In general, if the function $\mathbf{r}$ has real numbers for inputs, and n-dimensional vectors for outputs, we could write

$$\mathbf{r} : D \longrightarrow \mathbb{R}^n$$

and

$$\mathbf{r} : t \longmapsto \mathbf{r}(t) = \langle x_1(t), x_2(t), \ldots, x_n(t) \rangle.$$

If the outputs of a vector-valued function are 2- or 3-dimensional, then we can think of them as designating positions in a plane or in space, respectively. Here's how: if we represent a vector with its initial point (tail) at the origin, then the vector's components are exactly described by the coordinates of its terminal point (head). As the input t takes on values in the domain, the terminal points determined by $\mathbf{r}(t)$ trace out a curve in the plane or space (see Figure 13.1).

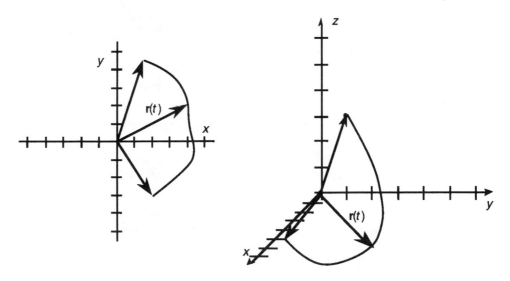

Figure 13.1 The terminal point of $\mathbf{r}(t)$ traces out a curve.

This curve gives us a way to visualize the behavior of the function. Indeed, much of the language we use to describe these functions is intimately tied to this visual idea of a curve.

If t is thought of as representing *time*, then we can think of the curve as the *path* of a moving object. This physical interpretation is also the source of much of the terminology we'll use.

EXAMPLE 1   The line in the plane passing through the point $(1, -2)$ and having direction $\langle 2, 3 \rangle$ can be parametrized by the equations

$$x = 2t + 1$$

$$y = 3t - 2$$

We can also represent this line as the curve defined by position function $\mathbf{r} : \mathbb{R} \longrightarrow \mathbb{R}^2$, where

$$\mathbf{r}(t) = \langle 2t + 1, 3t - 2 \rangle = (2t + 1)\mathbf{i} + (3t - 2)\mathbf{j}.$$

Figure 13.2 shows the *image curve* of $\mathbf{r}$, with the points corresponding to $\mathbf{r}(0)$ and $\mathbf{r}(1)$ indicated. We can say that the position function $\mathbf{r}$ **parametrizes** this line.   ■

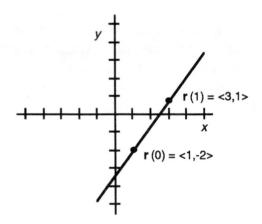

Figure 13.2 Line parametrized by $\mathbf{r}(t) = \langle 2t + 1, 3t - 2 \rangle$.

EXAMPLE 2   The circle of radius 1 centered at the origin has parametric equations

$$x = \cos t$$

$$y = \sin t$$

for $0 \le t \le 2\pi$.

The corresponding position function is $\mathbf{r} : [0, 2\pi] \longrightarrow \mathbb{R}^2$, where

$$\mathbf{r}(t) = \langle \cos t, \sin t \rangle = (\cos t)\mathbf{i} + (\sin t)\mathbf{j}.$$

See Figure 13.3.

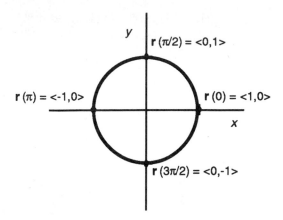

Figure 13.3 Circle parametrized by $\mathbf{r}(t) = (\cos t)\mathbf{i} + (\sin t)\mathbf{j}$.

The same circle is traced out by the function $\mathbf{u} : [0, 2\pi] \longrightarrow \mathbb{R}^2$, where

$$\mathbf{r}(t) = \langle \sin t, \cos t \rangle = (\sin t)\mathbf{i} + (\cos t)\mathbf{j},$$

and also by the function $\mathbf{w} : [0, \pi] \longrightarrow \mathbb{R}^2$, where

$$\mathbf{w}(t) = \langle \cos 2t, \sin 2t \rangle = (\cos 2t)\mathbf{i} + (\sin 2t)\mathbf{j}.$$

☞ **Note that a single curve can be parametrized in many different ways. That is,** *different* **position functions can have the** *same* **image curve.**

Curves in space can also be parametrized by position functions.

EXAMPLE 3 The space curve with parametric equations

$$x = 1 - t$$
$$y = t^2 + 1$$
$$z = 2t^3/3 + 1$$

for $-1 \leq t \leq 1$ has position function

$$\mathbf{r} : [-1, 1] \longrightarrow \mathbb{R}^3,$$

where

$$\mathbf{r}(t) = \langle 1 - t, t^2 + 1, \frac{2t^3}{3} + 1 \rangle = (1 - t)\mathbf{i} + (t^2 + 1)\mathbf{j} + (\frac{2t^3}{3} + 1)\mathbf{k}.$$

Figure 13.4 shows the curve with the points corresponding to

$\mathbf{r}(-1)$, $\mathbf{r}(-1/2)$, $\mathbf{r}(0)$, $\mathbf{r}(1/2)$, and $\mathbf{r}(1)$.

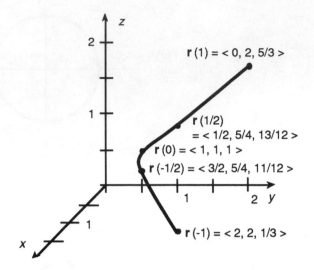

Figure 13.4 The space curve defined by $\mathbf{r}(t) = (1-t)\mathbf{i} + (t^2+1)\mathbf{j} + (\frac{2t^3}{3}+1)\mathbf{k}$ for $t \in [-1, 1]$.

Here's another example that we'll use for an illustration.

EXAMPLE 4   The space curve with parametric equations

$$x = \cos t$$

$$y = \sin t$$

$$z = t$$

for $0 \le t \le 2\pi$ has position function

$$\mathbf{r} : [0, 2\pi] \longrightarrow \mathbb{R}^3,$$

where

$$\mathbf{r}(t) = \langle \cos t, \sin t, t \rangle = (\cos t)\mathbf{i} + (\sin t)\mathbf{j} + (t)\mathbf{k}.$$

Figure 13.5 shows this spiral or *helix* with the points corresponding to $\mathbf{r}(0)$, $\mathbf{r}(\pi/2)$, $\mathbf{r}(\pi)$, $\mathbf{r}(3\pi/2)$, and $\mathbf{r}(2\pi)$.   ■

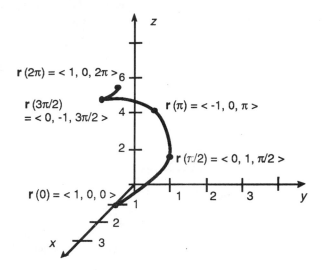

Figure 13.5 The space curve defined by $\mathbf{r}(t) = (\cos t)\mathbf{i} + (\sin t)\mathbf{j} + (t)\mathbf{k}$ for $t \in [0, 2\pi]$.

Figures 13.4 and 13.5 illustrate the importance of labelling of points in a two-dimensional representation of a curve in three-dimensional space. Without them, it is quite impossible to visualize exactly where these curves lie in space from the pictures alone.

Using machine graphics to visualize space curves

Even if you have a machine grapher that will plot a parametrized space curve quickly on a computer or calculator screen, this "flat" representation can leave much to be desired.

There is a way that you can use your graphing calculator or computer software to obtain additional information about the position of a curve in space by means of *multiple views*. In many design specifications for objects, three **principal views** are given. These usually are taken to be the *front view*, the *right-side view*, and the *top view*. In terms of our Cartesian coordinate system for space, we can think of these as corresponding to views obtained by looking toward the origin to see the xz-plane, the yz-plane, and the xy-plane, respectively, as shown in Figure 13.6.

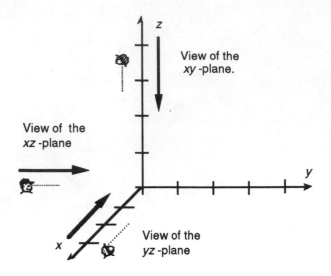

Figure 13.6 Three views of space.

You can obtain these three views easily, provided you have a two-dimensional parametric plotter, as found on many graphing calculators and graphing software. Just plot each possible pair out of the three parametric equations that define the coordinate functions. More specifically, if the space curve has position function

$$\mathbf{r}(t) = \langle f(t), g(t), h(t) \rangle,$$

then:

        1) Plot horizontal: $f(t)$ **Plot vertical:** $g(t)$

to obtain the view of the xy-plane;

        2) Plot horizontal: $f(t)$ **Plot vertical:** $h(t)$

to obtain the view of the xz-plane; and

        3) Plot horizontal: $g(t)$ **Plot vertical:** $h(t)$

to obtain the view of the yz-plane.

EXAMPLE 5   Use a two-dimensional parametric plotter to obtain three views of the space curve in Example 3.

Solution   Figure 13.7 illustrates the desired views of our first space curve example.

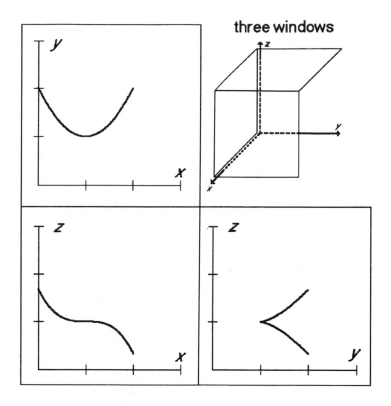

Figure 13.7 Three views of the curve defined by $\mathbf{r}(t) = (1 - t)\mathbf{i} + (t^2 + 1)\mathbf{j} + (\frac{2t^3}{3} + 1)\mathbf{k}$ for $t \in [-1,$

Notice the relationship between the three views as shown in Figure 13.7. You can think of each view as the picture seen through a clear window placed parallel to each of the three coordinate planes:

The xy-view (top) corresponds to the parametric plot of

$$x = 1 - t \quad \text{and} \quad y = t^2 + 1.$$

The xz-view corresponds to the parametric plot of

$$x = 1 - t \quad \text{and} \quad z = 2t^3/3 + 1.$$

The yz-view corresponds to the parametric plot of

$$y = t^2 + 1 \quad \text{and} \quad z = 2t^3/3 + 1.$$

■

EXAMPLE 6   Use a two-dimensional parametric plotter to obtain three views of the space curve in Example 4.

Solution   Figure 13.8 shows the three views of the helix.

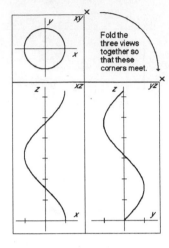

Figure 13.8 The space curve defined by $\mathbf{r}(t) = (\cos t)\mathbf{i} + (\sin t)\mathbf{j} + (t)\mathbf{k}$ for $t \in [0, 2\pi]$.

■

If you cut out the three views exactly as shown, and then folded them so that the corners indicated in Figure 13.8 aligned, then you have a three-dimensional model showing all three views (see Figure 13.9.)

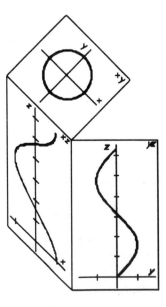

Figure 13.9 Folding three views to make a model of the helix.

Parametrized curves have orientation

Curves parametrized by position functions have an orientation given by "increasing t".

EXAMPLE 7 The orientation of the parametrized line in Example 1 is from lower left to upper right as t increases. The orientation of the parametrized unit circle in Example 2 is counterclockwise as t increases. If, instead, we parametrized the unit circle with the position function

$$\mathbf{r}(t) = \langle \cos(-t), \sin(-t) \rangle$$

for $0 \le t \le 2\pi$, then the orientation is reversed to be clockwise. The orientations of the parametrized space curves of Example 3 and Example 4 are upwards as t increases. ∎

Since machine graphics programs generally plot parametric curves dynamically as t increases, we are able to witness the orientation as we watch the plot progress.

EXERCISES

For each pair of parametric equations in exercises 1-8, graph the image curve of the indicated position function for the interval $0 \le t \le 2\pi$.

1. $\mathbf{r}(t) = 2\cos(t)\,\mathbf{i} + 3\sin(t)\,\mathbf{j}$.

2. $\mathbf{r}(t) = 2\cos(t)\,\mathbf{i} - 3\sin(t)\,\mathbf{j}$.

3. $\mathbf{r}(t) = -2\cos(t)\,\mathbf{i} + 3\sin(t)\,\mathbf{j}$.

4. $\mathbf{r}(t) = -2\cos(t)\,\mathbf{i} - 3\sin(t)\,\mathbf{j}$.

5. $\mathbf{r}(t) = 3\cos(t)\,\mathbf{i} + 2\sin(t)\,\mathbf{j}$.

6. $\mathbf{r}(t) = 3\cos(t)\,\mathbf{i} - 2\sin(t)\,\mathbf{j}$.

7. $\mathbf{r}(t) = -3\cos(t)\,\mathbf{i} + 2\sin(t)\,\mathbf{j}$.

8. $\mathbf{r}(t) = -3\cos(t)\,\mathbf{i} - 2\sin(t)\,\mathbf{j}$.

9. Describe the orientation (clockwise or counterclockwise) of each of the curves described by the position functions in exercises 1-8. What type of conic section is traced out by each of these position functions?

10. Describe as completely as possible, including orientation, the curve traced out by

$$\mathbf{r}(t) = a\cos(t)\mathbf{i} + b\sin(t)\mathbf{j}$$

for $0 \le t \le 2\pi$. Be specific regarding the effects of a and b on the shape and orientation of the curve.

For each pair of parametric equations in exercises 11-18, graph the curve for the interval $-2 \le t \le 2$.

11. $\mathbf{r}(t) = 2\cosh(t)\,\mathbf{i} + 3\sinh(t)\,\mathbf{j}$.

12. $\mathbf{r}(t) = 2\cosh(t)\,\mathbf{i} - 3\sinh(t)\,\mathbf{j}$.

13. $\mathbf{r}(t) = -2\cosh(t)\,\mathbf{i} + 3\sinh(t)\,\mathbf{j}$.

14. $\mathbf{r}(t) = -2\cosh(t)\,\mathbf{i} - 3\sinh(t)\,\mathbf{j}$.

15. $\mathbf{r}(t) = 3\cosh(t)\,\mathbf{i} + 2\sinh(t)\,\mathbf{j}$.

16. $\mathbf{r}(t) = 3\cosh(t)\,\mathbf{i} - 2\sinh(t)\,\mathbf{j}$.

17. $\mathbf{r}(t) = -3\cosh(t)\,\mathbf{i} + 2\sinh(t)\,\mathbf{j}$.

18. $\mathbf{r}(t) = -3\cosh(t)\,\mathbf{i} - 2\sinh(t)\,\mathbf{j}$.

19.   Describe the orientation of each of the curves described by the position functions in exercises 11-18. What type of conic section is traced out by each of these position functions?

20.   Describe as completely as possible, including orientation, the curve traced out by

$$\mathbf{r}(t) = a\cosh(t) + b\sinh(t)$$

for $-2 \le t \le 2$. Be specific regarding the effects of a and b on the shape and orientation of the curve.

For exercises 21-30, plot each of the three principal views of the space curve described by the indicated position function for $-3 \le t \le 3$. Then use your three views to sketch the curve on a 3-dimensional Cartesian coordinate system.

21. $\mathbf{r}(t) = (2t + 3)\,\mathbf{i} + (3t + 2)\,\mathbf{j} + (5 - t)\,\mathbf{k}$.

22. $\mathbf{r}(t) = (\sqrt{9 - t^2})\,\mathbf{i} + 2\mathbf{j} - t\,\mathbf{k}$.

23. $\mathbf{r}(t) = 2\mathbf{i} + (\sqrt{9 - t^2})\,\mathbf{j} + t\,\mathbf{k}$.

24. $\mathbf{r}(t) = t\,\mathbf{i} + 2\mathbf{j} - (\sqrt{9 - t^2})\,\mathbf{k}$.

25. $\mathbf{r}(t) = t\cos(t)\,\mathbf{i} + t\sin(t)\mathbf{j} - e^t\,\mathbf{k}$.

26. $\mathbf{r}(t) = t\,\mathbf{i} + (t^2/2)\mathbf{j} + (t^3/3)\,\mathbf{k}$.

27. $\mathbf{r}(t) = t\,\mathbf{i} + (1 - \cos t)\,\mathbf{j} + (\sin t)\,\mathbf{k}$.

28. $\mathbf{r}(t) = \left(\frac{1}{1+t^2}\right)\mathbf{i} + (\arctan t)\,\mathbf{j} + t^2\,\mathbf{k}$.

29. $\mathbf{r}(t) = 3\cos(2t)\,\mathbf{i} + 3\sin(2t)\mathbf{j} + 4t\,\mathbf{k}$.

30. $\mathbf{r}(t) = 2\cos(3t)\,\mathbf{i} + 2\sin(3t)\mathbf{j} + (t/3)\,\mathbf{k}$.

13.2 DIFFERENTIAL CALCULUS OF CURVES

The derivatives of a position function **r** have some natural geometric and physical interpretations when we imagine the *image curve* of **r** as describing the path of a moving object.

Limits of position functions

Derivatives are defined in terms of limits of difference quotients, so first we need to discuss what it means to find the limit of a vector-valued function.

Definition 1

The **limit** of the position vectors $\mathbf{r}(t)$ as $t \to a$ is

$$\lim_{t \to a} \mathbf{r}(t) = \mathbf{L},$$

provided that

$$\lim_{t \to a} \|\mathbf{r}(t) - \mathbf{L}\| = 0.$$

In other words, if the difference between the vectors $\mathbf{r}(t)$ and $\mathbf{L}$ approaches 0 as $t \to a$, then $\mathbf{L}$ is the limit vector. If no vector satisfies this requirement, we say that the limit *does not exist.*

The only way that we can have $\mathbf{r}(t) \to \mathbf{L}$ as $t \to a$ is if each component of $\mathbf{r}(t)$ approaches the corresponding component of $\mathbf{L}$. In other words, we can evaluate limits "component-wise."

EXAMPLE 8   Find $\lim_{t \to 0} \mathbf{r}(t)$, if $\mathbf{r}(t) = \langle \dfrac{\sin t}{t}, \quad t^2, \quad 5 - t \rangle$ for $t \neq 0$, and $\mathbf{r}(0) = 5\mathbf{k}$.

Solution   Note that the actual value $\mathbf{r}(0) = 5\mathbf{k}$ is ignored with regards to finding the limit of $\mathbf{r}(t)$ as $t \to 0$. Since

$$\mathbf{r}(t) = \left\langle \frac{\sin t}{t}, \quad t^2, \quad 5 - t \right\rangle \qquad \text{for } t \neq 0,$$

we have

$$\lim_{t \to 0} \mathbf{r}(t) = \langle \lim_{t \to 0} \frac{\sin t}{t}, \quad \lim_{t \to 0} t^2, \quad \lim_{t \to 0} 5 - t \rangle = \langle 1, 0, 5 \rangle.$$

L'Hôpital's Rule was used in evaluating the limit of the first component. ∎

Continuity of position functions

Continuity of position functions is defined using exactly the same criteria as for real-valued functions.

Definition 2

> The position vector **r** is **continuous** at $t = a$ provided that
>
> $$\lim_{t \to a} \mathbf{r}(t) = \mathbf{r}(a).$$
>
> If **r** is continuous at every value in its domain, then we say that **r** is a **continuous position function**.

EXAMPLE 9   The position function **r** of the previous example is *not continuous* at $t = 0$, because

$$\lim_{t \to 0} \mathbf{r}(t) = \langle 1, 0, 5 \rangle \neq \langle 0, 0, 5 \rangle = \mathbf{r}(0).$$

The function **r** is continuous at every other real value t. For instance, at $t = \pi$ we have

$$\lim_{t \to \pi} \mathbf{r}(t) = \langle \lim_{t \to \pi} \frac{\sin t}{t}, \quad \lim_{t \to \pi} t^2, \quad \lim_{t \to \pi} 5 - t \rangle = \langle 0, \pi^2, 5 - \pi \rangle = \mathbf{r}(\pi).$$

■

Derivatives of vector position functions

The derivative of a position function is defined as the limit of a difference quotient.

Definition 3

> The position function **r** is said to be **differentiable** at $t = a$ if and only if the limit
>
> $$\lim_{\Delta t \to 0} \frac{\mathbf{r}(a + \Delta t) - \mathbf{r}(a)}{\Delta t}$$
>
> exists. This value is called the **derivative** of **r** at $t = a$, and is denoted
>
> $$\mathbf{r}'(a) \qquad \text{or} \qquad \frac{d\mathbf{r}}{dt}\bigg|_{t=a}.$$
>
> If the limit does not exist, then we say that **r** is *not differentiable*.

Let's make a few comments about this definition. Suppose we have a position function

$$\mathbf{r} : \mathbb{R} \longrightarrow \mathbb{R}^3$$

$$\mathbf{r} : t \longmapsto \langle x(t), y(t), z(t) \rangle$$

and we examine the difference quotient

$$\frac{\mathbf{r}(a + \Delta t) - \mathbf{r}(a)}{\Delta t}.$$

First, note that the numerator of the difference quotient is a difference of two *vectors* $\mathbf{r}(a + \Delta t)$ and $\mathbf{r}(a)$ while the denominator is a scalar Δt. Since subtraction is performed component-wise, and scalar division is distributed to each of the components, this difference quotient can be written

$$\langle \frac{x(a + \Delta t) - x(a)}{\Delta t}, \quad \frac{y(a + \Delta t) - y(a)}{\Delta t}, \quad \frac{z(a + \Delta t) - z(a)}{\Delta t} \rangle.$$

Given the fact that limits of vector functions are also calculated component-wise, we can see that the derivative

$$\mathbf{r}'(a) = \lim_{\Delta t \to 0} \langle \frac{x(a + \Delta t) - x(a)}{\Delta t}, \quad \frac{y(a + \Delta t) - y(a)}{\Delta t}, \quad \frac{z(a + \Delta t) - z(a)}{\Delta t} \rangle$$

$$= \langle x'(a), y'(a), z'(a) \rangle.$$

Hence, to find the derivative $\mathbf{r}'(a)$, we simply differentiate each component.

EXAMPLE 10   If $\mathbf{r}(t) = \langle 2t + 1, 3t - 2 \rangle = (2t + 1)\mathbf{i} + (3t - 2)\mathbf{j}$, then

$$\frac{d\mathbf{r}}{dt} = \mathbf{r}'(t) = \langle 2, 3 \rangle$$

for all real values of t. Notice that the derivative gives us the direction vector for the line parametrized by this position function.   ■

EXAMPLE 11   If $\mathbf{r}(t) = \langle \cos t, \sin t \rangle = (\cos t)\mathbf{i} + (\sin t)\mathbf{j}$, for $0 \leq t \leq 2\pi$, then

$$\frac{d\mathbf{r}}{dt} = \mathbf{r}'(t) = \langle -\sin t, \cos t \rangle = (-\sin t)\mathbf{i} + (\cos t)\mathbf{j}$$

for $0 < t < 2\pi$. (Note that we do not assign a derivative value at the two endpoints.) For instance,

$$\mathbf{r}'(\pi/2) = \langle -\sin \pi/2, \cos \pi/2 \rangle = \langle -1, 0 \rangle,$$

and

$$\mathbf{r}'(\pi) = \langle -\sin \pi, \cos \pi \rangle = \langle 0, -1 \rangle.$$

In each instance, the vector value of the derivative points in the direction of the parametrized circle at that point. If you imagine an object constrained to travel in the circular path defined by this position function (like a swinging ball tied to the end of a rope), then the derivative vector gives us the direction that the object would travel if freed from the constraint at that point (the rope breaks). One way to illustrate this is to place the derivative vector so that its initial point coincides with the point corresponding to the value of the position vector (see Figure 13.10). ■

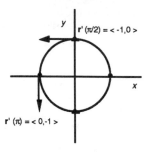

Figure 13.10 The derivative vector points in the direction of the curve.

EXAMPLE 12   Suppose

$$\mathbf{r} : [0, 2\pi] \longrightarrow \mathbb{R}^3,$$

where

$$\mathbf{r}(t) = \langle \cos t, \sin t, t \rangle = (\cos t)\mathbf{i} + (\sin t)\mathbf{j} + t\,\mathbf{k}.$$

Then the derivative is

$$\frac{d\mathbf{r}}{dt} = \mathbf{r}'(t) = \langle -\sin t, \cos t, 1 \rangle.$$

■

Graphical and physical interpretations of the derivative

If we imagine the image curve of $\mathbf{r}$ as representing the path of an object, then the vector represented by the difference

$$\mathbf{r}(a + \Delta t) - \mathbf{r}(a)$$

tells us the *displacement* (how far and in what direction the object moved) over the time interval of length Δt. If the object does not change direction suddenly at $t = a$, then for very small positive values of Δt, this vector also gives a close approximation to the actual path travelled by the object, and the length of this vector approximates the actual distance travelled by the object over this tiny time interval. If we divide by the scalar Δt to obtain the difference quotient

$$\frac{\mathbf{r}(a + \Delta t) - \mathbf{r}(a)}{\Delta t},$$

this new vector has the same direction, but now has a magnitude that represents the *average speed* of the object (distance covered divided by elapsed time). Since the difference quotient vector gives us both the direction travelled and the average speed, we call it the *average velocity vector* over the time interval $[a, a + \Delta t]$. Note that if Δt is negative, then the numerator of the difference quotient points in the opposite direction, but the denominator sets us right again. In other words, the difference quotient gives us the average velocity for both positive and negative values Δt.

As $\Delta t \to 0$, the limiting value of the difference quotient represents the *instantaneous velocity vector*. Its direction gives us the *instantaneous direction of movement*, and its magnitude gives us the *instantaneous speed*.

$$\mathbf{r}'(a) = \frac{d\mathbf{r}}{dt}\bigg|_{t=a} = \text{instantaneous velocity at } t = a,$$

$$\|\mathbf{r}'(a)\| = \|\frac{d\mathbf{r}}{dt}\bigg|_{t=a}\| = \text{instantaneous speed at } t = a.$$

We write $\mathbf{v}$ for the **velocity** vector function $\mathbf{r}'$. A curve is called **smooth** if the function $\mathbf{r}'$ is continuous and $\mathbf{r}'(t)$ is never 0.

EXAMPLE 13   Suppose

$$\mathbf{r} : [-1, 1] \longrightarrow \mathbb{R}^3,$$

where

$$\mathbf{r}(t) = \langle 1 - t, t^2 + 1, \frac{2t^3}{3} + 1 \rangle = (1 - t)\mathbf{i} + (t^2 + 1)\mathbf{j} + (\frac{2t^3}{3} + 1)\mathbf{k}.$$

Find the velocity and speed when $t = 1/2$.

Solution   The velocity is

$$\frac{d\mathbf{r}}{dt} = \mathbf{r}'(t) = \mathbf{v}(t)\langle -1, 2t, 2t^2 \rangle = -\mathbf{i} + 2t\mathbf{j} + 2t^2\mathbf{k}$$

for $-1 < t < 1$. The velocity at $t = 1/2$ is

$$\mathbf{r}'(1/2) = \mathbf{v}(t) = \langle -1, 1, 1/2 \rangle$$

and the speed is

$$\|\mathbf{v}(1/2)\| = \sqrt{(-1)^2 + 1^2 + (1/2)^2} = \sqrt{9/4} = 3/2.$$

Figure 13.11 shows the image curve and the velocity vector $\mathbf{v}(1/2)$. ■

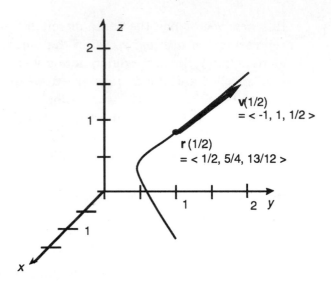

Figure 13.11 A velocity vector.

Tangent lines and unit tangent vectors

The velocity vector $\mathbf{r}'(a)$ gives us the tangent direction to the image curve at the point corresponding to $\mathbf{r}(a)$. Using this point and the velocity vector as a direction vector, we can parametrize the tangent line to the image curve at this point.

EXAMPLE 14   Find a parametrization for the tangent line to the image curve of $\mathbf{r}(t) = \mathbf{r}(t) = (1-t)\mathbf{i} + (t^2 + 1)\mathbf{j} + (\frac{2t^3}{3} + 1)\mathbf{k}$ at $t = 1/2$.

Solution   The velocity vector is

$$\mathbf{v}(t) = \mathbf{r}'(t) = -\mathbf{i} + 2t\,\mathbf{j} + 2t^2\,\mathbf{k}.$$

Hence, at $t = 1/2$, we have a direction vector $\mathbf{v}(1/2) = \langle -1, 1, 1/2 \rangle$. The point of tangency corresponds to $\mathbf{r}(1/2) = \langle 1/2, 5/4, 13/12 \rangle$, so the tangent line is parametrized by

$$\mathbf{r}(1/2) + t\mathbf{v}(1/2) = \langle 1/2 - t, 5/4 + t, 13/12 + t/2 \rangle.$$

Figure 13.12 illustrates this tangent line. ■

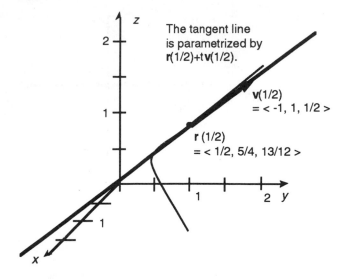

Figure 13.12 Tangent line to image curve of $\mathbf{r}(t)$ at $t = 1/2$.

If we are interested only in measuring the *direction* of the image curve at any point, then we can simply scale the velocity vector $\mathbf{r}'(t)$ to unit length by dividing by the speed $\|\mathbf{r}'(t)\|$. This is called the **unit tangent vector**, and is denoted

$$\mathbf{T}(t) = \frac{\mathbf{v}(t)}{\|\mathbf{v}(t)\|} = \frac{\mathbf{r}'(t)}{\|\mathbf{r}'(t)\|}.$$

EXAMPLE 15   Compute the unit tangent vector $\mathbf{T}(1/2)$ for the position vector function $\mathbf{r}$ of the previous example.

Solution   We have $\mathbf{v}(1/2) = \langle -1, 1, 1/2 \rangle$. Therefore,

$$T(1/2) = \frac{\langle -1, 1, 1/2 \rangle}{\|\langle -1, 1, 1/2 \rangle\|} = \frac{\langle -1, 1, 1/2 \rangle}{\sqrt{((-1)^2 + 1^2 + (1/2)^2}} = \frac{\langle -1, 1, 1/2 \rangle}{\sqrt{9/4}} \frac{2}{3}\langle -1, 1, 1/2 \rangle.$$

Figure 13.13 illustrates the unit tangent vector.   ∎

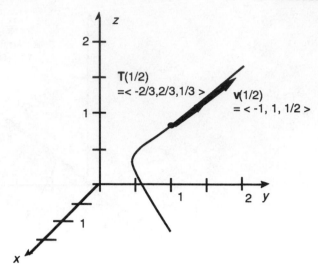

Figure 13.13 A unit tangent vector.

Higher order derivatives—acceleration

Of course, we can take higher order derivatives of position vector functions by simply computing the corresponding higher order derivatives of their component functions.

The second derivative of a position vector function is called the **acceleration** vector function, and is denoted

$$\mathbf{a}(t) = \mathbf{v}'(t) = \mathbf{r}''(t).$$

EXAMPLE 16   Compute the acceleration vector function $\mathbf{a}(t)$ for the position vector function $\mathbf{r}$ of the previous example. Use it to find the acceleration vector $\mathbf{a}(1/2)$.

Solution   We have $\mathbf{v}(t) = \mathbf{r}'(t) = -\mathbf{i} + 2t\,\mathbf{j} + 2t^2\,\mathbf{k}$. Therefore,

$$\mathbf{a}(t) = \mathbf{v}'(t) = \mathbf{r}''(t) = 2\,\mathbf{j} + 4t\,\mathbf{k}.$$

The acceleration vector at $t = 1/2$ is

$$\mathbf{a}(1/2) = 2\,\mathbf{j} + 4(1/2)\,\mathbf{k} = \langle 0, 2, 2\rangle.$$

The acceleration vector is illustrated in Figure 13.14.   ∎

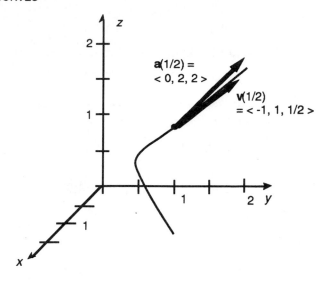

Figure 13.14 An acceleration vector.

EXERCISES

For exercises 1-5, consider the vector function for a curve in $\mathbb{R}^2$:

$$\mathbf{r}(t) = \langle 3t^2, \sqrt{t} \rangle \qquad \text{for } 1 \le t \le 3.$$

1. Find $\mathbf{v}(t)$ and $\mathbf{v}(2)$.
2. Find $\mathbf{a}(t)$ and $\mathbf{a}(2)$.
3. Find $\mathbf{T}(t)$ and $\mathbf{T}(2)$.
4. Find the speed at $t = 2$.
5. Find a parametrization of the tangent line to the curve at $\mathbf{r}(2)$.

For exercises 6-10, suppose a particle moves on the path

$$\mathbf{r} : [0, 2] \to \mathbb{R}^3$$

$$\mathbf{r}(t) = \langle \sin 3t, \cos 3t, 2t^{3/2} \rangle.$$

6. Find $\mathbf{v}(t)$ and $\mathbf{v}(1)$.
7. Find $\mathbf{a}(t)$ and $\mathbf{a}(1)$.
8. Find $\mathbf{T}(t)$ and $\mathbf{T}(1)$.

9.   Find the speed at $t = 1$.

10.   Find a parametrization of the tangent line to the curve at $\mathbf{r}(1)$.

For exercises 11-15, consider the path of a moving object described by the following position function:

$$\mathbf{r} : [0, 2] \mapsto \mathbb{R}^3$$

$$t \mapsto \langle \frac{t^2}{2}, \frac{4}{3}t^{3/2}, 2t \rangle.$$

11.   Find $\mathbf{v}(t)$ and $\mathbf{v}(1)$.

12.   Find $\mathbf{a}(t)$ and $\mathbf{a}(1)$.

13.   Find $\mathbf{T}(t)$ and $\mathbf{T}(1)$.

14.   Find the speed at $t = 1$.

15.   Find a parametrization of the tangent line to the curve at $\mathbf{r}(1)$.

For exercises 16-25, consider the path of a moving object described by the following position function:

$$\mathbf{r} : [\frac{1}{2}, e] \to \mathbb{R}^3$$

$$t \mapsto \langle \sin \pi t, \ln t, e^t \rangle.$$

16.   Find $\mathbf{v}(t)$ and $\mathbf{v}(1)$.

17.   Find $\mathbf{a}(t)$ and $\mathbf{a}(1)$.

18.   Find $\mathbf{T}(t)$ and $\mathbf{T}(1)$.

19.   Find the speed at $t = 1$.

20.   At what time t during the time interval $[\frac{1}{2}, e]$ is the object *farthest* from the origin?

21.   At what time t during the time interval $[\frac{1}{2}, e]$ does the object achieve *maximum speed*?

22.   At what time t during the time interval $[\frac{1}{2}, e]$ is the length of the acceleration vector at a maximum?

23.   At what times does the object touch one of the coordinate axes?

24.   During what times t is the object in the first octant?

25.   Find a parametrization of the tangent line to the curve at $\mathbf{r}(1)$.

13.3 PROPERTIES OF DERIVATIVES

Derivatives of position vector functions satisfy many properties similar to derivative properties of real-valued functions.

For example, we know that the derivative of a constant function

$$g : x \longmapsto c$$

is the zero function $g' : x \longmapsto 0$. Similarly, the constant position function

$$\mathbf{g} : t \longmapsto \mathbf{c}$$

(where $\mathbf{c} = \langle c_1, c_2, \ldots, c_n \rangle$ is a constant vector) has derivative

$$\mathbf{g}' : t \longmapsto \mathbf{0},$$

the zero position function. This follows from the observation that the derivative of each of the constant components is 0.

EXAMPLE 17   Find $\mathbf{r}'(t)$ if $\mathbf{r}(t) = \langle 1, -2, 3 \rangle$ for all t.

Solution   Since $\mathbf{r}(t)$ is a constant vector, $\mathbf{r}'(t) = \langle 0, 0, 0 \rangle$.   ■

We can also "factor out" constant (scalar) factors. That is, if a is a scalar, and $\mathbf{r}$ is a differentiable vector function, then

$$\frac{d}{dt}(a\mathbf{r}(t)) = a\mathbf{r}'(t).$$

EXAMPLE 18   Find $\mathbf{r}'(t)$ if $\mathbf{r}(t) = 5\langle \ln t, e^t \rangle$ for all $t > 0$.

Solution   Since $\mathbf{r}(t) = \langle 5\ln t, 5e^t \rangle$, we have

$$\mathbf{r}'(t) = \left\langle \frac{5}{t}, 5e^t \right\rangle = 5\left\langle \frac{1}{t}, e^t \right\rangle.$$

■

If $\mathbf{r}_1$ and $\mathbf{r}_2$ are any two differentiable position functions, then we have

$$\frac{d}{dt}[\mathbf{r}_1(t) + \mathbf{r}_2(t)] = \frac{d}{dt}\mathbf{r}_1(t) + \frac{d}{dt}\mathbf{r}_2(t).$$

This is verified by simply carrying out the computation indicated on both sides of the equation and noting that they yield the same result.

Taken together, we say that the scalar multiplication property and this addition property of vector-valued functions establish the *linearity properties* of the derivative: if a and b are any scalars, then

$$\frac{d}{dt}[a\mathbf{r}_1 + b\mathbf{r}_2] = a\frac{d\mathbf{r}_1}{dt} + b\frac{d\mathbf{r}_2}{dt}.$$

Product rules

The product rule for real-valued functions can be written

$$\frac{d}{dt}(fg) = \frac{df}{dt}g + f\frac{dg}{dt}.$$

There are different product rules for vector-valued functions corresponding to the different types of multiplication we can perform (scalar, dot, and cross). It is quite fortunate (and somewhat amazing) that the product rule in each case has essentially the same form as the familiar product rule for real-valued functions.

First, suppose we have a position vector function $\mathbf{r}$ and a *scalar-valued function f* of a real variable t. Then,

$$\frac{d}{dt}f(t)\mathbf{r}(t) = f'(t)\mathbf{r}(t) + f(t)\mathbf{r}'(t).$$

EXAMPLE 19   Suppose $\mathbf{r}(t) = \langle \cos t, \sin t, t \rangle$ and $f(t) = t^2$ for all t. Then

$$f(t)\mathbf{r}(t) = \langle t^2 \cos t, \quad t^2 \sin t, \quad t^3 \rangle,$$

and the derivative of this vector function is

$$\frac{d}{dt}f(t)\mathbf{r}(t) = \langle 2t \cos t - t^2 \sin t, 2t \sin t + t^2 \cos t, 3t^2 \rangle.$$

Note that $f'(t) = 2t$ and $\mathbf{r}'(t) = \langle -\sin t, \cos t, 1 \rangle$. If we apply the product rule, we have

$$f'(t)\mathbf{r}(t) + f(t)\mathbf{r}'(t) = \langle 2t \cos t, 2t \sin t, 2t^2 \rangle + \langle t^2(-\sin t), t^2(\cos t), t^2 \rangle$$

$$= \langle 2t \cos t - t^2 \sin t, \quad 2t \sin t + t^2 \cos t, \quad 3t^2 \rangle.$$

∎

We can verify the *product rule for scalar multiplication* by carrying out this same computation in general. If we have any 3-dimensional differentiable position function $\mathbf{r}(t) = \langle x(t), y(t), z(t) \rangle$ and any differentiable scalar-valued function $f(t)$, then

$$\frac{d}{dt} f\mathbf{r} = \frac{d}{dt} \langle fx, fy, fz \rangle$$

$$= \langle \frac{df}{dt}x + f\frac{dx}{dt}, \quad \frac{df}{dt}y + f\frac{dy}{dt}, \quad \frac{df}{dt}z + f\frac{dz}{dt} \rangle$$

$$= \frac{df}{dt} \langle x, y, z \rangle + f \langle \frac{dx}{dt}, \frac{dy}{dt}, \frac{dz}{dt} \rangle$$

$$= \frac{df}{dt}\mathbf{r} + f\frac{d\mathbf{r}}{dt}.$$

The verification for any other dimension is entirely similar.

As for derivatives of the dot product of two vector functions, assume that $\mathbf{r}_1$ and $\mathbf{r}_2$ are any differentiable position vector functions of t. The *dot product rule* states

$$\frac{d}{dt}(\mathbf{r}_1 \cdot \mathbf{r}_2) = \frac{d\mathbf{r}_1}{dt} \cdot \mathbf{r}_2 + \mathbf{r}_1 \cdot \frac{d\mathbf{r}_2}{dt}.$$

Again, we can verify this property by simply carrying out the computation indicated on both sides of the equation and noting that they yield the same result. If

$$\mathbf{r}_1(t) = \langle x_1(t), y_1(t), z_1(t) \rangle \quad \text{and} \quad \mathbf{r}_2(t) = \langle x_2(t), y_2(t), z_2(t) \rangle$$

are any two differentiable 3-dimensional position functions, then

$$\mathbf{r}_1(t) \cdot \mathbf{r}_2(t) = x_1(t)x_2(t) + y_1(t)y_2(t) + z_1(t)z_2(t).$$

Note that this is a *scalar* function. Applying the usual differentiation rules gives us

$$\frac{d}{dt}(\mathbf{r}_1 \cdot \mathbf{r}_2) = (x_1'x_2 + x_1x_2') + (y_1'y_2 + y_1y_2') + (z_1'z_2 + z_1z_2')$$

$$= (x_1'x_2 + y_1'y_2 + z_1'z_2) + (x_1x_2' + y_1y_2' + z_1z_2')$$

$$= \frac{d\mathbf{r}_1}{dt} \cdot \mathbf{r}_2 + \mathbf{r}_1 \cdot \frac{d\mathbf{r}_2}{dt}.$$

The verification for other dimensions is similar.

For vector functions having outputs in $\mathbb{R}^3$, we also have the *cross product rule*:

$$\frac{d}{dt}(\mathbf{r}_1(t) \times \mathbf{r}_2(t)) = \frac{d\mathbf{r}_1(t)}{dt} \times \mathbf{r}_2(t) + \mathbf{r}_1(t) \times \frac{d\mathbf{r}_2}{dt},$$

which can also be verified by direct computation of both sides of this equation.

EXAMPLE 20   Suppose

$$\mathbf{r}_1(t) = \langle \arctan(t), \ln(t), e^t \rangle \qquad \text{and} \qquad \mathbf{r}_2(t) = \langle t, t^2, t^3 \rangle$$

for $0 \leq t \leq 1$. Find $\mathbf{r}_1(t) \cdot \mathbf{r}_2(t)$ and $\mathbf{r}_1(t) \times \mathbf{r}_2(t)$. Then, compute

$$\frac{d}{dt}(\mathbf{r}_1(t) \cdot \mathbf{r}_2(t)) \qquad \text{and} \qquad \frac{d}{dt}(\mathbf{r}_1(t) \times \mathbf{r}_2(t)),$$

both directly and by use of the dot and cross product rules.

Solution   The dot product of the two functions is

$$\mathbf{r}_1(t) \cdot \mathbf{r}_2(t) = t \arctan t + t^2 \ln t + t^3 e^t$$

while the cross product can be computed using the "determinant" method:

$$\mathbf{r}_1(t) \times \mathbf{r}_2(t) = \begin{vmatrix} \mathbf{i} & \mathbf{j} & \mathbf{k} \\ \arctan(t) & \ln t & e^t \\ t & t^2 & t^3 \end{vmatrix}$$

$$= (t^3 \ln t - t^2 e^t)\, \mathbf{i} - (t^3 \arctan(t) - t e^t)\, \mathbf{j} + (t^2 \arctan(t) - t \ln t)\, \mathbf{k}.$$

The derivatives of the two given vector functions are

$$\mathbf{r}_1'(t) = \langle \frac{1}{1+t^2}, \frac{1}{t}, e^t \rangle \qquad \text{and} \qquad \mathbf{r}_2(t) = \langle 1, 2t, 3t^2 \rangle.$$

Direct computation of the derivative of the dot product gives us

$$\frac{d}{dt}(\mathbf{r}_1(t) \cdot \mathbf{r}_2(t)) = \frac{d}{dt}(t \arctan t + t^2 \ln t + t^3 e^t)$$

$$= t\frac{1}{1+t^2} + \arctan t + t^2 \frac{1}{t} + 2t \ln t + t^3 e^t + 3t^2 e^t$$

$$= (\frac{t}{1+t^2} + t + t^3 e^t) + (\arctan t + 2t \ln t + 3t^2 e^t).$$

On the other hand, the dot product rule gives us

$$\frac{d\mathbf{r}_1}{dt} \cdot \mathbf{r}_2(t) + \mathbf{r}_1(t) \cdot \frac{d\mathbf{r}_2}{dt}$$

$$= \langle \frac{1}{1+t^2}, \frac{1}{t}, e^t \rangle \cdot \langle t, t^2, t^3 \rangle + \langle \arctan(t), \ln(t), e^t \rangle \cdot \langle 1, 2t, 3t^2 \rangle$$

$$= (\frac{t}{1+t^2} + t + t^3 e^t) + (\arctan t + 2t \ln t + 3t^2 e^t).$$

We note that the result is the same in either case.

Direct computation of the derivative of the cross product gives us

$$\frac{d}{dt}(\mathbf{r}_1(t) \times \mathbf{r}_2(t))$$

$$= \frac{d}{dt}\left((t^3 \ln t - t^2 e^t)\,\mathbf{i} - (t^3 \arctan(t) - te^t)\,\mathbf{j} + (t^2 \arctan(t) - t \ln t)\,\mathbf{k}\right)$$

$$= \left(3t^2 \ln t + t^3 \frac{1}{t} - 2te^t - t^2 e^t\right)\mathbf{i}$$

$$- \left(3t^2 \arctan(t) + t^3 \frac{1}{1+t^2} - e^t - te^t\right)\mathbf{j}$$

$$+ \left(2t \arctan(t) + t^2 \frac{1}{1+t^2} - \ln t - 1\right)\mathbf{k}.$$

On the other hand, the cross product rule yields

$$\frac{d\mathbf{r}_1}{dt} \times \mathbf{r}_2(t) + \mathbf{r}_1(t) \times \frac{d\mathbf{r}_2}{dt}$$

$$= \begin{vmatrix} \mathbf{i} & \mathbf{j} & \mathbf{k} \\ \frac{1}{1+t^2} & \frac{1}{t} & e^t \\ t & t^2 & t^3 \end{vmatrix} + \begin{vmatrix} \mathbf{i} & \mathbf{j} & \mathbf{k} \\ \arctan(t) & \ln t & e^t \\ 1 & 2t & 3t^2 \end{vmatrix}$$

$$= \left[\left(t^3 \frac{1}{t} - t^2 e^t\right)\mathbf{i} - \left(t^3 \frac{1}{1+t^2} - te^t\right)\mathbf{j} + \left(t^2 \frac{1}{1+t^2} - 1\right)\mathbf{k}\right]$$

$$+ \left[(3t^2 \ln t - 2te^t)\,\mathbf{i} - (3t^2 \arctan(t) - e^t)\,\mathbf{j} + (2t \arctan(t) - \ln t)\,\mathbf{k}\right].$$

You can check that this matches the result of the direct computation. ■

  The order of the functions is important in the cross product rule because the cross product is *not* commutative.

Chain rule

For a composition of real-valued functions $g \circ f$, the chain rule provides us with the derivative formula:

$$(g \circ f)'(t) = g'(f(t))f'(t).$$

Using the Leibniz notation, and writing $u = g(w)$ and $w = f(t)$, we have the equivalent formulation

$$\frac{du}{dt} = \frac{du}{dw}\frac{dw}{dt}.$$

Now, suppose $\mathbf{u} = \mathbf{r}(w)$ is a *vector position function* of w, where in turn, $w = f(t)$ is a scalar-valued function of t. We can then ask what the velocity vector of $\mathbf{u}$ is with respect to t. The answer is again supplied by the *chain rule*:

$$(\mathbf{r} \circ f)'(t) = \mathbf{r}(f(t))f'(t),$$

or, using Leibniz notation:

$$\frac{d\mathbf{u}}{dt} = \frac{d\mathbf{u}}{dw}\frac{dw}{dt}.$$

This chain rule is verified by direct computation and use of the chain rule for real-valued functions. For a 3-dimensional vector position function $\mathbf{u} = \mathbf{r}(w) = \langle x(w), y(w), z(w)\rangle$, we have

$$\frac{d\mathbf{u}}{dt} = \frac{d}{dt}\left(\langle x(f(t)), y(f(t)), z(f(t))\rangle\right)$$

$$= \langle x'(f(t))f'(t), y'(f(t))f'(t), z'(f(t))f'(t)\rangle$$

$$= f'(t)\langle x'(f(t)), y'(f(t)), z'(f(t))\rangle$$

$$= f'(t)\mathbf{r}'(f(t)).$$

EXAMPLE 21   Suppose $\mathbf{u} = \mathbf{r}(w) = \langle \cos w, \sin w, w\rangle$ and $w = f(t) = t^2$. Find $\dfrac{d\mathbf{u}}{dt}$, both directly and by using the chain rule.

Solution   We have $\mathbf{u} = \langle \cos t^2, \sin t^2, t^2\rangle$, so

$$\frac{d\mathbf{u}}{dt} = \langle -\sin t^2(2t), \cos(t^2)2t, 2t\rangle.$$

Using the chain rule, we have

$$\frac{d\mathbf{u}}{dt} = \frac{d\mathbf{u}}{dw}\frac{dw}{dt} = \langle -\sin w, \cos w, 1\rangle(2t).$$

Substituting t^2 for w and multiplying each component by the scalar $2t$ gives us the same result as the direct computation.   ■

━━━━━━━   **EXERCISES**

For exercises 1-10, consider the vector function for a curve in $\mathbb{R}^2$:

$$\mathbf{r}(t) = \langle 3t^2, \sqrt{t}\rangle \qquad \text{for } 0 \leq t \leq 1$$

and the scalar function $f : t \longmapsto 1 - t^2$ for $0 \leq t \leq 1$.

1.   Find $\mathbf{r}'(t)$ and $\mathbf{r}'(1/2)$.

2.   Find $\mathbf{r}''(t)$ and $\mathbf{r}''(1/2)$.

3.   Find the angle between the velocity and acceleration vectors when $t = 1/2$.

4. Is there any time $t \in [0, 1]$ at which the velocity and acceleration vectors are orthogonal?

5. Find $\mathbf{T}(t)$ and $\mathbf{T}(1/2)$.

6. Let $u(t) = \|\mathbf{r}(t)\|$ and find $\dfrac{du}{dt}$.

7. Find the speed $\|\mathbf{r}'(t)\|$.

8. True or false: The length of the derivative is the same as the derivative of the length. (Compare your answers to the previous two exercises.)

9. Find $\dfrac{d(f\mathbf{r})}{dt}$ and verify the scalar product rule for the given functions.

10. Find $\dfrac{d(\mathbf{r} \circ f)}{dt}$ and verify the chain rule for the given functions.

For exercises 11-20, consider the two paths

$$\mathbf{r}_1 : [0, 2] \to \mathbb{R}^3$$

$$\mathbf{r}(t) = \langle \sin 3t, \cos 3t, 2t^{3/2} \rangle.$$

$$\mathbf{r}_2 : [0, 2] \mapsto \mathbb{R}^3$$

$$r_2(t) = \left\langle \frac{t^2}{2}, \frac{4}{3}t^{3/2}, 2t \right\rangle.$$

11. Find $\mathbf{v}_1(t)$ and $\mathbf{v}_2(t)$.

12. Find $\mathbf{a}_1(t)$ and $\mathbf{a}_2(t)$.

13. Find $\mathbf{T}_1(t)$ and $\mathbf{T}_2(t)$.

14. Find the velocity of the object whose path is described by $-2\mathbf{r}_1 + 3\mathbf{r}_2$.

15. Find the acceleration of the object whose position at time t is given by $\mathbf{r}_1(t^2)$.

16. Find the unit tangent vector of the object whose position at time t is given by $t^2\mathbf{r}_2(t)$.

17. Find $\mathbf{r}_1 \cdot \mathbf{r}_2$ and then compute $\dfrac{d(\mathbf{r}_1 \cdot \mathbf{r}_2)}{dt}$ both directly and by the dot product rule.

18. Find $\mathbf{r}_1 \times \mathbf{r}_2$ and then compute $\dfrac{d(\mathbf{r}_1 \times \mathbf{r}_2)}{dt}$ both directly and by the cross product rule.

19. Find $\dfrac{d}{dt}(\|\mathbf{r}_1\|)$.

20. Find $\dfrac{d}{dt}(\|\mathbf{r}_2\|)$.

21. Verify the cross product rule in general for two vector functions $r_1(t) = \langle x_1(t), y_1(t), z_1(t) \rangle$ and $r_2(t) = \langle x_2(t), y_2(t), z_2(t) \rangle$ by computing the derivative of $r_1 \times r_2$ both directly and by using the cross product rule.

22. Verify that if a vector position function has constant length (in other words, $\|r(t)\| = c$ for some scalar c), then $r'(t)$ is orthogonal to $r(t)$. (Hint: write $c = \|r(t)\| = \sqrt{r(t) \cdot r(t)}$ and differentiate.)

23. Find an example of a vector position function r such that $\|r(t)\|$ is constant, but $r'(t) \neq 0$.

24. Find a formula for the derivative of the triple scalar product $r_1 \cdot (r_2 \times r_3)$, where r_1, r_2, and r_3 are 3-dimensional differentiable vector functions.

25. Let $T(t) = r'(t)/\|r'(t)\|$ be the unit tangent vector function for a 3-dimensional differentiable vector function r. Verify that

$$\frac{dT}{dt} \times T(t) = \frac{r''(t)}{\|r'(t)\|} \times T(t).$$

13.4   INTEGRATION AND ARC LENGTH

Like derivatives, both definite and indefinite integrals of position vector functions are computed component-wise.

Indefinite integrals

In finding the antiderivative of a position vector function, it is important to realize that each component will involve its own arbitrary constant.

EXAMPLE 22   Find the general antiderivative of the vector function $v(t) = \langle t, t^2, t^3 \rangle$.

Solution   We write

$$\int \langle t, t^2, t^3 \rangle \, dt = \langle \frac{t^2}{2} + C_1, \frac{t^3}{3} + C_2, \frac{t^4}{4} + C_3 \rangle$$

$$= \langle \frac{t^2}{2}, \frac{t^3}{3}, \frac{t^4}{4} \rangle + C,$$

where $C = \langle C_1, C_2, C_3 \rangle$ is an arbitrary *constant vector*. Note that the constants C_1, C_2, C_3 may or may not have different values in specifying any particular antiderivative. ∎

Definite integrals

Definite integrals of vector functions are also computed component-wise.

EXAMPLE 23   Find $\int_1^2 \langle t, t^2, t^3 \rangle \, dt$.

Solution   We have

$$
\int_1^2 \langle t, t^2, t^3 \rangle \, dt = \langle \frac{t^2}{2}, \frac{t^3}{3}, \frac{t^4}{4} \rangle \bigg]_1^2
$$

$$
= \langle 2 - \frac{1}{2}, \frac{8}{3} - \frac{1}{3}, 4 - \frac{1}{4} \rangle
$$

$$
= \langle \frac{3}{2}, \frac{7}{3}, \frac{15}{4} \rangle.
$$

■

If the vector function we are integrating has a physical interpretation as a velocity vector function of time t, then the definite integral represents a *net vector change in position*. For example, if $\mathbf{v}(t) = \langle t, t^2, t^3 \rangle$ represents velocity of an object at time t, then

$$
\int_1^2 \langle t, t^2, t^3 \rangle \, dt = \left\langle \frac{3}{2}, \frac{7}{3}, \frac{15}{4} \right\rangle
$$

tells us that this object had a net change in position between $t = 1$ and $t = 2$ of $\frac{3}{2}$ in the x-direction, $\frac{7}{3}$ in the y-direction, and $\frac{15}{4}$ in the z-direction. The actual path travelled between the two times is determined by the antiderivative

$$
\mathbf{r}(t) = \left\langle \frac{t^2}{2}, \frac{t^3}{3}, \frac{t^4}{4} \right\rangle + \langle C_1, C_2, C_3 \rangle.
$$

Note that some initial condition is necessary to determine the actual values $\mathbf{r}(1)$ and $\mathbf{r}(2)$ (we don't know the value of the arbitrary constant vector). What we do know is that

$$
\mathbf{r}(2) - \mathbf{r}(1) = \left\langle \frac{3}{2}, \frac{7}{3}, \frac{15}{4} \right\rangle,
$$

regardless of the particular starting and ending points of the path in question.

Arc length

We have just discussed how a velocity vector function can be integrated to find the net change in position (also a vector). Unless the moving object always travels in the *same* direction along a straight line, this net change in position will not give us a true indication of the actual distance travelled. For example, consider an object moving in a circle of radius 3, whose path is parametrized by

$$\mathbf{r}(t) = \langle 3\cos t, 3\sin t \rangle$$

for $0 \le t \le 2\pi$. The starting point and ending point are

$$\mathbf{r}(0) = \langle 3\cos 0, 3\sin 0 \rangle = \langle 3, 0 \rangle = \langle 3\cos 2\pi, 3\sin 2\pi \rangle = \mathbf{r}(2\pi).$$

Hence, the net change in position is

$$\mathbf{r}(2\pi) - \mathbf{r}(0) = \langle 0, 0 \rangle = \mathbf{0},$$

the zero vector (after all, the object has travelled in a complete circle). Yet, the actual distance travelled by the object is 6π units, since the path is a circle of radius 3.

Let's think for a moment about a similar situation we encountered for real-valued functions. For an object travelling in one dimension (a straight line), integrating the *velocity* between time $t = a$ and $t = b$ gives us the *net distance* travelled during that time interval. If, instead, we integrate the *speed* (the absolute value of the velocity) between time $t = a$ and $t = b$, we obtain the *total distance* travelled.

To measure the actual total distance travelled by an object moving in a plane or in space with position function $\mathbf{r}$, the same approach may be valid. If the curve traced out by $\mathbf{r}(t)$ is **smooth** over the interval $[a, b]$, meaning that the velocity vector $\mathbf{v}(t) = \mathbf{r}'(t)$ is *continuous* and *nonzero* for all $t \in [a, b]$, then the object does not change direction suddenly, and we know the object's speed (the norm or magnitude $\|\mathbf{v}(t)\| = \|\mathbf{r}'(t)\|$) at each instant t. In this case, we can integrate the *scalar-valued* function $\|\mathbf{v}(t)\|$ between time $t = a$ and $t = b$ to find the *total distance* travelled:

$$\text{total distance travelled} \quad = \quad \int_a^b \|\mathbf{v}(t)\|\, dt \quad = \quad \int_a^b \|\mathbf{r}'(t)\|\, dt.$$

In terms of the image curve of a position vector function, we are saying that the **arc length** of the curve from $\mathbf{r}(a)$ to $\mathbf{r}(b)$ is given by $\int_a^b \|\mathbf{r}'(t)\|\, dt$. Note that $\|\mathbf{r}'\|$ is integrable on $[a, b]$ since we have already assumed that $\mathbf{r}'$ is continuous. We can justify this arc length formula for the path of a position function $\mathbf{r}$ in a similar way to our justification of the arc length formula for the graph of a function $y = f(x)$.

Consider the path $\mathbf{r}(t)$ traces out from $t = a$ to $t = b$, as shown in Figure 13.15.

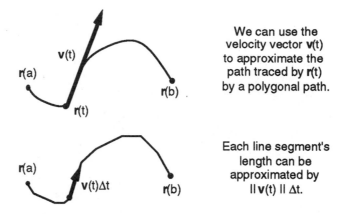

We can use the velocity vector **v**(t) to approximate the path traced by **r**(t) by a polygonal path.

Each line segment's length can be approximated by ‖ **v**(t) ‖ Δt.

Figure 13.15 Approximating an image curve with a polygonal path.

If we subdivide the time interval $[a, b]$ into n subintervals of length Δt, then the path of the object between time t and time $t + \Delta t$ can be approximated by a line segment of length

$$\|\mathbf{r}(t + \Delta t) - \mathbf{r}(t)\|,$$

which we could also write as

$$\frac{\|\mathbf{r}(t + \Delta t) - \mathbf{r}(t)\|\Delta t}{\Delta t},$$

by multiplying the numerator and denominator by Δt. For Δt very small, this is closely approximated by

$$\|\mathbf{v}(t)\|\Delta t.$$

The total length of the curve from $\mathbf{r}(a)$ to $\mathbf{r}(b)$ is found by summing up the lengths of the n line segments

$$\sum_{i=1}^{n} \frac{\|\mathbf{r}(t + \Delta t) - \mathbf{r}(t)\|}{\Delta t} \Delta t \approx \sum_{i=1}^{n} \|\mathbf{r}'(t)\| \, \Delta t = \sum_{i=1}^{n} \|\mathbf{v}(t)\| \, \Delta t.$$

As $\Delta t \to 0$, the approximation becomes better and better, with the limit providing the formula:

$$\text{arc length} \quad = \quad \int_{a}^{b} \|\mathbf{r}'(t)\| \, dt \quad = \quad \int_{a}^{b} \|\mathbf{v}(t)\| \, dt.$$

EXAMPLE 24   Use the arc length formula to find the distance travelled by an object whose position is given by

$$\mathbf{r}(t) = \langle 3\cos t, 3\sin t \rangle$$

for $0 \le t \le 2\pi$.

Solution   Since $\mathbf{r}'(t) = \langle -3\sin t, 3\cos t \rangle$, we have

$$\|\mathbf{r}'(t)\| = \sqrt{9\sin^2 t + 9\cos^2 t} = \sqrt{9(\sin^2 t + \cos^2 t)} = \sqrt{9} = 3,$$

and

$$\int_0^{2\pi} 3\,dt = 3t\,\Big]_{t=0}^{t=2\pi} = 6\pi.$$

This is precisely the circumference of a circle of radius 3.   ■

EXAMPLE 25   Find the arc length of the path traced out by an object whose position is given by

$$\mathbf{r}(t) = \langle 1 - t, t^2 + 1, 2t^3/3 + 1 \rangle$$

between $t = 0$ and $t = 1/2$.

Solution   The speed of the object is given by

$$\|\mathbf{r}'(t)\| = \|\langle -1, 2t, 2t^2 \rangle\|$$

$$= \sqrt{(-1)^2 + (2t)^2 + (2t^2)^2}$$

$$= \sqrt{1 + 4t^2 + 4t^4}$$

$$= \sqrt{(1 + 2t^2)^2}$$

$$= |1 + 2t^2| = 1 + 2t^2.$$

(The absolute value signs can be removed since $1 + 2t^2$ is positive for all t.) The arc length between $t = 0$ and $t = 1/2$ is given by

$$\int_0^{1/2} \|\mathbf{r}'(t)\| = \int_0^{1/2} (1 + 2t^2)\,dt = (t + 2t^3/3)\,\Big]_{t=0}^{t=1/2} = 7/12.$$

                                                                                     ■

Arc length function—parametrization by arc length

We can define an **arc length function**

$$s(t) = \int_a^t \|\mathbf{r}'(u)\| \, du$$

by integrating the speed from a fixed value $u = a$ and allowing the upper limit $u = t$ to vary. The arc length function output $s(t)$ tells us the arc length of the image curve of $\mathbf{r}$ from $\mathbf{r}(a)$ to $\mathbf{r}(t)$.

Now, the Second Fundamental Theorem of Calculus tells us that

$$\frac{ds}{dt} = \|\mathbf{r}'(t)\|.$$

This makes perfect sense—the rate of change of the distance travelled s with respect to time t is simply the speed!

One important consequence of this fact is simply that with respect to arc length, $\mathbf{r}$ has *constant unit speed*. In other words,

$$\left\| \frac{d\mathbf{r}}{ds} \right\| = 1.$$

To see why, note that by the chain rule we have

$$\mathbf{r}'(t) = \frac{d\mathbf{r}}{dt} = \frac{d\mathbf{r}}{ds}\frac{ds}{dt} = \frac{d\mathbf{r}}{ds}\|\mathbf{r}'(t)\|.$$

Now, if we take the norm of both sides of the equation, we have

$$\|\mathbf{r}'(t)\| = \left\| \frac{d\mathbf{r}}{ds} \right\| \|\mathbf{r}'(t)\|.$$

Dividing both sides by $\|\mathbf{r}'(t)\|$ gives us $\|d\mathbf{r}/ds\| = 1$.

EXAMPLE 26 Find the arc length function for the path traced out by an object whose position is given by

$$\mathbf{r}(t) = \langle 1 - t, t^2 + 1, 2t^3/3 + 1 \rangle$$

using $t = 0$ as the starting time.

Solution Using $\|\mathbf{r}'(t)\| = 1 + 2t^2$, we have

$$s(t) = \int_0^t (1 + 2u^2) \, du = (u + 2u^3/3) \Big]_{u=0}^{u=t} = t + 2t^3/3.$$

You can check that $\dfrac{ds}{dt} = 1 + 2t^2 = \|\mathbf{r}'(t)\|$. ∎

The arc length s is particularly useful as a "time-independent" parameter for describing a smooth curve. That is, we can use the arc length s

as our independent variable in parametrizing a curve by fixing a reference point on the curve, and then describing other positions in terms of the distance travelled from the reference point.

Given a parametrization $\mathbf{r}(t)$ of a smooth curve in terms of t, we can *reparametrize by arc length* s by choosing an initial time $t = a$ and

1) finding the resulting arc length function s in terms of t;

2) solving for t in terms of s; and

3) expressing the position function $\mathbf{r}$ in terms of s by substitution.

A few comments are in order. Because the curve is assumed to be *smooth*, the speed is always positive $(\mathbf{r}'(t) \neq 0)$ and the distance travelled is strictly increasing. This means each different value of t results in a different value s, and so t is determined uniquely by s. Theoretically, this allows us to solve for t in terms of s (although that might be quite difficult algebraically.) Here is an example illustrating the steps in reparametrizing a curve in terms of arc length.

EXAMPLE 27   Parametrize by arc length the curve traced out by an object whose position is given by

$$\mathbf{r}(t) = \langle \cos t, \sin t, t \rangle$$

for $0 \leq t \leq 2\pi$.

Solution   This curve is our helix example from earlier discussion. The first step is to find the arc length function. The velocity vector is given by

$$\mathbf{v}(t) = \mathbf{r}'(t) = \langle -\sin t, \cos t, 1 \rangle$$

and the speed is

$$\|\mathbf{v}(t)\| = \|\mathbf{r}'(t)\| = \sqrt{(-\sin t)^2 + (\cos t)^2 + 1^2} = \sqrt{2}.$$

The arc length function is

$$s(t) = \int_0^t \sqrt{2}\, du = \sqrt{2}u \Big]_{u=0}^{u=t} = \sqrt{2}t.$$

Solving for t in terms of s, we have

$$t = s/\sqrt{2}$$

and the parametrization by arc length is

$$\langle \cos(s/\sqrt{2}),\ \sin(s/\sqrt{2}),\ s/\sqrt{2} \rangle.$$

The domain of this new parametrization is $[0, 2\sqrt{2}\pi]$, since $s = \sqrt{2}t$ and $0 \leq t \leq 2\pi$. ∎

You can think of parametrizing a curve by arc length as "standardizing" our description of the curve, in the sense that the speed with respect to arc length is guaranteed to be the constant 1 unit.

EXAMPLE 28 In the previous example, the parametrization of the helix relative to arc length is

$$\langle \cos(s/\sqrt{2}), \sin(s/\sqrt{2}), s/\sqrt{2} \rangle.$$

Verify that the speed with respect to arc length is 1 unit.

Solution The velocity with respect to arc length is

$$\langle -\frac{\sin(s/\sqrt{2})}{\sqrt{2}}, \ \frac{\cos(s/\sqrt{2})}{\sqrt{2}}, \ \frac{1}{\sqrt{2}} \rangle,$$

and hence the speed with respect to arc length is

$$\sqrt{\left(-\frac{\sin(s/\sqrt{2})}{\sqrt{2}}\right)^2 + \left(\frac{\cos(s/\sqrt{2})}{\sqrt{2}}\right)^2 + \left(\frac{1}{\sqrt{2}}\right)^2}$$

$$= \sqrt{\frac{\sin^2(s/\sqrt{2}) + \cos^2(s/\sqrt{2})}{2} + \frac{1}{2}} = \sqrt{1} = 1.$$

∎

The fact that the speed is always one unit for a curve parametrized by arc length tells us that the velocity vector with respect to arc length is the same as the unit tangent vector. In other words, if $\mathbf{r}(s)$ is the position in terms of arc length s, then

$$\frac{d\mathbf{r}}{ds} = \mathbf{T}.$$

EXERCISES

For exercises 1-8, consider an object moving in $\mathbb{R}^2$ with velocity

$$\mathbf{v}(t) = \langle 3t^2, \sqrt{t} \rangle \qquad \text{for } 1 \le t \le 4.$$

1. Find the acceleration $\mathbf{a}(t)$ of the object.

2. Find $\displaystyle\int_2^3 \mathbf{v}(t)\,dt.$

3. Find the net distance travelled by the object between $t = 1$ and $t = 4$.

4. Find the total distance travelled by the object between $t = 1$ and $t = 4$.

5. If the object's location is $(-2, 3)$ at $t = 1$, find its position at $t = 4$.

6. If the object's location is $(-2, 3)$ at $t = 1$, find its position function $\mathbf{r}(t) = \langle x(t), y(t) \rangle$ for $1 \le t \le 4$.

7. Find the arc length function for $\mathbf{r}(t)$.

8. How far does the object travel between $t = 2$ and $t = 3$?

For exercises 9-14, suppose a particle moves on the path

$$\mathbf{r} : [0, 2] \to \mathbb{R}^3$$

$$\mathbf{r}(t) = \langle \sin 3t, \cos 3t, 2t^{3/2} \rangle.$$

9. Find $\displaystyle\int \mathbf{r}(t)\, dt$.

10. Find $\displaystyle\int_0^2 \mathbf{r}(t)\, dt$.

11. Find the net distance travelled by the particle.

12. Find the total distance travelled by the particle.

13. Find the arc length function s for $\mathbf{r}$.

14. Reparametrize by arc length the path traced by $\mathbf{r}$.

For exercises 15-18, consider the path of a moving object whose acceleration is described by:

$$\mathbf{a} : [0, 2] \mapsto \mathbb{R}^3$$

$$t \mapsto \langle t^2, t^3, 2t \rangle.$$

15. Find the velocity $\mathbf{v}(t)$ if the initial velocity is $\mathbf{v}(0) = \langle 1, 2, 3 \rangle$. What is the terminal velocity?

16. Find the position $\mathbf{r}(t)$ if $\mathbf{r}(0) = \langle -1, 0, -2 \rangle$. What is the terminal position?

17. Find the net distance travelled by the object.

18. Find the total distance travelled by the object.

---

19. Consider the path of a moving object described by the following position function:

$$\mathbf{r} : [0, 2\pi] \to \mathbb{R}^3$$

$$t \mapsto \langle 3\sin t, 3\cos t, 4t \rangle$$

Find the total distance travelled by the object and reparametrize its path by arc length.

---

20. Two of the ways to parametrize the unit circle in $\mathbb{R}^2$ starting at the point $(1, 0)$ are

$$\mathbf{r}_1(t) = \langle \cos(t/2), \sin(t/2) \rangle \qquad \text{for} \qquad 0 \le t \le 4\pi,$$

and

$$\mathbf{r}_2(t) = \langle \cos(2t), \sin(2t) \rangle \qquad \text{for} \qquad 0 \le t \le \pi.$$

The image curves are identical. The major difference between these two ways of parametrizing the unit circle is that $\mathbf{r}_2$ traces out the circle *four* times as fast (note the time interval $[0, \pi]$ is one-fourth as long as the time interval $[0, 4\pi]$ for $\mathbf{r}_1$). Show that starting with either r_1 or r_2, reparametrizing by arc length yields identical results.

13.5 CURVATURE AND THE PRINCIPAL UNIT NORMAL VECTOR

In this section, we'll take a look at how we can use calculus to describe more completely certain characteristics of the path traced out by a vector position function, including "how curved" the path is. First, we consider the notion of a *normal* vector to the curve.

Normal vectors to a curve

At each point along a smooth curve described by the differentiable position function $\mathbf{r}$, we have a tangent direction provided by the velocity vector $\mathbf{v}(t) = \mathbf{r}'(t)$. As we have discussed, if we scale this vector to have unit length, then we call the resulting vector the **unit tangent vector**

$$\mathbf{T}(t) = \frac{\mathbf{v}(t)}{\|\mathbf{v}(t)\|}.$$

We can use the direction of this vector to parametrize the tangent line to the curve at that point.

A vector *perpendicular* to the tangent vector is called a **normal** vector. If a normal vector has length 1 unit, we call it a **unit normal vector**. For a smooth curve in the plane $\mathbb{R}^2$, there are exactly two unit normal vectors at each point (having opposite directions). For a smooth curve in space $\mathbb{R}^3$, there are *infinitely* many unit normal vectors at each point, all lying in the plane normal to the tangent vector (see Figure 13.16).

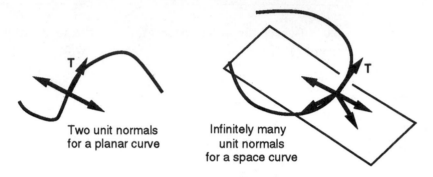

Figure 13.16 Unit normal vectors to a plane curve and a space curve.

It will be convenient to make a specific choice of unit normal at each point of the curve, which we will call the **principal unit normal** and denote by **N**.

EXAMPLE 29 In the plane we have two choices, and the principal unit normal is always located on the same side of the tangent line as the curve itself. If $\mathbf{T} = \langle u_1, u_2 \rangle$ then we must have $\mathbf{N} = \langle -u_2, u_1 \rangle$ or $\mathbf{N} = \langle u_2, -u_1 \rangle$, depending on whether we must choose the unit normal $90°$ counterclockwise or clockwise from the unit tangent vector.   ■

We agree to choose as the principal unit normal the vector

$$\mathbf{N} = \frac{d\mathbf{T}/dt}{||d\mathbf{T}/dt||}$$

where $d\mathbf{T}/dt$ is the derivative of the unit tangent vector **T** with respect to t.

We can see that this vector **N** has unit length (note that we are dividing $d\mathbf{T}/dt$ by its length), but it is not immediately clear why this vector is necessarily orthogonal to **T**. To see why, note that

$$\mathbf{T} \cdot \mathbf{T} = 1$$

since **T** is a *unit vector*. If we differentiate both sides with respect to t using the dot product rule, we see that

$$\frac{d\mathbf{T}}{dt} \cdot \mathbf{T} + \mathbf{T} \cdot \frac{d\mathbf{T}}{dt} = 0.$$

Using the fact that the dot product is commutative, we can rewrite this as

$$2\frac{d\mathbf{T}}{dt} \cdot \mathbf{T} = 0.$$

If we divide both sides by 2, we can see that $\dfrac{d\mathbf{T}}{dt} \cdot \mathbf{T} = 0$, so that $\dfrac{d\mathbf{T}}{dt}$ is orthogonal to $\mathbf{T}$. Hence, we indeed have

$$\mathbf{N} \cdot \mathbf{T} = \frac{d\mathbf{T}/dt}{||d\mathbf{T}/dt||} \cdot \mathbf{T} = 0.$$

EXAMPLE 30   Find the principal unit normal $\mathbf{N}$ to the curve described by the position function

$$\mathbf{r}(t) = \langle 3\cos t, 3\sin t \rangle$$

at $t = 3\pi/4$.

Solution   The velocity vector is

$$\mathbf{v}(t) = \langle -3\sin t, 3\cos t \rangle,$$

with speed $||\mathbf{v}(t)|| = \sqrt{9\sin^t + 9\cos^t} = \sqrt{9} = 3$. So, the unit tangent vector is

$$\mathbf{T}(t) = \langle -\sin t, \cos t \rangle$$

and at $t = 3\pi/4$, we have $\mathbf{T} = \langle -\sqrt{2}/2, -\sqrt{2}/2 \rangle$.

Since $\mathbf{r}$ is a plane curve, there are only two choices for $\mathbf{N}$:

$$\langle -\sqrt{2}/2, \sqrt{2}/2 \rangle \quad \text{or} \quad \langle \sqrt{2}/2, -\sqrt{2}/2 \rangle.$$

By inspecting the circle, it is clear that the correct choice is

$$\mathbf{N} = \langle \sqrt{2}/2, -\sqrt{2}/2 \rangle.$$

However, we can check this by computing

$$\frac{d\mathbf{T}}{dt} = \langle -\cos t, -\sin t \rangle.$$

At $t = 3\pi/4$, this vector is $\langle \sqrt{2}/2, -\sqrt{2}/2 \rangle$, and is the principal unit normal vector (it already has unit length).   ■

EXAMPLE 31   Find the principal unit normal $\mathbf{N}$ to the curve described by the position function

$$\mathbf{r}(t) = \langle 1 - t, t^2 + 1, 2t^3/3 + 1 \rangle$$

at $t = 1/2$.

Solution   The velocity vector is $\mathbf{r}'(t) = \langle -1, 2t, 2t^2 \rangle$, and the unit tangent vector is given by

$$\mathbf{T}(t) = \frac{\mathbf{r}'(t)}{\|\mathbf{r}'(t)\|} = \frac{\langle -1, 2t, 2t^2 \rangle}{1 + 2t^2} = \left\langle -\frac{1}{1 + 2t^2}, \frac{2t}{1 + 2t^2}, \frac{2t^2}{1 + 2t^2} \right\rangle.$$

Now we take the derivative of the unit tangent vector function with respect to t:

$$\frac{d\mathbf{T}}{dt} = \left\langle \frac{4t}{(1 + 2t^2)^2}, \frac{2 - 4t^2}{(1 + 2t^2)^2}, \frac{4t}{(1 + 2t^2)^2} \right\rangle.$$

At $t = 1/2$, we have

$$\frac{d\mathbf{T}}{dt}\bigg|_{t=1/2} = \langle \frac{8}{9}, \frac{4}{9}, \frac{8}{9} \rangle.$$

The length of this vector is

$$\sqrt{\frac{64}{81} + \frac{16}{81} + \frac{64}{81}} = \sqrt{\frac{144}{81}} = \frac{12}{9} = \frac{4}{3}.$$

Dividing by this scalar gives the principal unit normal

$$\mathbf{N}(1/2) = \left\langle \frac{2}{3}, \frac{1}{3}, \frac{2}{3} \right\rangle.$$

We note that $\mathbf{T}(1/2) = \left\langle -\frac{2}{3}, \frac{2}{3}, \frac{1}{3} \right\rangle$, and we verify that

$$\mathbf{N}(1/2) \cdot \mathbf{T}(1/2) = \left\langle \frac{2}{3}, \frac{1}{3}, \frac{2}{3} \right\rangle \cdot \left\langle -\frac{2}{3}, \frac{2}{3}, \frac{1}{3} \right\rangle = -\frac{4}{9} + \frac{2}{9} + \frac{2}{9} = 0,$$

showing that $\mathbf{N}(1/2)$ is indeed orthogonal to $\mathbf{T}(1/2)$. (You can also check that $\mathbf{N}(1/2)$ has length 1 unit.)   ■

EXAMPLE 32   Find the principal unit normal N to the curve described by the position function

$$\mathbf{r}(t) = \langle \cos t, \sin t, t \rangle$$

at $t = \pi$.

Solution   The velocity vector is $\mathbf{r}'(t) = \langle -\sin t, \cos t, 1 \rangle$, and the unit tangent vector is given by

$$\mathbf{T}(t) = \frac{\mathbf{r}'(t)}{\|\mathbf{r}'(t)\|} = \frac{\langle -\sin t, \cos t, 1 \rangle}{\sqrt{2}}.$$

Now we take the derivative of the unit tangent vector function with respect to t:

$$\frac{d\mathbf{T}}{dt} = \langle -\frac{\cos t}{\sqrt{2}}, -\frac{\sin t}{\sqrt{2}}, 0 \rangle,$$

and that gives us

$$\frac{d\mathbf{T}}{dt}\bigg|_{t=\pi} = \langle -\frac{\cos \pi}{\sqrt{2}}, -\frac{\sin \pi}{\sqrt{2}}, 0 \rangle = \langle \frac{1}{\sqrt{2}}, 0, 0 \rangle.$$

A unit vector in the same direction is

$$\mathbf{N}(\pi) = \langle 1, 0, 0 \rangle = \mathbf{i}.$$

Since $\mathbf{T}(\pi) = \langle 0, -1/\sqrt{2}, 1/\sqrt{2} \rangle$, we can check that

$$\mathbf{N}(\pi) \cdot \mathbf{T}(\pi) = \langle 1, 0, 0 \rangle \cdot \langle 0, -1/\sqrt{2}, 1/\sqrt{2} \rangle = 0 + 0 + 0 = 0,$$

showing that $\mathbf{N}(\pi)$ is orthogonal to $\mathbf{T}(\pi)$.   ■

The osculating plane

The plane that is determined by N and T at a particular point on a smooth curve is called the **osculating plane** to the curve at that point. In a very real sense, this plane "best fits" the curve at this point.

Let's clarify this statement a bit. Consider a specific time t_0 and the point corresponding to the position $\mathbf{r}(t_0)$ on the smooth curve described by r. Now, suppose we choose three *distinct* points in time t_1, t_2, t_3, and the three points on the curve corresponding to $\mathbf{r}(t_1)$, $\mathbf{r}(t_2)$, and $\mathbf{r}(t_3)$. Unless these three points are collinear, they determine a plane. As all three time values approach t_0, you can think of the osculating plane as representing the "limiting plane" determined by these three points.

EXAMPLE 33   Find the general equation of the osculating plane to the curve described by the position function

$$\mathbf{r}(t) = \langle 1 - t, t^2 + 1, 2t^3/3 + 1 \rangle$$

at $t = 1/2$.

Solution   We've already computed the unit tangent vector and the principal unit normal vector to the curve at $t = 1/2$. A vector giving the normal direction of the plane determined by these two vectors $\mathbf{N}(1/2)$ and $\mathbf{T}(1/2)$ is

$$\mathbf{N}(1/2) \times \mathbf{T}(1/2) = \begin{vmatrix} \mathbf{i} & \mathbf{j} & \mathbf{k} \\ \frac{2}{3} & \frac{1}{3} & \frac{2}{3} \\ -\frac{2}{3} & \frac{2}{3} & \frac{1}{3} \end{vmatrix} = \langle -\frac{1}{3}, -\frac{2}{3}, \frac{2}{3} \rangle$$

and contains the point corresponding to $\mathbf{r}(1/2) = \langle 1/2, 5/4, 13/12 \rangle$.

The general equation of the osculating plane is of the form

$$ax + by + cz + d = 0.$$

The vector $\langle a, b, c \rangle$ can be taken to be $\langle -\frac{1}{3}, -\frac{2}{3}, \frac{2}{3} \rangle$, so we have

$$-\frac{1}{3}x - \frac{2}{3}y + \frac{2}{3}z + d = 0.$$

To find d, we substitute $(1/2, 5/4, 13/12)$ for (x, y, z) and solve for d:

$$d = (\frac{1}{3})(\frac{1}{2}) + (\frac{2}{3})(\frac{5}{4}) - (\frac{2}{3})(\frac{13}{12}) = \frac{5}{18}.$$

Substituting $d = 5/18$, we have

$$-\frac{1}{3}x - \frac{2}{3}y + \frac{2}{3}z + \frac{5}{18} = 0$$

as an equation of the osculating plane.   ∎

Another property of the osculating plane is that the acceleration vector **a** must lie in this plane. Said in other words, we must have scalars p and q such that

$$\mathbf{a} = p\mathbf{T} + q\mathbf{N}.$$

That is, the acceleration vector must be a linear combination of the unit tangent vector and the principal unit normal vector.

EXAMPLE 34   Verify that the acceleration vector is a linear combination of the unit tangent vector and the principal unit normal vector to the curve described by

$$\mathbf{r}(t) = \langle 1 - t, t^2 + 1, 2t^3/3 + 1 \rangle$$

at $t = 1/2$.

Solution   The acceleration vector is

$$\mathbf{a}(1/2) = \langle 0, 2, 2 \rangle.$$

The question calls for us to verify that there are scalars p and q such that

$$\langle 0, 2, 2 \rangle = p\langle -\frac{2}{3}, \frac{2}{3}, \frac{1}{3} \rangle + q\langle \frac{2}{3}, \frac{1}{3}, \frac{2}{3} \rangle = \langle \frac{2q - 2p}{3}, \frac{2p + q}{3}, \frac{p + 2q}{3} \rangle.$$

We note that this is indeed the case if we have $p = q = 2$.   ■

Measuring curvature for curves in a plane

We can think of *curvature* as a measure of how "fast" the direction of a curve changes. Of course, the speed of change in direction depends in part on simply how quickly a curve is traversed, and that depends entirely on the particular parametrization we choose for the curve. Certainly, we want any measure of curvature to depend intrinsically on the shape of the curve, and not on some arbitrary choice of the infinitely many ways that a curve can be parametrized.

A natural and objective choice to make is to insist on parametrization by arc length for purposes of measuring curvature. The unit tangent vector **T** at any point tells us the direction of the curve at that point, so we define the **curvature** at the point to be

$$\kappa = ||d\mathbf{T}/ds||,$$

the length of the derivative of the unit tangent vector with respect to arc length. The *vector* $d\mathbf{T}/ds$ is sometimes called the **curvature vector**, but it is its *scalar* length that gives us the measure of curvature.

Exactly what does this value κ tell us? If we were to find a circle that "best fits" the curve at a particular point on a curve, then the curvature κ at that point represents the *reciprocal of the radius* of this circle (see Figure 13.17).

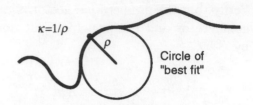

Figure 13.17 The relationship between curvature and radius of curvature.

This idea makes sense—if a circle has a small radius, its curvature is great (it turns in a tight space); if a circle has a large radius, its curvature is small (a person can walk in a large circle while believing that the path is straight). The radius ρ of this circle of best fit is called the **radius of curvature** and as we have stated, it has an inverse relationship to the curvature κ:

$$\kappa = 1/\rho \quad \text{and} \quad \rho = 1/\kappa.$$

If the curve is a straight line (or the part of the curve including the point in question is straight), then $\kappa = 0$ (zero curvature) and we agree to say the radius of curvature is infinite and write $\rho = \infty$.

It may be difficult to parametrize a curve relative to arc length, but fortunately, there are several alternative, but equivalent, formulas for measuring curvature. For example, by the chain rule, we have

$$\frac{d\mathbf{T}}{dt} = \frac{d\mathbf{T}}{ds}\frac{ds}{dt}.$$

We can solve this equation for $d\mathbf{T}/ds$, and noting that

$$\frac{ds}{dt} = \|\mathbf{v}(t)\|$$

we have

$$\frac{d\mathbf{T}}{ds} = \frac{d\mathbf{T}/dt}{\|\mathbf{v}(t)\|}.$$

We can see that this tells us that *the curvature vector points in the same direction as the principal unit normal* (recall that $\mathbf{N} = (d\mathbf{T}/dt)/\|d\mathbf{T}/dt\|$). We now have an another way to express curvature:

$$\kappa = \|d\mathbf{T}/ds\| = \frac{\|d\mathbf{T}/dt\|}{\|\mathbf{v}(t)\|}.$$

EXAMPLE 35   Compute the curvature κ and the radius of curvature ρ for the curve defined by

$$\mathbf{r}(t) = \langle 1 - t, t^2 + 1, 2t^3/3 + 1\rangle$$

at $t = 1/2$.

Solution   As we have computed before, the derivative of the unit tangent vector with respect to t is

$$\frac{d\mathbf{T}}{dt} = \langle \frac{4t}{(1 + 2t^2)^2}, \frac{2 - 4t^2}{(1 + 2t^2)^2}, \frac{4t}{(1 + 2t^2)^2}\rangle,$$

from which we have

$$\left. \frac{d\mathbf{T}}{dt} \right|_{t=1/2} = \langle \frac{8}{9}, \frac{4}{9}, \frac{8}{9}\rangle.$$

The length of this vector is

$$\|d\mathbf{T}/dt\| = 4/3.$$

The speed is $\|\mathbf{v}(t)\| = 1 + 2t^2$, from which we have

$$\|\mathbf{v}(1/2)\| = 3/2.$$

We can now evaluate the curvature at time $t = 1/2$:

$$\kappa(1/2) = \frac{4/3}{3/2} = \frac{8}{9}$$

and the radius of curvature at time $t = 1/2$:

$$\rho(1/2) = \frac{1}{\kappa} = \frac{9}{8}.$$

■

EXAMPLE 36   Compute the curvature $\kappa(t)$ and the radius of curvature $\rho(t)$ for the curve defined by

$$\mathbf{r}(t) = \langle \cos t, \sin t, t\rangle$$

for any value t.

Solution   The derivative of the unit tangent vector with respect to t is

$$\frac{d\mathbf{T}}{dt} = \langle -\frac{\cos t}{\sqrt{2}}, -\frac{\sin t}{\sqrt{2}}, 0\rangle,$$

whose length is

$$\|d\mathbf{T}/dt\| = \sqrt{\frac{\cos^2 t}{2} + \frac{\sin^2 t}{2} + 0} = 1/\sqrt{2}$$

for any value t. The speed is $\|\mathbf{v}(t)\| = \sqrt{2}$ for any value t, so we have

$$\kappa = \frac{1/\sqrt{2}}{\sqrt{2}} = \frac{1}{2} \quad \text{and} \quad \rho = 2$$

for any value t. This helix is a curve with *constant curvature* κ.   ■

Relation between curvature and $d^2\mathbf{r}/ds^2$

If we have a parametrization by arc length, the position vector $\mathbf{r}(s)$ has constant unit speed (in other words, $\|d\mathbf{r}/ds\| = 1$). That means any change in the velocity vector $\mathbf{r}'(s)$ is purely *directional*. The acceleration vector $\mathbf{r}''(s)$ measures the rate of change in the velocity vector, so it makes sense that the magnitude of the acceleration vector $\|\mathbf{r}''(s)\|$ would be related in some way to the curvature κ. Indeed, we have

$$\kappa = \|\frac{d^2\mathbf{r}}{ds^2}\| = \|\mathbf{r}''(s)\|.$$

To see this, note that

$$\frac{d^2\mathbf{r}}{ds^2} = \frac{d}{ds}\left(\frac{d\mathbf{r}}{ds}\right) = \frac{d}{ds}\left(\frac{d\mathbf{r}/dt}{ds/dt}\right) = \frac{d}{ds}\left(\frac{\mathbf{v}(t)}{\|\mathbf{v}(t)\|}\right) = \frac{d\mathbf{T}}{ds},$$

which is precisely the curvature vector.

Restated, the curvature κ can be considered as the length of the second derivative of position with respect to arc length $d^2\mathbf{r}/ds^2$.

Curvature of space curves—an alternative formula

For space curves, we have an equivalent formulation of arc length that in practice is usually more directly computable (in the sense that we can avoid the intermediary step of finding the unit tangent vector function). This formula is

$$\kappa = \frac{\|\mathbf{r}''(t) \times \mathbf{r}'(t)\|}{\|\mathbf{r}'(t)\|^3} = \frac{\|\mathbf{a}(t) \times \mathbf{v}(t)\|}{\|\mathbf{v}(t)\|^3}.$$

So, for the special case of a curve in $\mathbb{R}^3$, the curvature κ is the length of the cross product of acceleration and velocity, divided by the cube of the speed. We postpone verifying this formula to a set of exercises at the end of this section. However, let's check it out on our previous examples.

EXAMPLE 37 Compute the curvature κ for the curve defined by

$$\mathbf{r}(t) = \langle 1 - t, t^2 + 1, 2t^3/3 + 1 \rangle$$

at $t = 1/2$.

Solution The velocity vector is $\mathbf{v}(t) = \langle -1, 2t, 2t^2 \rangle$, from which we can compute

$$\mathbf{v}(1/2) = \langle -1, 1, 1/2 \rangle \quad \text{and} \quad \|\mathbf{v}(1/2)\| = \sqrt{9/4} = 3/2.$$

The acceleration vector is $\mathbf{a}(t) = \langle 0, 2, 4t \rangle$, from which we can compute

$$\mathbf{a}(1/2) = \langle 0, 2, 2 \rangle.$$

The cross product of the acceleration and velocity vectors is

$$\mathbf{a}(1/2) \times \mathbf{v}(1/2) = \begin{vmatrix} \mathbf{i} & \mathbf{j} & \mathbf{k} \\ 0 & 2 & 2 \\ -1 & 1 & 1/2 \end{vmatrix} = \langle -1, -2, 2 \rangle,$$

so our alternative formula for curvature yields

$$\kappa = \frac{\|\mathbf{a}(1/2) \times \mathbf{v}(1/2)\|}{\|\mathbf{v}(1/2)\|^3} = \frac{\sqrt{(-1)^2 + (-2)^2 + 2^2}}{(3/2)^3} = \frac{\sqrt{9}}{27/8} = \frac{8}{9},$$

matching our previous computation. ∎

EXAMPLE 38 Compute the curvature $\kappa(t)$ for the curve defined by

$$\mathbf{r}(t) = \langle \cos t, \sin t, t \rangle$$

for any value t.

Solution The velocity and acceleration vectors are

$$\mathbf{v}(t) = \langle -\sin t, \cos t, 1 \rangle \quad \text{and} \quad \mathbf{a}(t) = \langle -\cos t, -\sin t, 0 \rangle.$$

The speed at any time t is

$$\|\mathbf{v}(t)\| = \sqrt{(-\sin t)^2 + (\cos t)^2 + 1} = \sqrt{2},$$

and

$$\mathbf{a}(t) \times \mathbf{v}(t) = \begin{vmatrix} \mathbf{i} & \mathbf{j} & \mathbf{k} \\ -\cos t & -\sin t & 0 \\ -\sin t & \cos t & 1 \end{vmatrix} = \langle -\sin t, \cos t, -1 \rangle.$$

Hence, at any time t we have curvature

$$\kappa = \frac{\|\mathbf{a}(t) \times \mathbf{v}(t)\|}{\|\mathbf{v}(t)\|^3} = \frac{\sqrt{(-\sin t)^2 + (\cos t)^2 + 1}}{(\sqrt{2})^3} = \frac{\sqrt{2}}{2\sqrt{2}} = \frac{1}{2},$$

the same constant curvature we computed before. ∎

Curvature of plane curves—an alternative formula

Another formula for κ if $\mathbf{r}(t) = \langle x(t), y(t) \rangle$ is simply

$$\kappa = \frac{|x'y'' - y'x''|}{((x')^2 + (y')^2)^{3/2}}.$$

This can be verified by first noting that

$$\mathbf{T} = \frac{\mathbf{v}}{||\mathbf{v}||} = \langle \frac{x'}{\sqrt{(x')^2 + (y')^2}}, \frac{y'}{\sqrt{(x')^2 + (y')^2}}$$

and then computing

$$\frac{d\mathbf{T}}{ds} = \frac{d\mathbf{T}/dt}{||\mathbf{v}(t)||} = \frac{d\mathbf{T}/dt}{\sqrt{((x')^2 + (y')^2)}}.$$

The length of the resulting vector is precisely the expression above.

Curvature of function graphs and the osculating circle

At each point $(x_0, f(x_0))$ of the graph of $y = f(x)$, we can use the first derivative to find the slope $f'(x_0)$ of the graph. We can use this information to derive the equation of the tangent line (graph of the best linear approximation) to f:

$$y = y_0 + m(x - x_0)$$

where $y_0 = f(x_0)$ and $m = f'(x_0)$.

The **curvature** of the graph of $y = f(x)$ at the point $(x_0, f(x_0))$ is

$$\kappa = \frac{|f''(x_0)|}{(1 + [f'(x_0)]^2)^{3/2}}.$$

EXAMPLE 39   Calculate the curvature κ and the radius of curvature r at the point $(1, -1)$ on the graph of $y = f(x) = x^3 - 2x$.

Solution   The first derivative is $f'(x) = 3x^2 - 2$ and the second derivative is $f''(x) = 6x$. Using the formula for κ with $x_0 = 1$, we have

$$\kappa = \frac{6}{(1 + [1]^2)^{3/2}} = \frac{6}{2^{3/2}} \approx 2.12$$

and the radius of curvature is $r = 1/\kappa \approx 0.4714$.   ■

Note that the formula for curvature κ depends both on the concavity, as measured by the second derivative f'', as well as the slope, as measured by f'.

If the definition of radius of curvature is to be reasonable, then it should match our usual notion of radius when applied to a circle.

EXAMPLE 40   Find the curvature and radius of curvature at the point $(3, 4)$ on a circle of radius 5 centered at the origin.

Solution   The top "half" of the circle can be represented as the graph of the function

$$f : x \longmapsto \sqrt{25 - x^2}$$

over the interval $[-5, 5]$. The first derivative is

$$f' : x \longmapsto \frac{-x}{\sqrt{25 - x^2}}$$

and the second derivative is

$$f'' : x \longmapsto \frac{-25}{(25 - x^2)^{3/2}}.$$

From these derivatives, we have $f'(3) = -3/4$ and $f''(3) = -25/64$. The curvature is

$$\kappa = \frac{|-25/64|}{(1 + [-3/4]^2)^{3/2}} = \frac{25/64}{(25/16)^{3/2}} = \frac{25/64}{125/64} = \frac{1}{5},$$

and hence the radius of curvature is $r = 1/\kappa = 1/(1/5) = 5$, the same as the radius of the original circle.   ∎

We can use the radius of curvature to define the **osculating circle** (or **circle of curvature** for a graph $y = f(x)$. The osculating circle for f at the point $(x_0, f(x_0))$ is a circle whose tangent at that point is the same as the tangent line to the function graph, and whose radius is the same as the radius of curvature of the function graph at that point. There are two such circles satisfying these requirements (they "kiss" at the point of tangency). THE osculating circle is the one on the same side of the tangent line as the function graph. Figure 13.18 illustrates the osculating circle at the point $(1, -1)$ for the graph of $y = f(x) = x^3 - 2x$.

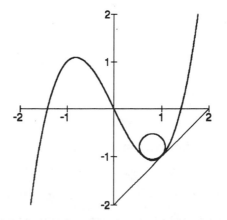

Figure 13.18 Osculating circle and tangent line.

EXERCISES

For exercises 1-3, consider the vector function for a curve in $\mathbb{R}^2$:

$$\mathbf{r}(t) = \langle 3t^2, \sqrt{t} \rangle \qquad \text{for } 1 \le t \le 3.$$

1. Find $\mathbf{T}(t)$ and $\mathbf{T}(2)$.

2. Find $\mathbf{N}(t)$ and $\mathbf{N}(2)$.

3. Find the curvature κ and the radius of curvature ρ at $t = 2$.

For exercises 4-7, suppose a particle moves on the path

$$\mathbf{r} : [0, 2] \to \mathbb{R}^3$$

$$\mathbf{r}(t) = \langle \sin 3t, \cos 3t, 2t^{3/2} \rangle.$$

4. Find $\mathbf{T}(t)$ and $\mathbf{T}(1)$.

5. Find $\mathbf{N}(t)$ and $\mathbf{N}(1)$.

6. Find the curvature κ and the radius of curvature ρ at $t = 1$.

7. Find the equation of the osculating plane at $t = 1$.

For exercises 8-11, consider the path of a moving object described by the following position function:

$$\mathbf{r} : [0, 2] \mapsto \mathbb{R}^3$$

$$\mathbf{r}(t) = \left\langle \frac{t^2}{2}, \frac{4}{3}t^{3/2}, 2t \right\rangle.$$

8. Find $\mathbf{T}(t)$ and $\mathbf{T}(1)$.

9. Find $\mathbf{N}(t)$ and $\mathbf{N}(1)$.

10. Find the curvature κ and the radius of curvature ρ at $t = 1$.

11. Find the equation of the osculating plane at $t = 1$.

For exercises 12-15, consider the path of a moving object described by the following position function:

$$\mathbf{r} : [\tfrac{1}{2}, e] \to \mathbb{R}^3$$

$$\mathbf{r}(t) = \langle t^2, \ln t, e^t \rangle.$$

12. Find $\mathbf{T}(t)$ and $\mathbf{T}(1)$.

13. Find $\mathbf{N}(t)$ and $\mathbf{N}(1)$.

14. Find the curvature κ and the radius of curvature ρ at $t = 1$.

15. Find the equation of the osculating plane at $t = 1$.

The **torsion** of a space curve at a particular point is a measure of the "twist" in the curve at that instant. To find the torsion, we first need to define the **unit binormal vector**

$$\mathbf{B}(t) = \mathbf{N}(t) \times \mathbf{T}(t).$$

Notice that the unit binormal vector is mutually orthogonal to both unit tangent vector and the principal unit normal vector, and has unit length (why?). As such, we know that

$$\mathbf{B} \cdot \mathbf{T} = 0 \quad \text{and} \quad \mathbf{B} \cdot \mathbf{B} = 1.$$

16. Differentiate $\mathbf{B} \cdot \mathbf{T} = 0$ with respect to s, and use the fact that $d\mathbf{T}/ds$ has the same direction as $\mathbf{N}$ to explain why $d\mathbf{B}/ds$ must be orthogonal to $\mathbf{T}$.

17. Differentiate $\mathbf{B} \cdot \mathbf{B} = 1$ with respect to s and use that to explain why $d\mathbf{B}/ds$ must be orthogonal to $\mathbf{B}$.

If $d\mathbf{B}/ds$ is orthogonal to both $\mathbf{T}$ and $\mathbf{B}$, then it must have the same or opposite direction as $\mathbf{N}$. This means that there is a scalar a such that

$$d\mathbf{B}/ds = a\mathbf{N}.$$

The **torsion** is $\tau = -a$.

18. Find the unit binormal vector $\mathbf{B}(1)$ for each of the three space curves given in the instructions for exercises 4-15.

19. Find $d\mathbf{B}/ds$ for each of the three space curves given in the instructions for exercises 4-15.

20. Find the torsion τ at $t = 1$ for each of the three space curves given in the instructions for exercises 4-15.

In exercises 21-30, find the curvature κ, radius of curvature ρ, and the equation of the osculating circle to the given function f at the specified point x_0. Then graph both the circle and the original function in an appropriate viewing window centered at $(x_0, f(x_0))$.

21. $f : x \longmapsto x^2 + x + 1; \quad x_0 = 1.6$

22. $f : x \longmapsto 2x^2 - 7x + 3; \quad x_0 = 3/2$

23. $f : x \longmapsto x^3 - 12x; \quad x_0 = -2$

24. $f : x \longmapsto \dfrac{3x - 17}{5}; \quad x_0 = 4$

25. $f : x \longmapsto x^5; \quad x_0 = 1$

26. $f : x \longmapsto 3; \quad x_0 = 2$

27. $f : x \longmapsto \sin^2 x; \quad x_0 = \pi/3$

28. $f : x \longmapsto \tan x; \quad x_0 = 3\pi/4$

29. $f : x \longmapsto \sin x^2; \quad x_0 = 0$

30. $f : x \longmapsto \sqrt{25 - x^2}; \quad x_0 = 4$

---

31. Verify the curvature formula for plane curves

$$\kappa(t) = \frac{|x'y'' - y'x''|}{((x')^2 + (y')^2)^{3/2}}$$

by considering a position function $\mathbf{r}(t) = \langle x(t), y(t) \rangle$ and computing $\kappa = \dfrac{d\mathbf{T}/dt}{\|\mathbf{r}'(t)\|}$ directly.

32. A function graph $y = f(x)$ can be parametrized simply by the position function $\mathbf{r}(t) = \langle t, f(t) \rangle$ for all t in the domain of the function f. Use this parametrization and the formula from the previous exercise to compute the curvature κ and verify the special formula for function graphs.

For exercises 32-38, let $\mathbf{r}$ be a smooth 3-dimensional position function and let

$$v(t) = \|\mathbf{v}(t)\| = \|\mathbf{r}'(t)\|.$$

33. Show that $\mathbf{v}(t) = v(t)\mathbf{T}(t)$.

34. Differentiate both sides of $\mathbf{v}(t) = v(t)\mathbf{T}(t)$ with respect to t to verify that

$$\mathbf{a}(t) = \frac{dv}{dt}\mathbf{T}(t) + v(t)\frac{d\mathbf{T}}{dt}.$$

35. If $\kappa(t)$ is the curvature at time t, show that $d\mathbf{T}/dt = \kappa(t)v(t)\mathbf{N}(t)$. (Hint: substitute the definitions for $\kappa(t)$, $v(t)$ and $\mathbf{N}(t)$.)

36. Use the results of exercise 34 and 35 to verify that

$$\mathbf{a}(t) = \frac{dv}{dt}\mathbf{T}(t) + \kappa(t)v^2(t)\frac{d\mathbf{T}}{dt}.$$

37. Substitute the result of exercise 36 for $\mathbf{a}(t)$ and substitute the result of exercise 31 for $\mathbf{v}(t)$ and compute

$$\mathbf{a}(t) \times \mathbf{v}(t) = \kappa(t)v^3(t)\mathbf{N}(t) \times \mathbf{T}(t).$$

38. Use exercise 37 to verify that

$$\|\mathbf{a}(t) \times \mathbf{v}(t)\| \doteq \kappa(t)v(t)^3.$$

Dividing both sides by $v(t)^3$ gives us the curvature formula for space curves.

39. Use exercise 37 to verify that

$$\frac{\mathbf{a} \times \mathbf{v}}{v^3} = \kappa \mathbf{B}$$

where $\mathbf{B}$ is the unit binormal vector.

40. Suppose $\|\mathbf{a}(t)\| = a$ and $\|\mathbf{v}(t)\| = b$ for all t, where a and b are constants. Show that the curvature κ is also constant, and express it in terms of a and b.

14

Fundamentals of Multivariable Functions

So far, we have focused virtually all our attention on *functions of a single variable*—those functions having a single real number input as independent variable.

In *Volume I*, we studied single variable functions of the form

$$y = f(x),$$

having a single real number as output. We saw how calculus can be used to measure and investigate the behavior of these functions. Differentiation and integration have both geometrical interpretations in terms of slopes and areas, as well as physical interpretations in terms of rates of change and net effect.

In the last chapter, we studied single variable functions

$$\mathbf{r}(t)$$

having *vectors* as outputs. Thinking of the independent variable t as a time parameter, and the vector output as a position, we were able to associate a path with each such function. Calculus again allowed us to investigate both geometrical properties of the path's image curve as well as the physical notions such as velocity and acceleration.

In the remainder of this book, we proceed to consider *multivariable functions*, that is, functions having more than one independent variable. In this chapter, we'll concentrate on *scalar-valued* multivariable functions. These are often called **scalar fields**. We'll discuss some techniques for visualizing and interpreting their graphical representations, and we'll examine the special examples of linear and quadratic multivariable functions in detail. In the next two chapters, we'll show how derivatives and integrals can be extended as tools for analyzing multivariable functions.

14.1　MULTIVARIABLE FUNCTIONS—EXAMPLES AND TERMINOLOGY

In this first section, we'll introduce some examples of multivariable functions and some of the language and notation we'll need to discuss them.

Functions of several variables arise naturally everywhere we look. Indeed, it is more common for us to find that a quantity depends on the values of several other variable quantities rather than just one. Many formulas from elementary geometry can be thought of as defining functions of several variables.

EXAMPLE 1　The Pythagorean Theorem's statement that

$$z = \sqrt{x^2 + y^2}$$

expresses the length of the hypotenuse z in terms of x and y, the lengths of the legs. By varying x and y, we can think of z as a function of two variables. To emphasize this functional relationship, we write

$$z(x, y) = \sqrt{x^2 + y^2}$$

(see Figure 14.1). ■

The very first functions you ever studied were all two-input functions: addition, subtraction, multiplication, and division!

EXAMPLE 2　The volume V of a box (rectangular parallelepiped) can be written

$$V(w, \ell, h) = w\ell h,$$

where w is the width, ℓ the length, and h the height of the box. Thus, the volume V is a function of three variables (see Figure 14.1). ■

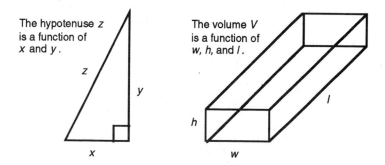

The hypotenuse z is a function of x and y.

The volume V is a function of w, h, and l.

Figure 14.1　Multivariable functions: $z = \sqrt{x^2 + y^2}$ and $V = w\ell h$.

In all of these examples, the output is a single number that is uniquely determined by the values of more than one input.

The term **scalar field** is often used in the context of physics to refer to a scalar-valued multivariable function. If we think of three inputs x, y, and z as specifying the location (x, y, z) of an object in space, then there are a variety of scalar quantities that we may associate with this location. These include temperature, humidity, barometric pressure, light intensity, etc. The values of each of these quantities may well depend on the precise location of the object. Each of these quantities defines a *field* of scalar values. For example, we think of a temperature field as "tagging" a specific temperature

$$T(x, y, z)$$

to each position in space. In other words, the temperature at a point is a function of the three variables that specify its position.

Notation for multivariable functions

The inputs to a multivariable function can specify an ordered pair (x, y) (for 2 variables), an ordered triple (x, y, z), (for 3 variables), or, in general an n-tuple $(x_1, x_2, \ldots, x_n)$ (for n variables), depending on how many real number inputs are required. As such, we think of the domain of a multivariable function as a subset of $\mathbb{R}^n$, where n denotes the number of independent variables. If the domain of a multivariable function f includes all real numbers for each of its n inputs, we can write

$$f : \mathbb{R}^n \longrightarrow \mathbb{R}.$$

If the domain D is some subset of $\mathbb{R}^n$, then we write

$$f : D \longrightarrow \mathbb{R}.$$

EXAMPLE 3   Because of the physical circumstances involved, only positive values make sense as inputs in the first two examples above. In other words, to indicate their domains, we could write

$$z : \{(x, y) : x > 0, y > 0\} \longrightarrow \mathbb{R}$$

and

$$V : \{(w, \ell, h) : w > 0, \ell > 0, h > 0\} \longrightarrow \mathbb{R}$$

for these two functions.   ■

Graph of $z = f(x, y)$—surfaces

Because we live in three spatial dimensions, we'll devote a good deal of our attention to scalar-valued functions of two variables (two inputs and one output), for they still provide a ready means of visualization. If we denote the two independent variables as x and y, and the dependent variable as z, then the graph of the function f is the set of points (x, y, z) in space satisfying the equation

$$z = f(x, y).$$

Generally, we can think of this graph as describing a *surface* in space (see Figure 14.2).

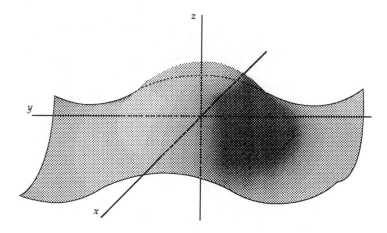

Figure 14.2 The graph of $z = f(x, y)$ is a surface.

For each ordered pair (x, y) in the domain of the function f, we plot the point

$$(x, y, z = f(x, y))$$

on the surface. In $\mathbb{R}^3$, we can think of locating the input pair (x, y) in the xy-plane (actually the point $(x, y, 0)$), and then move vertically up or down to the appropriate level for $z = f(x, y)$ to plot the point on the graph.

EXAMPLE 4   The multivariable function

$$f : \mathbb{R}^2 \longrightarrow \mathbb{R}$$

$$z = f(x, y) = \sqrt{x^2 + y^2}$$

has a graph in the shape of a cone with vertex at the origin and opening upwards. One point on the graph is $(3, 4, 5)$, since

$$f(3, 4) = \sqrt{3^2 + 4^2} = \sqrt{25} = 5.$$

Figure 14.3 illustrates how this point is located on the graph.   ■

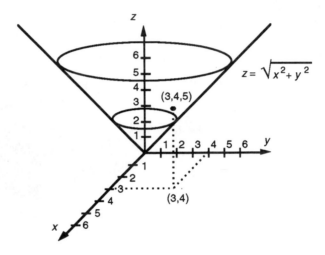

Figure 14.3 Locating a point on the graph of $f(x, y) = \sqrt{x^2 + y^2}$.

 The surface representing the graph of a function of two variables will satisfy the "vertical line test." That is, any vertical line (parallel to the z-axis) will pierce the surface in at most one point.

Although we "run out" of dimensions for pictorially representing the graph of a function with more than two inputs, we can still use the idea of a surface to motivate our thinking. Some of the other visualization techniques we'll discuss in the next section extend well to functions of any number of inputs, while others work well only for two-input functions.

Examples of surfaces

Let's start out by taking a look at several examples of surfaces. By way of comparison, let's note that a *curve* in the Cartesian plane is often described as the solution set of an equation in two variables x and y. While this equation may represent a functional relationship between the two variables, such as

$$y = x^2,$$

an equation in two variables can also represent some non-functional relationship, such as

$$x^2 + y^2 = 9,$$

whose graph is a circle of radius 3 centered at the origin.

Similarly, a surface in space can be described as the solution set to an equation in the three variables x, y, and z.

EXAMPLE 5   Since the distance between any point (x, y, z) in space and the origin is $\sqrt{x^2 + y^2 + z^2}$, the solution set of the equation

$$x^2 + y^2 + z^2 = 4$$

describes a *sphere* of radius 2, centered at the origin (see Figure 14.4).

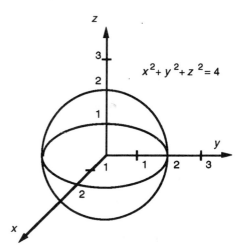

Figure 14.4 A sphere of radius 2 centered at the origin.

Note that this sphere is not the graph of a function of two variables x and y, since there can be two different values z corresponding to a single input pair (x, y). If we solve the equation for z, we have

$$z = \pm\sqrt{4 - x^2 - y^2}.$$

The graph of $z = \sqrt{4 - x^2 - y^2}$ is the upper hemisphere and the graph of $z = -\sqrt{4 - x^2 - y^2}$ is the lower hemisphere.

On the other hand, we could view the left-hand side of the original equation as a function of all three variables x, y, and z:

$$f(x, y, z) = x^2 + y^2 + z^2.$$

The sphere of radius 2 centered at the origin is now exactly the set of *all* inputs in $\mathbb{R}^3$ that produce the output

$$f(x, y, z) = 4.$$

■

EXAMPLE 6   The solution set of the linear equation

$$x + 2y + 3z = 6$$

describes a *plane*. Three points in this plane that we can identify readily are the *intercepts* with each of the three coordinate axes:

$$(6, 0, 0) \qquad (0, 3, 0) \qquad (0, 0, 2).$$

The triangle determined by these three points is shown in Figure 14.5.

This plane can be thought of as the graph of the function

$$z = f(x, y) = \frac{6 - x - 2y}{3}.$$

■

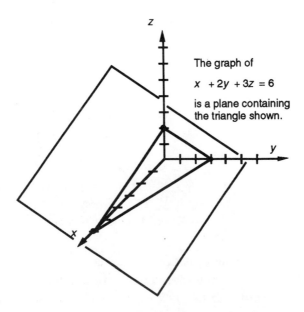

Figure 14.5 Three coordinate axis intercepts of the plane $x + 2y + 3z = 6$.

EXAMPLE 7   The graph of an equation in three variables could consist of a single point. The only point on the graph of $x^2 + y^2 + z^2 = 0$ is the origin $(0, 0, 0)$. A graph could even be empty, as is the case with the graph of $x^2 + y^2 + z^2 = -1$, which has no solution in real numbers x, y, and z.   ■

Cylindrical surfaces

A cylinder can be thought of as the surface generated by translating or sliding a circle along an axis. In general, a *cylindrical surface* is generated by translating a curve along some line. The graph of an equation involving only 2 of the 3 variables x, y, and z will be a cylindrical surface in space. We simply graph the curve in the appropriate coordinate plane and then visualize the surface traced out as we translate it parallel to the axis of the missing variable.

EXAMPLE 8   The graph of $x^2 + y^2 = 4$ is an infinite cylinder of radius 2 with the z-axis running down its center. (Note that z is not involved in the equation.) The graph is depicted in Figure 14.6.   ■

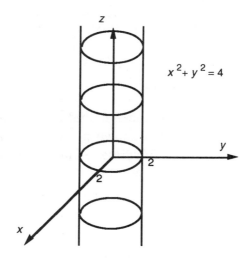

Figure 14.6 The infinite cylinder $x^2 + y^2 = 4$.

EXAMPLE 9   The graph of $z = y^2$ is the infinite "trough" generated by translating the appropriate parabola parallel to the x-axis. (Note that x is not involved in the equation.) Part of the graph is shown in Figure 14.7.   ■

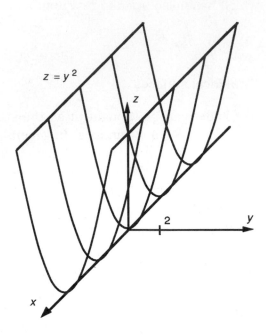

Figure 14.7 The infinite "trough" $z = y^2$.

Quadric surfaces

The sphere, the cylinder, and the parabolic "trough" all belong to a special class of surfaces known as the **quadric surfaces**. A quadric surface is the graph of a quadratic equation in three variables, that is, an equation involving only terms of degree 2 and lower.

You are familiar with quadratic equations in two variables. Their graphs in the Cartesian plane are known as **conic sections**: *ellipses* (including circles), *hyperbolas*, and *parabolas*. Each of the conic sections is a curve with very special geometric properties.

Quadric surfaces provide an important set of examples, because upon close inspection, many surfaces behave very much like quadric surfaces. Below, we look at several quadric surfaces. Take special note of the *cross-sections*, that is, the curves representing the intersection of the surface with a plane parallel to one of the coordinate planes. The cross-sections of quadric surfaces are also conic sections.

EXAMPLE 10 As already mentioned, a **sphere** is a quadric surface. In general, a sphere of radius R centered at the origin has equation

$$x^2 + y^2 + z^2 = R^2.$$

All cross-sections of a sphere are circles (see Figure 14.8). ■

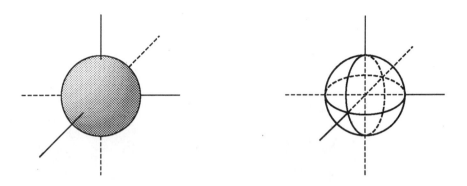

Figure 14.8 The sphere $x^2 + y^2 + z^2 = R^2$ and its cross-sections.

EXAMPLE 11 The sphere is a special case of an **ellipsoid**. An ellipsoid centered at the origin has an equation of the form

$$\frac{x^2}{a^2} + \frac{y^2}{b^2} + \frac{z^2}{c^2} = 1,$$

where a, b, and c are the distances from the origin to the intercepts with the x-axis, y-axis, and z-axis, respectively. As the name might suggest, the cross-sections of an ellipsoid are ellipses (see Figure 14.9). ■

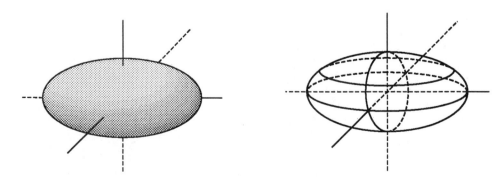

Figure 14.9 The ellipsoid $\frac{x^2}{a^2} + \frac{y^2}{b^2} + \frac{z^2}{c^2} = 1$ and its cross-sections.

EXAMPLE 12   An **elliptic cylinder** having the z-axis running down its center has an equation of the form

$$\frac{x^2}{a^2} + \frac{y^2}{b^2} = 1.$$

Each horizontal cross-section is an ellipse (see Figure 14.10).   ■

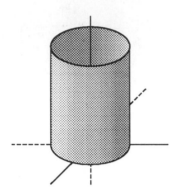 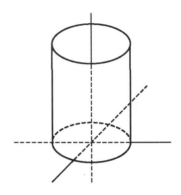

Figure 14.10 The elliptic cylinder $\dfrac{x^2}{a^2} + \dfrac{y^2}{b^2} = 1$ and its cross-sections.

EXAMPLE 13   An **elliptic cone** having the z-axis running down its center has an equation of the form

$$\frac{z^2}{c^2} = \frac{x^2}{a^2} + \frac{y^2}{b^2}.$$

Again, note that each horizontal cross-section is an ellipse (see Figure 14.11). The vertical cross-sections are hyperbolas, with the exception of the xz-plane and the yz-plane. In these two cases, the vertical cross-sections are two intersecting lines.   ■

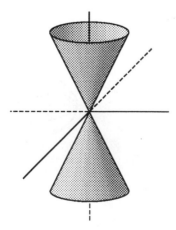 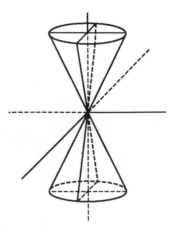

Figure 14.11 The elliptic cone $\dfrac{z^2}{c^2} = \dfrac{x^2}{a^2} + \dfrac{y^2}{b^2}$ and its cross-sections.

EXAMPLE 14   A **hyperboloid of one sheet** having the z-axis running down its center has an equation of the form

$$\frac{z^2}{c^2} = \frac{x^2}{a^2} + \frac{y^2}{b^2} - 1,$$

and a **hyperboloid of two sheets** has an equation of the form

$$\frac{z^2}{c^2} = \frac{x^2}{a^2} + \frac{y^2}{b^2} + 1.$$

The horizontal cross-sections of both hyperboloids are ellipses, while the vertical cross-sections are hyperbolas (see Figures 14.12 and 14.13).   ■

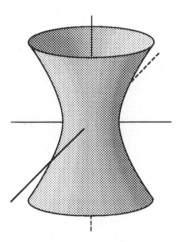 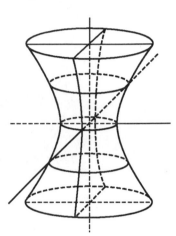

Figure 14.12 The hyperboloid of one sheet $\dfrac{z^2}{c^2} = \dfrac{x^2}{a^2} + \dfrac{y^2}{b^2} - 1$ and its cross-sections.

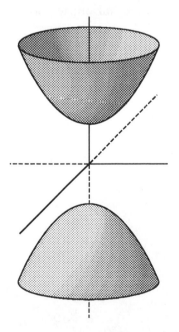 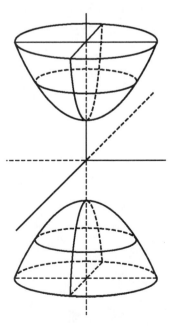

Figure 14.13 The hyperboloid of two sheets $\dfrac{z^2}{c^2} = \dfrac{x^2}{a^2} + \dfrac{y^2}{b^2} + 1$ and its cross-sections.

EXAMPLE 15   An **elliptic paraboloid** having the z-axis running down its center has an equation of the form

$$cz = \frac{x^2}{a^2} + \frac{y^2}{b^2}.$$

The horizontal cross-sections are ellipses, while the vertical cross-sections are parabolas. The sign of c determines whether the surface is "bowl up" ($c > 0$) or "bowl down" ($c < 0$). Figure 14.14 illustrates an elliptic paraboloid with $c > 0$.   ■

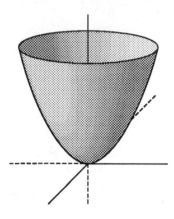
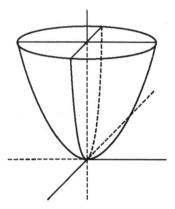

Figure 14.14 The elliptic paraboloid $cz = \dfrac{x^2}{a^2} + \dfrac{y^2}{b^2}$ $(c > 0)$ and its cross-sections.

EXAMPLE 16   By changing a sign in the previous equation, we obtain the equation of a **hyperbolic paraboloid**:

$$cz = \frac{x^2}{a^2} - \frac{y^2}{b^2}.$$

Now, the horizontal cross-sections are hyperbolas, while the vertical cross-sections are parabolas. If $c < 0$, cross-sections parallel to the xz-plane open down while cross-sections parallel to the yz-plane open up (see Figure 14.15). If $c > 0$, the situation is reversed. In either case, the surface has the shape of a "saddle."   ■

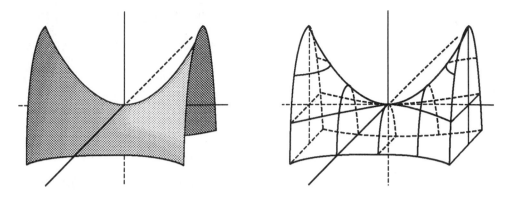

Figure 14.15 The hyperbolic paraboloid $cz = \dfrac{x^2}{a^2} - \dfrac{y^2}{b^2}$ ($c < 0$) and its cross-sections.

The origin $(0, 0, 0)$ is a key point for many of the quadric surfaces described above. The sphere and ellipsoid have the origin at their centers. The elliptic cone has the origin at its vertex, as does the elliptic paraboloid. The origin is a point of symmetry for both hyperboloids, and is the "saddle point" of the hyperbolic paraboloid.

Given any point (x_0, y_0, z_0), we can translate any of these surfaces so that (x_0, y_0, z_0) acts as the new "origin," simply by replacing x by $x - x_0$, y by $y - y_0$, and z by $z - z_0$. The new "axes" all pass through this point, but are still parallel to the original coordinate axes.

EXAMPLE 17 What is the equation of the ellipsoid whose center is $(1, -2, 3)$, and whose three "radii" are 4, 5, and 2 in the directions parallel to the x-axis, y-axis, and z-axis, respectively?

Solution We have $a = 4$, $b = 5$, $c = 2$, and $(x_0, y_0, z_0) = (1, -2, 3)$. The equation of the ellipsoid is

$$\frac{(x-1)^2}{16} + \frac{(y+2)^2}{25} + \frac{(z-3)^2}{4} = 1.$$

∎

After expanding and collecting terms, we can arrive at an equivalent equation for this ellipse:

$$25x^2 + 16y^2 + 100z^2 - 50x + 64y - 600z + 589 = 0.$$

There are still no mixed terms (xy, xz, or yz) involved. Quadric surfaces whose equations include mixed terms can have axes with directions not parallel to any of the coordinate axes.

EXERCISES

1. The graphs of the addition and subtraction functions

$$z = a(x, y) = x + y \quad \text{and} \quad z = s(x, y) = x - y$$

are two planes, each containing the origin $(0, 0, 0)$. Describe the line of intersection of these two planes.

2. The graph of the multiplication function

$$z = m(x, y) = xy$$

also contains the origin $(0, 0, 0)$. Is this graph also a plane?

3. What is the domain of the division function

$$z = d(x, y) = x/y,$$

that is, what is the set of ordered pairs (x, y) that are acceptable inputs for this function?

Sketch each of the cylindrical surfaces whose equations are given in exercises 4-9.

4. $xy = 1$.

5. $xz = 1$.

6. $yz = 1$.

7. $x^2 + z^2 = 9$.

8. $\dfrac{y^2}{9} + \dfrac{z^2}{16} = 1$.

9. $|x| + |y| = 1$.

---

10. When is the graph of a function $z = f(x, y)$ a cylindrical surface? Give an example.

11. What is the equation of a sphere of radius 7, centered at the point $(-2, 4, -7)$?

12. What is the equation of a circular cylinder of radius 7, with a vertical axis containing the point $(-2, 4, -7)$?

13. For what values a, b, and c is the ellipsoid of Figure 14.9 the same as the sphere of Figure 14.8?

For each of the quadric surfaces described in exercises 14-20, indicate whether or not it could be the graph of a function

$$z = f(x, y)$$

by using the vertical line test.

14. The ellipsoid shown in Figure 14.9.

15. The elliptic cylinder shown in Figure 14.10.

16. The elliptic cone shown in Figure 14.11.

17. The hyperboloid of one sheet shown in Figure 14.12.

18. The hyperboloid of two sheets shown in Figure 14.13.

19. The elliptic paraboloid shown in Figure 14.14.

20. The hyperbolic paraboloid shown in Figure 14.15.

Each of the quadric surfaces whose equation is given in exercises 21-26 has the z-axis as its "central axis." In each exercise, make the necessary change to the equation so that the type of quadric surface is the same, but the central axis is now the x-axis. Then make the change so that the central axis is the y-axis.

21. The elliptic cylinder shown in Figure 14.10.

22. The elliptic cone shown in Figure 14.11.

23. The hyperboloid of one sheet shown in Figure 14.12.

24. The hyperboloid of two sheets shown in Figure 14.13.

25. The elliptic paraboloid shown in Figure 14.14.

26. The hyperbolic paraboloid shown in Figure 14.15.

14.2 VISUALIZING AND INTERPRETING MULTIVARIABLE FUNCTIONS

All of our visual communications media (written material, computer and calculator screens, and even the retinal surfaces of our eyes, for that matter) present at most two dimensions, so multivariable functions require more effort to be represented completely and unambiguously. In this section we will explore a number of methods for graphically representing multivariable functions and for interpreting these graphs. Although no single technique can capture all of a multivariable function's behavior, together they provide an important aid to understanding.

Plotting individual points in a haphazard or random manner is usually not a good way to graph an equation in three variables (because it's so hard to figure out how to "connect the dots"). It is much more sensible to graph these equations in some systematic manner. Let's turn now to some strategies for graphing multivariable functions of the form

$$z = f(x, y).$$

Slicing

Suppose we *freeze* one of the inputs of $z = f(x, y)$. The result can be thought of as a single-input function that can be graphed, visualized, and interpreted like any other function of one variable. The graph of this new function is a cross-sectional *slice* of the surface graph of the original function.

EXAMPLE 18   Find and graph the slices of the two-input function

$$f(x, y) = x^2 + y^2$$

obtained by freezing the second input at $y = -2, -1, 0, 1$, and 2.

Solution   If we *freeze* the second input at $y = -2$, the result is a function of x alone:

$$f(x, -2) = x^2 + 4.$$

If instead, we freeze the second input at $y = -1$, the result is a different, but related function of x:

$$f(x, -1) = x^2 + 1.$$

Continuing in this manner, we also have

$$f(x, 0) = x^2, \qquad f(x, 1) = x^2 + 1, \qquad f(x, 2) = x^2 + 4.$$

The graphs of these "slice functions" are shown in Figure 14.16.   ∎

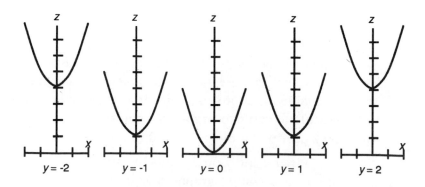

Figure 14.16 Graphs of slice functions.

The function, $z = f(x, c)$, for each value $y = c$, corresponds to restricting the domain of f to the set of inputs

$$\{(x, y) : x \in \mathbb{R}, y = c\}.$$

The graphs of these slice functions can now be placed in the appropriate plane, corresponding to the value of the frozen input. The graph of $f(x, -2) = x^2 + 4$ is placed in the vertical plane $y = -2$, the graph of $f(x, -1) = x^2 + 1$ is placed in the vertical plane $y = -1$, and so on.

Figure 14.17 shows the relative positions of the slices in three dimensions. Each slice is a parabola with the vertex height depending on the value y.

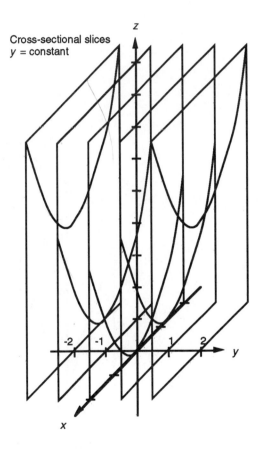

Figure 14.17 Visualizing the slices (for y frozen) of $z = x^2 + y^2$ in $\mathbb{R}^3$.

You can think of the vertical planes $y = c$ as panes of glass (or transparency slides) with the graphs drawn on them. The graph of a slice gives us an image along a single direction of the shape of the graph of $z = f(x, y)$ in the same way that the edge of a slice of bread gives us an image (along one direction) of the shape of the loaf.

Of course, we can also slice the same surface with the first input x frozen. Now we plot the function of the form $z = f(c, y)$ for some fixed value $x = c$. This corresponds to restricting the domain of f to the set of inputs

$$\{(x, y) : y \in \mathbb{R}, x = c\}.$$

Each slice can then be placed in the vertical plane $x = c$.

EXAMPLE 19   Figure 14.18 illustrates the slices of the two-input function

$$z = f(x, y) = x^2 + y^2$$

obtained by freezing the first input at $x = -2, -1, 0, 1,$ and 2. In order from smallest value of x to largest, these graphs are of

$$z = 4 + y^2, \quad z = 1 + y^2, \quad z = y^2, \quad z = 1 + y^2, \quad z = 4 + y^2.$$

■

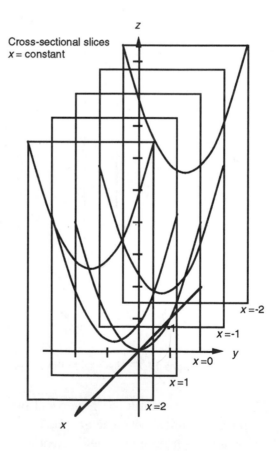

Figure 14.18 Visualizing the slices (for x frozen) of $z = x^2 + y^2$ in $\mathbb{R}^3$.

If we are especially interested in the "shape" of a function graph near a particular point, it is often useful to graph each of the cross-sectional slices at that point. This is analogous to first slicing a loaf of bread cross-wise and then length-wise to examine the shape of the loaf near the intersection.

Repeating this process of setting one of the variables equal to a constant, graphing the result and placing the result in the appropriate plane for several values of both inputs can give us a good visual image of the graph of the surface $z = f(x, y)$. Figure 14.19 shows several slices of $z = x^2 + y^2$ for both x and y. The "fishnet" has a shape known as a *paraboloid of revolution*, since the graph can be generated by rotating a parabola about the z-axis.

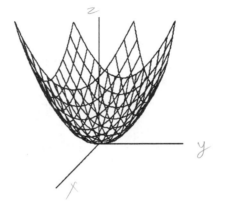

Figure 14.19 Several cross-sections of the paraboloid $z = x^2 + y^2$.

Slicing functions with more than two inputs

We cannot graph a function of three variables

$$w = f(x, y, z)$$

because we need four dimensions (three inputs, one output). However, slicing is one of the few methods that also works reasonably well for functions with three or more inputs.

We freeze all but one of the inputs and graph the resulting single-input function, repeating this for various choices of the frozen inputs. It is helpful to arrange the resulting graphs in an orderly manner reflecting the choice of frozen inputs.

EXAMPLE 20   Graph the y-slices of the function

$$w = f(x, y, z) = 2.62 + 0.6x + 0.4y + z - x^2 - y^2 - z^2$$

over the region $[0, 0.6] \times [0, 0.6] \times [0, 0.6]$ obtained by freezing both x and z at all possible combinations of $x = 0$, 0.2, 0.4, 0.6 and $z = 0$, 0.2, 0.4, 0.6 and graphing the resulting single variable functions of y.

Solution   The region $[0, 0.6] \times [0, 0.6] \times [0, 0.6]$ represents all points (x, y, z) satisfying

$$0 \leq x, y, z \leq 0.6$$

(the cube with one corner at the origin $(0, 0, 0)$ and the opposite corner at $(0.6, 0.6, 0.6)$). Figure 14.20 shows the sixteen different slices arranged by the values of x and z.

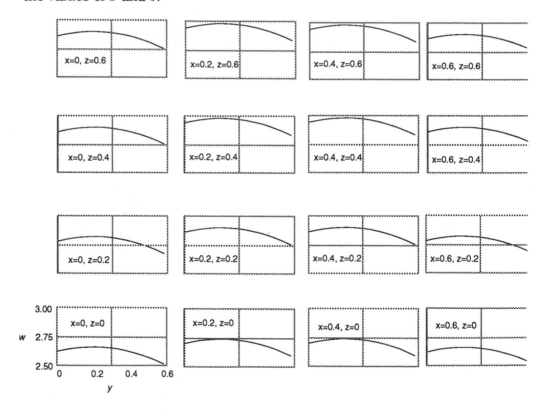

Figure 14.20 Slices of a function graph $w = f(x, y, z)$ with respect to y.

Note the scaling on each window (shown on the lower left-hand window): along the horizontal, the y-range runs from 0 to 0.6; along the vertical, the w-range runs from 2.5 to 3.0. The dotted "crosshairs" provide a visual reference and do *not* represent the y- and w-axes.

To illustrate, we find the output value w corresponding to

$$(x, y, z) = (0.4, 0.6, 0.2)$$

in the third picture of the third row from the top. At the far right hand edge of the window, we can see that

$$f(0.4, 0.6, 0.2) \approx 2.75,$$

and you can check that the actual function value at this point is 2.74. ■

A scientist may use this method when studying a quantity that depends on several variables. By fixing the values of all the variables except one, she can isolate the effect of varying the lone remaining independent variable on the value of the dependent variable. This process of replacing all but one of the inputs to a function with fixed values to produce a single-input slice function is one important enough to be named. It is called *Currying* in honor of the mathematician Haskel Curry.

EXAMPLE 21   Use the slices in Figure 14.20 to determine approximately where the relative maximum output of f occurs over the region $[0, 0.6] \times [0, 0.6] \times [0, 0.6]$.

Solution   In just the few y-slices taken, the very highest w values appear to be approximately 3.0 in four different windows, namely the second and third windows of the top two rows. The highest point in each of these four windows appears to occur near $y = 0.2$. In fact, the highest point in each of the sixteen windows appears to occur near $y = 0.2$. The top two rows correspond to $z = 0.6$ and $z = 0.4$, while the middle two columns correspond to $x = 0.2$ and $x = 0.4$. Averaging the z-values and x-values, we might make a reasonable guess that a maximum occurs near $(x = 0.3, y = 0.2, z = 0.5)$.

You can verify that

$$2.62 + 0.6x + 0.4y + z - x^2 - y^2 - z^2 = (1 - (x - .3)^2) + (1 - (y - .2)^2) + (1 - (z - .5)^2)$$

by expanding out the right-hand side and collecting like terms. Using this form of the function, we can see that the maximum value of $w = 3$ actually does occur exactly at $(x = 0.3, y = 0.2, z = 0.5)$. ■

Contour plots

A variant of the slicing method is widely used in graphing elevation (that is, height above sea level) of regions of the earth. A **contour** is a "line of constant elevation." A contour may also be called a **level curve**. A contour map of a region shows and labels several level curves.

Figure 14.21 shows a landscape with two tall hills and a dry lake bed nearby.

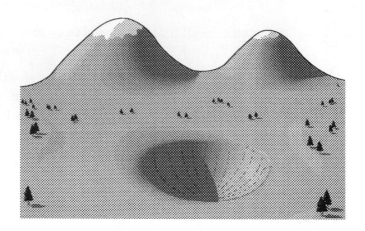

Figure 14.21 A landscape.

A contour map of the region pictured in the previous illustration is shown in Figure 14.22.

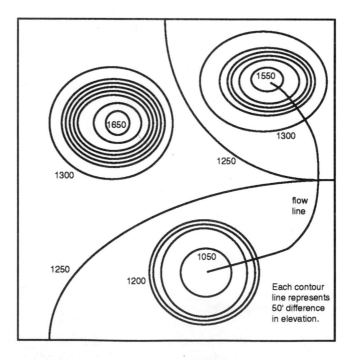

Figure 14.22 Contour map: each line represents 50 feet of elevation change.

Near the hilltops there are small, roughly circular contours. Similarly, near the bottom of the lake bed are more roughly circular contours. Between two given contour lines, the land is steepest where the curves are closest together. The direction perpendicular to the contour is called the "fall line" or "flow line," and represents the direction of steepest ascent/descent. One flow line is shown in the contour map above. It runs from the top of one of the hills onto the plain, and then down a gully to the bottom of the lake bed.

Instead of the cross-sectional slices made by vertical planes, contours represent the level curves of a surface sliced along horizontal planes. For a function

$$z = f(x,y),$$

a level curve is found by setting $z = c$ for a constant value c and graphing the result. This curve (or curves) represent the set of inputs (x,y) that produce the output

$$z = f(x,y) = c.$$

For three-dimensional graphing, the level curve can be placed in the horizontal plane $z = c$.

EXAMPLE 22   Graph the level curves of $z = f(x,y) = x^2 + y^2$ corresponding to $z = 0$, 1, 2, 3, and 4. (The graph has no points corresponding to negative values of z.)

Solution   The level curve corresponding to $z = 0$ is a single point $(0,0)$, since $x = 0$ and $y = 0$ is the only pair of real values satisfying

$$0 = x^2 + y^2.$$

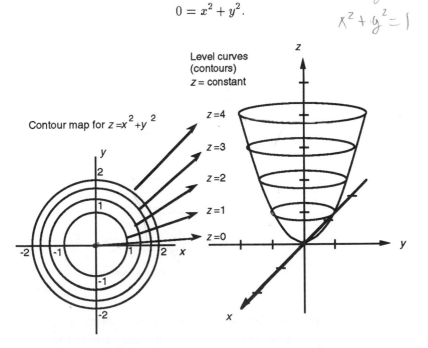

Figure 14.23 Graph of $z = x^2 + y^2$ showing level curves.

The other level curves are circles with centers at $(0,0)$. The contours are shown in Figure 14.23, along with their relationship to the graph of the surface.   ■

 When viewed from above, contours of a function graph

$$c = f(\dot{x}, y)$$

representing different values $z = c$ cannot overlap, because only one output is associated with any given input pair (x, y).

The "mountaintops" and the "lake bottoms" of a surface represent its *local maxima* and *local minima*. For the paraboloid $z = x^2 + y^2$, there is a single local extremum (a minimum) at $(0,0)$. In fact, the absolute minimum output is 0, and there is no maximum output.

EXAMPLE 23   The contours of a function were plotted over the domain $[0,3] \times [0,3]$ for values

$$c = \ldots, \; -1.0, \; -0.5, \; 0, \; 0.5, \; 1.0, \; 1.5, \; \ldots$$

and are shown in Figure 14.24. Use them to determine the approximate locations of the extrema (maxima and minima) and the location of the steepest region of the graph over this region.

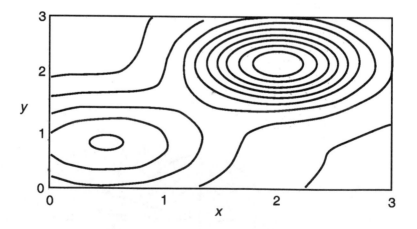

Figure 14.24 Contour plot of the function f.

Solution   From this picture we see that the extrema must be located near $(0.5, 1)$ and $(2, 2)$. The steep regions surround these extrema, and the flattest regions are off in the opposite corners. (To identify each extremum as either a relative maximum or minimum, we need to know the actual value c associated to each level curve shown.) ■

The cross-sections of a function graph $z = f(x,y)$ are usually easy to graph with most graphing software or graphing calculators. Just substitute the appropriate frozen value $x = c$ or $y = c$ and graph the one-input function (of the other variable) that results.

Level curves can be more difficult to plot, because the graph of

$$c = f(x,y)$$

is a more general equation in two variables. In the exercises, we discuss a way you can use slope fields to obtain contour plots.

Level surfaces

For a function of three variables $w = f(x,y,z)$, we have **level surfaces** determined by setting the output w equal to a constant. The level surface $c = f(x,y,z)$ gives us the set of inputs (x,y,z) in space that all produce the same output $w = c$.

EXAMPLE 24 A heat source is placed at the origin. After some time elapses, the temperature at point (x,y,z) is given by the function

$$T(x,y,z) = \frac{300}{1 + x^2 + y^2 + z^2},$$

where the temperature is 300 degrees Kelvin at the heat source. Describe the level surfaces of the function T for temperatures $c \le 300$.

Solution Setting $T(x,y,z) = c$, we have

$$\frac{300}{1 + x^2 + y^2 + z^2} = c,$$

or equivalently,

$$\frac{300}{c} - 1 = x^2 + y^2 + z^2.$$

The level surfaces are concentric spheres centered at the origin. The set of points having temperature $T = c$ lie on the sphere of radius $(300/c) - 1$ for $c < 300$ (and just the origin $(0,0,0)$ for $T = 300$). ■

Perspective

Imagine taking the graph of a single-input function, drawn on a sheet of paper, standing it on edge, and observing it from various locations in the room. As you move away from it, the image you see gets smaller. As you move your head up, the image seems to move down, as you move left, the image moves right. These simple observations are the basis for graphing a two-input function in *perspective*. The technique of drawing in perspective was discovered and explored in the Renaissance, and quickly became an important element of graphical artistry.

If we take the set of graphs produced by slicing the graph of a function of two variables $z = f(x, y)$, and shrink them in proportion to their distance from an imaginary "eye point," offsetting them left-to-right and up and down in the same manner, and then combine the resulting images, we will have produced a *perspective view* of the graph of the function.

EXAMPLE 25 Figure 14.25 illustrates a collection of slices on the right (near to far from bottom to top). On the left, these slices have first been stacked and offset vertically to show their order, and then shrunk in proportion to their distance from the eye point.

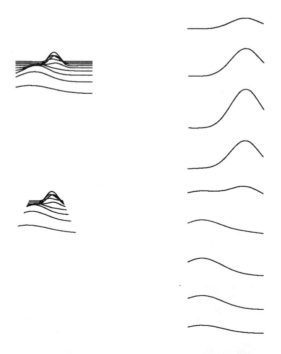

Figure 14.25 Putting slices in perspective.

While this is a simple combination of the pictures produced by slicing, its three-dimensional effect can be of benefit in grasping the shape of a surface. ∎

Ash

Wireframe plotting

A variant of the perspective graphing technique is quite simple but often adds a considerable amount of realism to the picture. First we make a rectangular grid in the domain of the function. Then, we plot the graph points corresponding to the grid intersections. Graph points plotted from adjacent grid points are connected in space with straight line segments.

Instead of plotting the smooth curves of the several slices corresponding to the grid lines, we are essentially plotting their *piecewise linear* approximations. The result resembles a piece of wire screening molded into the shape of the surface defined by the function's graph. For this reason, it is often called a *wireframe plot* of the function. If the grid is not too fine, then a machine can generate a wireframe plot fairly quickly, since there are relatively few "true" graph points to be computed. However, even a coarse grid can often give us a reasonably good feel for the shape of the graph.

Figure 14.26 shows the grid used to plot part of the paraboloid

$$z = x^2 + y^2.$$

Less than 300 graph points were actually computed. (To compare, most graphing calculators plot well over 100 points to graph a function of one variable $y = f(x)$.) The straight line segments connecting these points are short enough that they give us a feeling for the shape of the paraboloid.

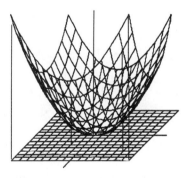

Figure 14.26 Points on a grid determine this wireframe plot of $z = x^2 + y^2$.

EXAMPLE 26   Figure 14.27 shows a wireframe plot of the function

$$f(x,y) = \frac{1 - x(x-1)(x-2)(x-3)}{(y-1)^2 + 1}$$

as well as its slices plotted in perspective over the input region $[0,3] \times [0,3]$.

Figure 14.27 $f(x,y) = \dfrac{1 - x(x-1)(x-2)(x-3)}{(y-1)^2 + 1}$ plotted over $[0,3] \times [0,3]$.

The graph of f has two peaks: one near $(0.5, 1)$ and one near $(2.5, 1)$. Between these two, at approximately $(1.5, 1)$, is an input at which the graph of the function resembles a "mountain pass" —locally flat, downhill in two directions, uphill in two other directions. Such a point is called a **saddle point**, so named because of the saddle-shape of the surface. ∎

EXERCISES

In exercises 1-5 graphically determine the approximate locations of all extrema of the given function in the input region $[0,3] \times [0,3]$.

1. $f(x,y) = (x-1)^2 + (x-0.3)(y-0.7) + (y-1)^2$.

2. $f(x,y) = \dfrac{x^3 + x^2 y - y^2}{(x-1.5)^2 + (y-2.3)^2}$.

3. $f(x,y) = x^{(y+1)^2}$.

4. $f(x,y) = x + \ln(x^2 + y^2 + 1)$.

5. $f(x,y) = 3.5x - 7.2y$.

In exercises 6-13 describe in your own words the shape of the graph of the given function near the input $(0,0)$. (Words you may consider using are: "bowl up," "bowl down," "saddle," "trough," or "flat," etc.)

6. $f(x,y) = x^2 + y^2$.

7. $f(x,y) = x^3 + y^3$.

8. $f(x,y) = x^3 - y^2$.

9. $f(x,y) = x^4 - y^4$.

10. $f(x,y) = xy$.

11. $f(x,y) = x^2 y$.

12. $f(x, y) = xy^3$.

13. $f(x, y) = x^2 y^2$.

Here's a way to use slope fields to get an idea of the shape of the contours of a function of two variables. We will illustrate the technique with the following example:

$$f(x, y) = x^2 + \sin(\pi y)x + y.$$

First, note that the level curves have an equation of the form

$$f(x, y) = c.$$

For our example, the level curves have equations of the form

$$x^2 + \sin(\pi y)x + y = c.$$

Recall that we can use implicit differentiation to find the slope of such a curve. For our example, we treat y implicitly as a function of x, and differentiate with respect to x:

$$2x + \pi y' \cos(\pi y)x + \sin(\pi y) + y' = 0.$$

Now we solve for y' in terms of both x and y:

$$y' = -\frac{2x + \sin(\pi y)}{1 + \pi \cos(\pi y)}.$$

We plot the slope field generated by this slope function of two variables in the input region $[0, 3] \times [0, 3]$, and obtain the window illustrated below:

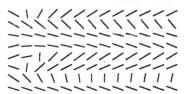

Using this we can rough sketch the shape of some of the level curves:

Note that this method of plotting level curves gives us no information about what level $z = c$ is represented by each contour. For this, we need to sample values to identify the level curves. Otherwise, there is no reason to believe that the curves we sketch correspond to *equally spaced* values of c, so be careful in using the closeness of the contours to judge how steep the surface is at any given input. Try the method out on exercises 6-13 above.

An ant is standing on a hill whose height is given by

$$H(x, y) = x^4 + y^4 - 4x^2y^2.$$

Take "North" to be the direction of the positive y-axis and "East" to be the direction of the positive x-axis and use graphical methods to answer exercises 14-24.

14. If the ant is standing at the origin and begins walking North, will it soon be going uphill, downhill, or neither?

15. If the ant is standing at the origin and begins walking South, will it soon be going uphill, downhill, or neither?

16. If the ant is standing at the origin and begins walking East, will it soon be going uphill, downhill, or neither?

17. If the ant is standing at the origin and begins walking West, will it soon be going uphill, downhill, or neither?

18. If the ant is standing at the origin and begins walking North-East, will it soon be going uphill, downhill, or neither?

19. If the ant is standing at the origin and begins walking North-West, will it soon be going uphill, downhill, or neither?

20. If the ant is standing at the origin and begins walking South-East, will it soon be going uphill, downhill, or neither?

21. If the ant is standing at the origin and begins walking South-West, will it soon be going uphill, downhill, or neither?

Suppose now that the height of the hill is given by

$$H(x, y) = x/y.$$

22. If the ant is standing at $x = 1$, $y = 1$, in which directions can it walk staying at the same height?

23. If the ant is standing at $x = 0$, $y = 1$, in which directions can it walk staying at the same height?

24. If the ant is standing at $x = -1$, $y = 1$, in which directions can it walk staying at the same height?

---

*The **potential** of a charged test particle at a particular position in an electric field is the amount of work it would take to move the particle from its position out to "infinity," that is, to a region far away where the field is negligible. Such a particle will tend to move "downhill," that is, in the direction of most rapidly decreasing potential. Two charged objects with charge 2 and −1 units, respectively, located in the plane at locations $(0.2, 0.01)$ and $(0.8, 0.01)$ produce a potential function on the plane given by*

$$P(x, y) = \frac{2}{\sqrt{(x - 0.2)^2 + (y - 0.01)^2}} + \frac{-1}{\sqrt{(x - 0.8)^2 + (y - 0.01)^2}}.$$

25. Use graphical techniques to determine if there is any point on the line given by $y = 1$ with $-1 < x < 1$ where the test particle would tend to move directly towards the x-axis? If so, determine the approximate location; if not, explain your conclusion.

26. Use graphical techniques to determine if there is any point on the line given by $y = 1$ with $-1 < x < 1$ where the test particle would tend not to move at all. If so, determine the approximate location; if not, explain your conclusion.

Suppose that the field-producing objects had charge 2 and 1 instead of 2 and −1. The potential would then be given by

$$P(x, y) = \frac{2}{\sqrt{(x - 0.2)^2 + (y - 0.01)^2}} + \frac{1}{\sqrt{(x - 0.8)^2 + (y - 0.01)^2}}.$$

27. Use graphical techniques to determine if there is any point on the line given by $y = 1$ with $-1 < x < 1$ where the test particle would tend to move directly away from the x-axis? If so, determine the approximate location; if not, explain your conclusion.

28. Use graphical techniques to determine if there is any point on the line given by $y = 0.01$ with $0.2 < x < 0.8$ where the test particle would tend not to move at all. If so, determine the approximate location; if not, explain your conclusion.

---

29. Graph the function $f(x, y) = \dfrac{x^2 \sin(xy)}{x^2 + y^2 + 1}$ on the region $[0, 3] \times [0, 3]$ by *freezing* the second input at $0, 0.3, 0.6, \ldots, 3.0$ and graphing the resulting slices. Then use these to approximate the inputs for f where $f(x, y)$ achieves its maximum and minimum values on this region.

30. Use the graphical slicing technique to determine whether the input $(1, -3, 2)$ to the function $f(x, y, z) = x^2 y z^2 - 3xyz + y^2$ is a maximum, minimum, or neither.

14.3 LINEAR AND QUADRATIC MULTIVARIABLE FUNCTIONS

In this section, we take time to discuss two simple but very important types of multivariable functions in more detail.

Linear functions

A linear function of one variable has a straight line for a graph and a formula of the form

$$y = f(x) = mx + b,$$

where m is the slope of the line, and b is the y-intercept.

A two-input function f is called a **linear function of two variables** provided its formula can be put in the form

$$f(x, y) = m_1 x + m_2 y + c.$$

If f has such a formula, then its graph is a *plane* with z-intercept $c = f(0,0)$ (see Figure 14.28).

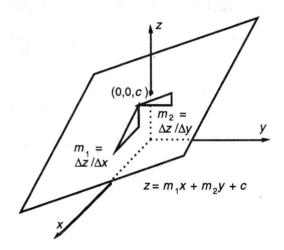

Figure 14.28 The graph of a linear two-input function is a plane.

Planes are important surfaces to study because upon close inspection, many surfaces also look planar (just as close-up, many curves look straight). Indeed, for centuries the human race assumed that the world was flat because people could visually grasp only relatively tiny regions of the earth.

We can think of m_1 and m_2 as slopes in much the same way we think of m representing the slope of the line $y = mx + b$.

If we pick any two input pairs (x_1, y_0) and (x_2, y_0) having the same y-coordinate y_0, and calculate the corresponding outputs

$$z_1 = f(x_1, y_0), \quad \text{and} \quad z_2 = f(x_2, y_0),$$

then

$$m_1 = \frac{f(x_2, y_0) - f(x_1, y_0)}{x_2 - x_1} = \frac{z_2 - z_1}{x_2 - x_1}.$$

In other words, m_1 is the "rise over run" of the graph plane where we measure the rise vertically in the z-direction and the run in the x-direction. We write

$$m_1 = \frac{\Delta f}{\Delta x} = \frac{\Delta z}{\Delta x}$$

to denote that m_1 is the ratio of the change in function output $z = f(x, y)$ to the change in input x, and we call m_1 the **slope with respect to** x.

Similarly, if we pick any two input pairs (x_0, y_1) and (x_0, y_2) having the same x-coordinate x_0, and calculate the corresponding outputs

$$z_1 = f(x_0, y_1), \quad \text{and} \quad z_2 = f(x_0, y_2),$$

then we have

$$m_2 = \frac{f(x_0, y_2) - f(x_0, y_1)}{y_2 - y_1} = \frac{z_2 - z_1}{y_2 - y_1}.$$

In other words, m_2 is the "rise over run" of the graph plane where we measure the rise vertically in the z-direction and the run in the y-direction.

We write

$$m_2 = \frac{\Delta f}{\Delta x} = \frac{\Delta z}{\Delta y}$$

to denote that m_2 is the ratio of the change in function output $z = f(x, y)$ to the change in input y, and we call m_2 the **slope with respect to** y.

The cross-sectional slices of the graph of a linear function will be straight lines. Figure 14.29 shows several slices of a linear function taken at equally spaced intervals along the y-direction, each representing a fixed value of y.

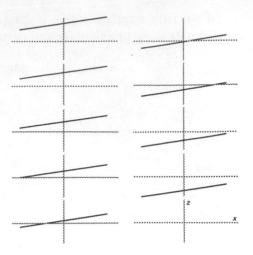

Figure 14.29 Slices of a linear function $z = f(x, y)$ with y fixed.

Notice that each slice is a straight line, all with the same slope with respect to x, and each is translated up or down by the same amount from the previous slice. Similarly, any slice for a fixed value of x will have the same slope with respect to y.

EXAMPLE 27   Consider the graph of the linear function $f(x, y) = 3x + 7y + 1$. Find the slope with respect to x, the slope with respect to y, and the z-intercept.

Solution   We can see that the slope with respect to x is

$$m_1 = 3,$$

the slope with respect to y is

$$m_2 = 7,$$

and the z-intercept is $c = 1$.

Let's verify these computations directly. First, we choose any two distinct ordered pairs with the same y-coordinate, say $(1, 2)$ and $(4, 2)$. We can use these to calculate the slope with respect to x:

$$\frac{\Delta f}{\Delta x} = \frac{f(4, 2) - f(1, 2)}{4 - 1} = \frac{(12 + 14 + 1) - (3 + 14 + 1)}{3} = \frac{9}{3} = 3.$$

To calculate the slope with respect to y, we choose any two distinct ordered pairs with the same x-coordinate, say $(-2, 3)$ and $(-2, -5)$:

$$\frac{\Delta f}{\Delta y} = \frac{f(-2, -5) - f(-2, 3)}{-5 - 3} = \frac{(-6 - 35 + 1) - (-6 + 21 + 1)}{-8} = \frac{-56}{-8} = 7.$$

Finally, the z-intercept is $f(0, 0) = 3(0) + 7(0) + 1 = 1.$ ■

EXAMPLE 28   Find the slope with respect to x, the slope with respect to y, and the z-intercept of the plane with equation

$$x + 2y + 3z = 6.$$

Solution   The equation of this plane is not written in a form expressing z as a function of x and y. If we solve for z, we obtain

$$z = \frac{-x - 2y + 6}{3},$$

from which we can see that the slope with respect to x is $-1/3$ and the slope with respect y is $-2/3$, and the z-intercept is $6/3 = 2$.   ■

EXAMPLE 29   A *horizontal* plane has equation $z = c$ for some constant c. The x-slope and y-slope of a horizontal plane are both 0, and the z-intercept is c.   ■

EXAMPLE 30   A *vertical* plane has an equation involving only x and y, and therefore cannot represent the graph of a function $z = f(x, y)$. Examples of equations of vertical planes include

$$x = 2, \qquad y = -3, \qquad \text{and} \qquad 2x - 3y = 5.$$

The x-slope and y-slope of a vertical plane are *undefined*. The notion of z-intercept does not make sense either, for a vertical plane will either contain the entire z-axis, or not intersect it at all.   ■

Taylor form of a linear function

The point-slope form of the equation for a line having slope m and passing through the point (x_0, y_0) is

$$y - y_0 = m(x - x_0).$$

Written in function notation (with $x_0 = a$, $y = f(x)$ and $y_0 = f(a)$), and rearranging the terms gives us the **Taylor form** for the linear function f about $x = a$:

$$f(x) = f(a) + m(x - a).$$

Similarly, there is a Taylor form for a linear function in two variables about a point (a, b):

$$f(x, y) = f(a, b) + m_1(x - a) + m_2(y - b),$$

where (a, b) is a specific input pair, m_1 is the x-slope, and m_2 is the y-slope.

EXAMPLE 31   Find the Taylor form of the linear function $f(x, y) = 3x + 7y + 1$ about the point $(4, -2)$.

Solution   Since $f(4, -2) = 3(4) + 7(-2) + 1 = -1$, we have

$$f(x, y) = 3(x - 4) + 7(y + 2) - 1$$

as the Taylor form of f.   ■

A linear function of three variables has the form

$$w = f(x, y, z) = m_1 x + m_2 y + m_3 z + c,$$

where

$$m_1 = \frac{\Delta w}{\Delta x}, \qquad m_2 = \frac{\Delta w}{\Delta y}, \qquad m_3 = \frac{\Delta w}{\Delta z}$$

are the three slopes with respect to x, y, and z, respectively. To compute the x-slope, choose two distinct ordered triples (x_1, y_0, z_0) and (x_2, y_0, z_0) with the same y and z coordinates and compute the difference quotient

$$m_1 = \frac{\Delta w}{\Delta x} = \frac{f(x_2, y_0, z_0) - f(x_1, y_0, z_0)}{x_2 - x_1}.$$

This value m_1 is the slope of every straight line slice obtained by freezing the values of y and z. The other two slopes are computed in a similar manner: m_2 using two points with the same x and z coordinates; m_3 using two points with the same x and y coordinates.

In general, a linear function of n variables has the form

$$f(x_1, x_2, \ldots, x_n) = m_1 x_1 + m_2 x_2 + \cdots + m_n x_n + c,$$

where each m_i $(1 \le i \le n)$ is the slope with respect to x_i, and represents the "rise/run" of the slice obtained by freezing the values of all the other variables and plotting the resulting line.

Quadratic functions

A **quadratic function** has a polynomial form involving only terms of degree 2 or less. (Linear functions involve only terms of degree 1 or less.) A quadratic function of two variables has the form

$$f(x, y) = c_{11} x^2 + c_{12} xy + c_{22} y^2 + c_1 x + c_2 y + c_0.$$

Note that we use the double-subscripted coefficients c_{11}, c_{12}, and c_{22} for the second-degree terms x^2, xy, and y^2, respectively, and the coefficients c_1, c_2, and c_0 for the linear and constant terms.

The graph of a quadratic function will be a quadric surface, of which several examples were discussed in the first section of the chapter (see Figures 14.8 through 14.15). Here we want to take a close look at how the coefficients of the quadratic affect the shape of the graph of

$$z = f(x, y).$$

We can consider the function $z = f(x, y)$ as the sum of a *homogeneous quadratic* (all terms of degree two) and a linear function:

$$f(x, y) = (c_{11}x^2 + c_{12}xy + c_{22}y^2) \quad + \quad (c_1x + c_2y + c_0).$$

We already know that the graph of $z = c_1x + c_2y + c_0$ is a *plane*. Hence, each point on the graph of $z = f(x, y)$ has the same height above or below this plane as given by

$$z = c_{11}x^2 + c_{12}xy + c_{22}y^2$$

for the corresponding input (x, y).

To concentrate on the effects of the quadratic coefficients, let's assume that the linear and constant coefficients are all zero:

$$c_1 = c_2 = c_0 = 0.$$

Let's look at some examples.

EXAMPLE 32   Examine the vertical cross-sections at the input $(0, 0)$ for each of the following quadratic functions:

$$f(x, y) = 2x^2 + y^2,$$

$$g(x, y) = -x^2 - 2y^2,$$

$$h(x, y) = y^2 - 2x^2,$$

$$j(x, y) = 2xy - x^2 - y^2.$$

Determine which has a strict maximum at $(0, 0)$, which has a strict minimum, and which must have neither. Describe the graph of each quadratic function.

Solution   The slices determined by setting $y = 0$ and $x = 0$ have equations:

$$f(x, 0) = 2x^2 \qquad f(0, y) = y^2$$

$$g(x, 0) = -x^2 \qquad g(0, y) = -2y^2$$

$$h(x, 0) = -2x^2 \qquad h(0, y) = y^2$$

$$j(x, 0) = -x^2 \qquad j(0, y) = -y^2$$

We can see in Figure 14.30 that f *could* have a minimum at $(0,0)$ since both $f(x,0)$, and $f(0,y)$ have a minimum there. Indeed, f *must* have a strict minimum at $(0,0)$, since $2x^2 + y^2 > 0$ for all other ordered pairs (x,y). The graph of $z = f(x,y)$ is a "bowl up" elliptic paraboloid.

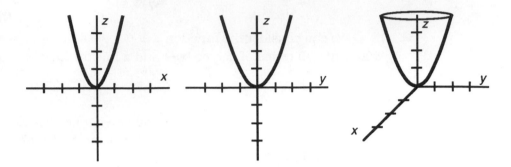

Figure 14.30　Vertical slices and graph of $z = f(x,y)$.

On the other hand, the slices of g suggest that a maximum occurs at $(0,0)$ (see Figure 14.31). It is the case that g has a strict maximum at $(0,0)$, since $-x^2 - y^2 < 0$ except when $x = 0$ and $y = 0$. The graph of $z = g(x,y)$ is a "bowl down" elliptic paraboloid.

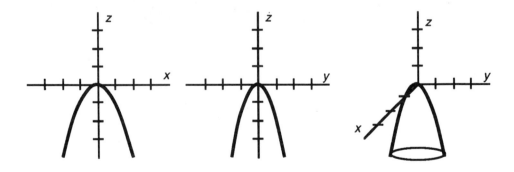

Figure 14.31　Vertical slices and graph of $z = g(x,y)$.

The function h must have neither a maximum nor a minimum at $(0,0)$, since it has a maximum along one slice but a minimum along the other! The graph of $z = h(x,y)$ is a "saddle" or hyperbolic paraboloid, and the origin is called the saddle point (see Figure 14.32).

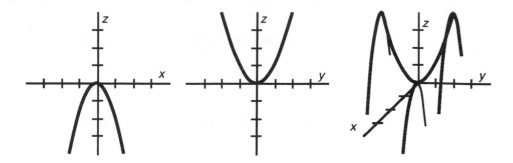

Figure 14.32 Vertical slices and graph of $z = h(x, y)$.

Finally, the slices of j also suggest that a maximum occurs at $(0, 0)$, but this is not a strict maximum, since

$$j(x, y) = 2xy - x^- y^2 = -(x - y)^2,$$

which means $j(x, y) = 0$ whenever $x = y$. The graph of j is an inverted "trough" whose crest runs along the line $x = y$ in the xy-plane (see Figure 14.33).

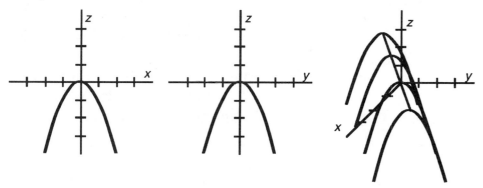

Figure 14.33 Vertical slices and graph of $z = j(x, y)$.

■

Slicing $z = Ax^2 + 2Bxy + Cy^2$ at the origin

The behavior we have just seen in these examples is typical of homogeneous quadratic functions. To justify this statement, we'll find it convenient to write the quadratic coefficients as follows:

$$c_{11} = A, \qquad c_{12} = 2B, \qquad c_{22} = C,$$

so that our quadratic has the form

$$z = Ax^2 + 2Bxy + Cy^2,$$

whose graph contains the origin $(0, 0, 0)$. Here's the main observation:

 A vertical plane containing the origin will intersect the surface

$$z = Ax^2 + 2Bxy + Cy^2$$

in either a parabola with its vertex at the origin, or in a horizontal line lying in the xy-plane.

For instance, the intersection of the xz-plane (the plane $y = 0$) and the surface $z = Ax^2 + 2Bxy + Cy^2$ is

$$z = Ax^2.$$

If $A \neq 0$, this is a parabola with vertex at the origin. If $A = 0$, the intersection is simply the x-axis.

On the other hand, the intersection of the yz-plane (the plane $x = 0$) and the surface $z = Ax^2 + 2Bxy + Cy^2$ is

$$z = Cy^2,$$

which is either a parabola with vertex at the origin ($C \neq 0$) or simply the y-axis ($C = 0$).

Any other vertical plane containing the origin has an equation of the form $y = mx$ ($m \neq 0$). Substituting mx for y, we see that the intersection of this plane with the surface has equation

$$z = Ax^2 + 2Bxy + Cy^2 = Ax^2 + 2Bx(mx) + C(mx)^2 = (A + 2Bm + Cm^2)x^2.$$

So, again, we can see that this is either a parabola with vertex at the origin (if $A + 2Bm + Cm^2 \neq 0$) or the *line* $y = mx$ lying in the xy-plane (if $A + 2Bm + Cm^2 = 0$).

With these observations in mind, we can list the possibilities:

1) If *all* the cross-sections at the origin are parabolas opening up, then the shape of the surface will be "bowl up" (an elliptic paraboloid with minimum at the origin).

2) If all the cross-sections at the origin are parabolas opening down, the shape of the surface will be "bowl down" (an elliptic paraboloid with maximum at the origin).

3) If there is a mix of parabolas, some opening up and some opening down, then the shape is a "saddle" (hyperbolic paraboloid with saddle point at the origin).

4) It is also possible that the graph could be a parabolic "trough" with its bottom (or crest if inverted) running along a line in the xy-plane.

5) The extreme case is the horizontal plane $z = 0$ (if $A = B = C = 0$).

The discriminant $D = AC - B^2$

Fortunately, there is often an easy way to detect the shape of the graph of $z = Ax^2 + 2Bxy + Cy^2$ by examining the coefficients.

Using some algebra, we can rewrite the form of the quadratic function $z = Ax^2 + 2Bxy + Cy^2$ in a way that gives us useful information about the shape of the graph. Assuming $A \neq 0$, we can rewrite the equation in the form

$$z = \frac{1}{A}[(Ax + By)^2 + (AC - B^2)y^2].$$

The quantity

$$D = AC - B^2$$

is called the **discriminant** of this quadratic function, and substituting D into the equation gives us

$$z = \frac{1}{A}[(Ax + By)^2 + Dy^2].$$

The *sign* of the discriminant D tells us much about the shape of the graph.

First, note that if $D < 0$, then the quantity in the square brackets can be either negative (substitute $x = -B/A$ and $y = 1$) or positive (substitute $x = 1$ and $y = 0$). Hence, the vertical cross-sections of the graph at the origin will include a mix of open-up and open-down parabolas with the origin at the vertex. Therefore, the graph must be *saddle-shaped* at the origin.

Now, if $D > 0$, then the quantity in the square brackets is positive at every point (x, y) except the origin. Hence, the factor $1/A$ determines whether all the cross-sections are open-up ($A > 0$) or open-down ($A < 0$) parabolas with vertex at the origin. In either case, the graph must be *bowl-shaped* with a strict minimum or strict maximum at the origin.

If $D = 0$, then the quantity in the square brackets is simply $(Ax + By)^2$, which is 0 all along the line $Ax + By = 0$, and positive off this line. The graph is *trough-shaped* (opening up or down depending on the sign of A).

All this analysis was done assuming $A \neq 0$. However, if $A = 0$, then the original function simplifies to

$$z = 2Bxy + Cy^2$$

and the discriminant simplifies to $D = -B^2$. If $B \neq 0$, then $D < 0$ and the graph is again saddle-shaped. To see this, note that the cross-section with the plane $x = 0$ is $z = Cy^2$, while the cross-section with the plane $x = -Cy/B$ is $z = -Cy^2$. If $C \neq 0$, then these two slices are parabolas opening in opposite directions. If $C = 0$, then the graph of $z = 2Bxy$ is still saddle-shaped. Finally, if both $A = 0$ and $B = 0$, then $D = 0$ and the graph of $z = Cy^2$ is either trough-shaped ($C \neq 0$), or a horizontal plane ($C = 0$).

We summarize these observations in a theorem.

Theorem 14.1

> **Hypothesis:** Suppose $z = Ax^2 + 2Bxy + Cy^2$. Let $D = AC - B^2$.
>
> **Case 1:** $D < 0$.
> **Conclusion:** A saddle point occurs at $(0,0)$.
>
> **Case 2:** $D > 0$ and $A > 0$.
> **Conclusion:** A strict minimum occurs at $(0,0)$.
>
> **Case 3:** $D > 0$ and $A < 0$.
> **Conclusion:** A strict maximum occurs at $(0,0)$.
>
> **Case 4:** $D = 0$.
> **Conclusion:** None. No strict maximum, minimum, or saddle point occurs at $(0,0)$.

$\square$

EXAMPLE 33   Use the theorem to analyze the five quadratic functions:

$$f(x,y) = 2x^2 + y^2,$$

$$g(x,y) = -x^2 - 2y^2,$$

$$h(x,y) = y^2 - 2x^2,$$

$$j(x,y) = 2xy - x^2 - y^2,$$

$$k(x,y) = 4xy.$$

Solution   All of these functions satisfy the hypothesis of the theorem, so for each we evaluate the discriminant D and make the conclusion indicated.

For $f(x,y) = 2x^2 + y^2$, we have $A = 2$, $B = 0$, and $C = 1$, so the discriminant is $AC - B^2 = 2 > 0$. Since $A > 0$, the function has a *strict minimum* at $(0,0)$.

For $g(x,y) = -x^2 - 2y^2$, we have $A = -1$, $B = 0$, and $C = -2$, so the discriminant is $D = AC - B^2 = 2 > 0$. Since $A < 0$, the function has a *strict maximum* at $(0,0)$.

For $h(x,y) = y^2 - 2x^2$, we have $A = -2$, $B = 0$, and $C = 1$, so the discriminant is $D = AC - B^2 = -2 < 0$. The function has a *saddle point* at $(0,0)$.

For $j(x, y) = 2xy - x^2 - y^2$, we have $A = -1$, $B = 1$, and $C = -1$, so the discriminant is $D = AC - B^2 = 0$. We make no strict conclusion regarding the behavior at $(0, 0)$. Recall that this function has no strict minimum, maximum, or saddle point at $(0, 0)$.

For $k(x, y) = 4xy$, we have $A = 0$, $B = 2$, and $C = 0$, so the discriminant is $D = AC - B^2 = -4$. The function has a *saddle point* at $(0, 0)$. ∎

If we replace x by $x - x_0$, y by $y - y_0$, and z by $z - z_0$ in any equation of three variables, the geometric effect on the graph is the same as relabelling the origin (x_0, y_0, z_0). Hence, the discriminant D can also be used to judge the behavior of a quadric surface

$$z - z_0 = A(x - x_0)^2 + 2B(x - x_0)(y - y_0) + C(y - y_0)^2$$

at the point (x_0, y_0, z_0).

Taylor form for quadratic functions

If

$$f(x, y) = c_{11}x^2 + c_{12}xy + c_{22}y^2 + c_1 x + c_2 y + c_0,$$

then the Taylor form of f about the point (a, b) is

$$f(x, y) = f(a, b) + d_1(x - a) + d_2(y - b) + c_{11}(x - a)^2 + c_{12}(x - a)(y - b) + c_{22}(y - b)^2.$$

Note that the quadratic coefficients are the same, but the linear and constant coefficients may change. In particular, the constant term is $f(a, b)$.

EXAMPLE 34   Write the quadratic function

$$f(x, y) = x^2 - 2xy - y^2 + 2x + 4y - 5$$

in Taylor form about the point $(-1, 2)$.

Solution   Since $f(-1, 2) = 1 + 4 - 4 - 2 + 8 - 5 = 2$, we can write

$$f(x, y) = 2 + d_1(x + 1) + d_2(y - 2) + (x + 1)^2 - 2(x + 1)(y - 2) - (y - 2)^2,$$

and we need only determine d_1 and d_2.

Perhaps the easiest way to accomplish this is to substitute values for (x, y) into both the original form and the Taylor form and set them equal to each other.

For instance, if we substitute $(0, 2)$ into the original form, we have

$$f(0, 2) = 0 - 0 - 4 + 0 + 8 - 5 = -1.$$

If we substitute the same point into the Taylor form we have

$$f(0,2) = 2 + d_1 + 0 + 1 - 0 - 0 = 3 + d_1.$$

Equating these, we conclude that $d_1 = -4$.

We chose the point $(0,2)$ for no other reason than to simplify the computation— each term in the Taylor form having $(y - 2)$ as a factor has value 0 at this point. Now, if we substitute $(-1, 0)$ into the original form, we have

$$f(-1,0) = 1 - 0 - 0 - 2 + 0 - 5 = -6,$$

and in the Taylor form:

$$f(-1,0) = 2 + 0 - 2d_2 + 0 + 0 - 4 = -2d_2 - 2.$$

Equating these, we conclude that $d_2 = 2$.

Now we can write the complete Taylor form of f about the point $(-1, 2)$:

$$f(x,y) = 2 - 4(x + 1) + 2(y - 2) + (x + 1)^2 - 2(x + 1)(y - 2) - (y - 2)^2.$$

As one more check, we substitute $(0,0)$ into the Taylor form and find $f(0,0) = 2 - 4 - 4 + 1 + 4 - 4 = -5$, which is the correct value of the function at this point.   ■

The Taylor form is useful when we are interested in studying a function's behavior around a specific point (a, b). In particular, if the Taylor form of a quadratic function has no linear terms, then we can use the discriminant to judge the shape of the graph at that point.

EXAMPLE 35   The Taylor form of a quadratic function f about the point $(3, -4)$ is found to have the form

$$z = f(x,y) = -7 + 5(x - 3)^2 - 2(x - 3)(y + 4) + 6(y + 4)^2.$$

Characterize the behavior of the graph at $(3, -4, -7)$.

Solution   There are no linear terms. Note that the equation can be written

$$z + 7 = 5(x - 3)^2 - 2(x - 3)(y + 4) + 6(y + 4)^2,$$

so we can make use of the discriminant.

In this form, we have $A = 5$, $B = -1$, and $C = 6$, so the discriminant is

$$D = AC - B^2 = 30 - (-1)^2 = 29 > 0.$$

We also have $A > 0$, so the function f must have a *strict minimum* at $(x, y) = (3, -4)$. The graph of $z = f(x, y)$ is a bowl-up elliptic paraboloid with a strict minimum at $(3, -4, -7)$. ■

EXERCISES

Suppose the graph of a two-input linear function $z = f(x, y)$ is a plane containing the three points $P = (1, 1, 1)$, $Q = (-1, 1, -3)$, and $R = (1, 3, 2)$. Exercises 1-5 refer to f.

1. Use the points P and Q to compute m_1, the x-slope of the plane.

2. Use the points P and R to compute m_2, the y-slope of the plane.

3. Express the equation of the plane in the form $z = m_1 x + m_2 y + c$.

4. Find the points at which the plane intersects each coordinate axis.

5. Find the Taylor form of f at $(-3, 2)$.

For each of the planes in exercises 6-10, find the slope with respect to x, the slope with respect to y, and the z-intercept, if they exist. If the plane is the graph of a function, find its Taylor form at $(-1, 4)$.

6. $z = 1 - 5y + 2x$.
7. $z = y$.
8. $z = \sqrt{17}$.
9. $x = y$.
10. $2x - 5y + 8z - 9 = 0$.

For each of the quadratic functions in exercises 11-15, describe the behavior of the graph at the origin.

11. $f(x, y) = 3x^2 - y^2$.
12. $g(x, y) = -x^2 + 6xy - 2y^2$.
13. $h(x, y) = x^2 - 4xy + 4y^2$.
14. $j(x, y) = 3xy - 5x^2 - y^2$.
15. $k(x, y) = 4xy + (x - y)^2 - (x + y)^2$.

For each of the quadratic functions in exercises 16-20, find its Taylor form about the point $(-1, 3)$.

16. $z = x^2 + xy + y^2 + x + y + 1$.

17. $z = x^2 - y^2 + 5$.

18. $z = xy$.

19. $z = x^2 + 2y^2 + 2x - 12y + 7$.

20. $z = -8$.

---

21. A quadratic function is found to have the Taylor form

$$z = 17 - (x + 1)^2 + 8(x + 1)(y - 3) - 4(y - 3)^2$$

about the point $(-1, 3)$. Characterize the behavior of the graph at the point $(-1, 3, 17)$.

22. What is the Taylor form of

$$z = 17 - (x + 1)^2 + 8(x + 1)(y - 3) - 4(y - 3)^2$$

at $(0, 0)$?

23. We know that three noncollinear points determine a unique plane in space. What is the least number of points on the graph of the quadratic function

$$z = c_{11}x^2 + c_{12}xy + c_{22}y^2 + c_1 x + c_2 y + c_0$$

that uniquely determine its equation?

Given each of the descriptions in exercises 24-30, find an example of a function whose graph fits the description.

24. The graph is a bowl-up elliptic paraboloid with a strict minimum at the point $(1, -2, 3)$.

25. The graph is a bowl-down elliptic paraboloid with a strict maximum at the point $(0, 1, -5)$.

26. The graph is a hyperbolic paraboloid with a saddle point at $(-1, 3, -5)$.

27. The graph is an open-up parabolic trough whose bottom is the intersection of the vertical plane $2x - y + 3 = 0$ and the horizontal plane $z = 5$.

28. The graph is an inverted parabolic trough whose crest is the y-axis.

29. The graph is a plane that intersects the y-axis and the z-axis, but not the x-axis.

30. Every vertical cross-section made parallel to the xz-plane or yz-plane is a straight line, but the graph is not a plane.

14.4 CYLINDRICAL AND SPHERICAL COORDINATES

(Note: This section may be delayed until after Section 16.2 if desired.)

Polar coordinates are a useful alternative to rectangular coordinates for describing certain curves in the plane, especially those involving circular arcs. There are also alternative coordinate systems for three-dimensional space, and they are the subject of this section.

Cylindrical coordinates

Given a point expressed in rectangular coordinates (x, y, z), we obtain its representation in **cylindrical coordinates** (r, θ, z) by simply converting x and y to polar form:

$$x = r \cos\theta \qquad y = r \sin\theta.$$

In cylindrical coordinates,

r is the horizontal distance from the z-axis;

θ is the angle measured counterclockwise from the positive x-axis side of the xz-plane;

z is still measured vertically from the xy-plane.

Given a point expressed in cylindrical coordinates (r, θ, z), we can imagine locating it on a vertical cylinder of radius r with the z-axis running down its center (see Figure 14.34).

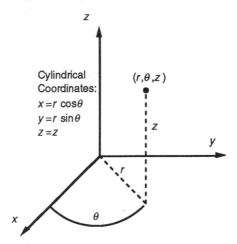

Figure 14.34 Cylindrical coordinates.

While we sometimes have occasion to consider $r < 0$ and arbitrary angles θ when using polar coordinates, we make the stipulation

$$r \geq 0 \qquad \text{and} \qquad 0 \leq \theta < 2\pi$$

for cylindrical coordinates. This gives each point in space except those on the z-axis exactly one representation in cylindrical coordinates.

EXAMPLE 36 A point in space has cylindrical coordinates $(r, \theta, z) = (4, \frac{\pi}{3}, -3)$. Find its rectangular coordinates.

Solution We have

$$x = 4\cos\frac{\pi}{3} = 4 \cdot \frac{1}{2} = 2$$

$$y = r\sin\frac{\pi}{3} = 4 \cdot \frac{\sqrt{3}}{2} = 2\sqrt{3}$$

$$z = -3.$$

Therefore, $(x, y, z) = (2, 2\sqrt{3}, -3)$. ∎

To convert from rectangular coordinates to cylindrical coordinates, we make use of the formula

$$r = \sqrt{x^2 + y^2}$$

to determine r. The value θ is determined by

$$\theta = \begin{cases} \arctan(y/x) & \text{if } x > 0 \text{ and } y \geq 0 \\ \pi + \arctan(y/x) & \text{if } x < 0 \\ 2\pi + \arctan(y/x) & \text{if } x > 0 \text{ and } y < 0. \end{cases}$$

and if $x = 0$, then we can take $\theta = \pi/2$ for $y > 0$ or $\theta = 3\pi/2$ for $y < 0$.

EXAMPLE 37 A point in space has rectangular coordinates $(x, y, z) = (3, 4, -2)$. Find its cylindrical coordinates.

Solution We have

$$r = \sqrt{x^2 + y^2} = \sqrt{3^2 + 4^2} = \sqrt{25} = 5,$$

and

$$\theta = \arctan(4/3).$$

Therefore, $(r, \theta, z) = (5, \arctan(4/3), -2)$. ∎

Some surfaces have particularly simple equations in terms of cylindrical coordinates. A cylinder of radius c and z-axis running down its center has equation

$$r = c.$$

A vertical "half-plane" with edge along the z-axis has equation

$$\theta = c,$$

and a horizontal plane still has equation

$$z = c.$$

Spherical coordinates

If we imagine locating points on a sphere centered at the origin, we have the basic idea of **spherical coordinates**.

For a point expressed in spherical coordinates (ρ, θ, φ):

ρ is the distance from the origin;

θ is the angle measured counterclockwise from the positive x-axis side of the xz-plane;

φ is the angle measured down from the positive z-axis.

Figure 14.35 illustrates the location of a point by spherical coordinates.

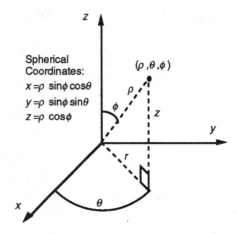

Figure 14.35 Spherical coordinates.

To give each point in space except those on the z-axis a unique representation in spherical coordinates, we make the stipulations

$$\rho \geq 0, \qquad 0 \leq \theta < 2\pi, \qquad 0 \leq \varphi \leq \pi.$$

To convert from rectangular coordinates (x, y, z) to spherical coordinates, we make use of the formula

$$\rho = \sqrt{x^2 + y^2 + z^2}$$

to determine the spherical radius ρ. The coordinate θ is determined in the same way as for cylindrical coordinates. Using $r = \sqrt{x^2 + y^2}$, we can write

$$\varphi = \begin{cases} \arctan(r/z) = \arctan(\sqrt{x^2 + y^2}/z) & \text{for } z > 0 \\ \pi + \arctan(r/z) = \pi + \arctan(\sqrt{x^2 + y^2}/z) & \text{for } z < 0 \\ \pi/2 & \text{for } z = 0 \end{cases}$$

for points not on the z-axis. We have $\varphi = 0$ for points on the positive z-axis, and $\varphi = \pi$ for points on the negative z-axis.

EXAMPLE 38   A point in space has rectangular coordinates $(x, y, z) = (3, 4, -2)$. Find its spherical coordinates (ρ, θ, φ).

Solution   We have

$$\rho = \sqrt{x^2 + y^2 + z^2} = \sqrt{3^2 + 4^2 + (-2)^2} = \sqrt{29}.$$

The value θ is calculated as it was in the previous example to be

$$\theta = \arctan(4/3).$$

Finally, since $r = \sqrt{3^2 + 4^2} = \sqrt{25} = 5$, we have

$$\varphi = \pi + \arctan(5/(-2)).$$

Therefore, the spherical coordinates are

$$(\rho, \theta, \varphi) = (\sqrt{29}, \arctan(4/3), \pi + \arctan(-5/2)).$$

■

We can convert directly from cylindrical coordinates (r, θ, z) to spherical coordinates by noting that

$$\rho = \sqrt{r^2 + z^2}.$$

The coordinate θ is the same, and φ can be calculated as discussed above using r and z.

EXAMPLE 39   A point in space has cylindrical coordinates $(r, \theta, z) = (4, \frac{\pi}{3}, -3)$. Find its spherical coordinates.

Solution   We have

$$\rho = \sqrt{r^2 + z^2} = \sqrt{4^2 + (-3)^2} = \sqrt{25} = 5.$$

Now, $\theta = \frac{\pi}{3}$ and

$$\varphi = \operatorname{arccot}(z/r) = \operatorname{arccot}(-3/4).$$

Thus, the point has spherical coordinates $(\rho, \theta, \varphi) = (5, \frac{\pi}{3}, \operatorname{arccot}(-3/4))$.   ∎

To change from spherical coordinates (ρ, θ, φ) to cylindrical coordinates (r, θ, z), we note that

$$r = \rho \sin \varphi,$$

$$\theta = \theta,$$

$$z = \rho \cos \varphi.$$

To change from spherical coordinates to rectangular, first convert to cylindrical coordinates, and then convert these to rectangular coordinates:

$$x = \rho \sin \varphi \cos \theta,$$

$$y = \rho \sin \varphi \sin \theta,$$

$$z = \rho \cos \varphi.$$

EXAMPLE 40   A point has spherical coordinates $(\rho, \theta, \varphi) = (4, \pi/4, 2\pi/3)$. Express the point in both cylindrical and rectangular coordinates.

Solution   For cylindrical coordinates, we have

$$r = \rho \sin \varphi = 4 \sin(2\pi/3) = 4(\sqrt{3}/2) = 2\sqrt{3},$$

$$\theta = \pi/4,$$

$$z = \rho \cos \varphi = 4 \cos(2\pi/3) = 4(-1/2) = -2.$$

Now, converting to rectangular coordinates, we have

$$x = \rho \sin \varphi \cos \theta = r \cos \theta = (2\sqrt{3}) \cos(\pi/4) = (2\sqrt{3})(\sqrt{2}/2) = \sqrt{6},$$

$$y = \rho \sin \varphi \sin \theta = r \sin \theta = (2\sqrt{3}) \cos(\pi/4) = (2\sqrt{3})(\sqrt{2}/2) = \sqrt{6},$$

$$z = \rho \cos \varphi = -2.$$

Hence, the cylindrical coordinates of the point are $(r, \theta, z) = (2\sqrt{3}, \pi/4, -2)$ and the rectangular coordinates are $(x, y, z) = (\sqrt{6}, \sqrt{6}, -2)$.   ∎

Some surfaces have particularly simple equations in terms of spherical coordinates. A sphere of radius c has equation

$$\rho = c.$$

Again, a vertical "half-plane" with edge along the z-axis has equation

$$\theta = c,$$

and a circular cone (actually the top or bottom half of a quadric surface) with vertex at the origin: $\varphi = c$.

EXERCISES

Fill in the table for exercises 1-12.

	rectangular	cylindrical	spherical
1.	$(1, 2, -2)$		
2.		$(4, 5\pi/4, -3)$	
3.			$(5, 2\pi/3, 2\pi/3)$
4.	$(-3, -3, -3)$		
5.		$(7, 3\pi/4, 6)$	
6.			$(5, 11\pi/6, \pi/2)$
7.	$(0, 0, -5)$		
8.		$(2, 0, 0)$	
9.			$(4, \pi/2, \pi/2)$
10.	$(0, 6, 0)$		
11.		$(2, 330°, -1)$	
12.			$(6, 256°, 127°)$

13. What points in space have exactly the same coordinates in both the rectangular and cylindrical systems (angles measured in radians)?

14. What points in space have exactly the same coordinates in both the rectangular and spherical systems (angles measured in radians)?

15. What points in space have exactly the same coordinates in both the cylindrical and spherical systems (angles measured in radians)?

16. What points in space have exactly the same coordinates in all three systems (angles measured in radians)?

17. What points in space have more than one representation in the cylindrical and spherical coordinate systems? Why?

An equation in terms of the coordinates of one system can be written in terms of another by substitution, using the conversion formulas. For example, the equation of the plane

$$x + 2y - 3z = 6$$

in rectangular coordinates has the equation

$$r\cos\theta + 2r\sin\theta - 3z = 6$$

in cylindrical coordinates, and the equation

$$\rho\cos\theta\sin\varphi + 2\rho\sin\theta\sin\varphi - 3\rho\cos\varphi = 6$$

in spherical coordinates. In exercises 18-30, make the conversion of the given equation into cylindrical and spherical coordinates.

18. The elliptic paraboloid $z = x^2 + y^2$.

19. The hyperbolic paraboloid $z = x^2 - y^2$.

20. The sphere $x^2 + y^2 + z^2 = 16$.

21. The plane $z = 5$.

22. The plane $y = x$.

23. The cylinder $x^2 + y^2 = 9$.

24. The cone $z^2 = x^2 + y^2$.

25. The parabolic trough $z = y^2$.

26. The hyperbolic paraboloid $z = 2xy$.

27. The surface $z = x/y$.

28. The hyperboloid of one sheet $z^2 = x^2 + y^2 - 1$.

29. The hyperboloid of two sheets $z^2 = x^2 + y^2 + 1$.

30. The surface $z = \arctan(y/x)$.

15

Differential Calculus of Multivariable Functions

In *Volume I* we saw how the derivative can a powerful tool for analyzing functions of a single variable.

The derivative

$$f'(x) = \frac{dy}{dx}$$

has both geometrical and physical interpretations.

1) Geometrically, the derivative tells us the local slope of the graph of $y = f(x)$.

2) Physically, the derivative tells us the instantaneous rate of change of output $f(x)$ per unit change in input x.

The derivative allows us to detect the relative extrema (maxima and minima) of a function, and determines the *best linear approximation* (first degree Taylor polynomial) at any input $x = a$. The graph of this best linear approximation is the tangent line to the graph. Higher order derivatives allow us to measure concavity and detect inflection points and determine the best quadratic (second degree Taylor polynomial) as well as higher degree approximations of the function.

In Chapter 13, we saw how to extend the idea of derivative to study *vector-valued* functions $\mathbf{r}(t)$ of a single variable t. In this chapter, we'll extend the idea of derivative to scalar-valued functions of several variables (scalar fields).

A function of several variables has several derivatives, one for each input. These *partial derivatives* can be used to study the various vertical slices of a multivariable function graph in much the same way that the derivative of single variable function is used.

Taken together, the partial derivatives provide a *total derivative*, which describes the locally linear behavior of a multivariable function. One form of total derivative we'll find particularly useful is called the *gradient* vector, and we can use it to measure rate of change, detect relative extrema, and form the best linear approximation of a scalar field.

Higher order partial derivatives give us additional information regarding the shape of the graph of a multivariable function, and can be used to find the best quadratic approximation of a scalar field. The best quadratic approximation, in turn, affords a powerful test for classifying the critical points of a function of two variables.

15.1 PARTIAL DERIVATIVES

The derivative of a single variable function $y = f(x)$ is defined as

$$f'(x) = \frac{dy}{dx} = \lim_{\Delta x \to 0} \frac{\Delta y}{\Delta x} = \lim_{\Delta x \to 0} \frac{f(x + \Delta x) - f(x)}{\Delta x}.$$

The difference quotient $\frac{\Delta y}{\Delta x}$ measures the *average rate of change*, while its limiting value $\frac{dy}{dx}$ tells us the *instantaneous rate of change* of the output y per unit change in input x.

For a multivariable function, we can consider the rate of change of output with respect to *each* input variable. For example, suppose we have a function of two variables

$$z = f(x, y).$$

If we hold the input variable y fixed, then we can study the rate of change of output z with respect to change in x alone. The difference quotient

$$\frac{\Delta z}{\Delta x} = \frac{f(x + \Delta x, y) - f(x, y)}{\Delta x}$$

gives us the ratio of change in output z to change in the input x (the two points $(x + \Delta x, y)$ and (x, y) have the same y-coordinate). We call $\frac{\Delta z}{\Delta x}$ the *average rate of change of z with respect to x.*

Similarly,

$$\frac{\Delta z}{\Delta y} = \frac{f(x, y + \Delta y) - f(x, \Delta y)}{\Delta y}$$

gives us the ratio of change in output z to change in the input y (the two points $(x, y + \Delta y)$ and (x, y) have the same x-coordinate). We call $\frac{\Delta z}{\Delta y}$ the *average rate of change of z with respect to y.*

The limiting values of these difference quotients are called **partial derivatives.**

Definition 1

The **partial derivative of** $z = f(x,y)$ **with respect to** x is

$$\frac{\partial z}{\partial x} = \lim_{\Delta x \to 0} \frac{\Delta z}{\Delta x} = \lim_{\Delta x \to 0} \frac{f(x + \Delta x, y) - f(x,y)}{\Delta x},$$

and the **partial derivative of** $z = f(x,y)$ **with respect to** y is

$$\frac{\partial z}{\partial y} = \lim_{\Delta y \to 0} \frac{\Delta z}{\Delta y} = \lim_{\Delta y \to 0} \frac{f(x, y + \Delta y) - f(x,y)}{\Delta y},$$

provided these limits exist.

Notice that the partial derivative symbol ∂ is different than the Leibniz d used for single variable functions. Other notations for $\dfrac{\partial z}{\partial x}$ include

$$\frac{\partial f}{\partial x} \qquad f_x \qquad z_x \qquad D_x f \qquad D_x z.$$

Similarly, other notations for $\dfrac{\partial z}{\partial y}$ include

$$\frac{\partial f}{\partial y} \qquad f_y \qquad z_y \qquad D_y f \qquad D_y z.$$

The partial derivatives $\dfrac{\partial f}{\partial x}$ and $\dfrac{\partial f}{\partial y}$ give us the *instantaneous rate of change* of $z = f(x,y)$ with respect to x and y, respectively.

For a function of three variables $w = f(x,y,z)$ we have three **partial derivatives**:

$$\frac{\partial f}{\partial x} = \lim_{\Delta x \to 0} \frac{f(x + \Delta x, y, z) - f(x,y,z)}{\Delta x}$$

$$\frac{\partial f}{\partial y} = \lim_{\Delta y \to 0} \frac{f(x, y + \Delta y, z) - f(x,y,z)}{\Delta y}$$

$$\frac{\partial f}{\partial z} = \lim_{\Delta z \to 0} \frac{f(x, y, z + \Delta z) - f(x,y,z)}{\Delta z}.$$

In general, a function of n variables has n partial derivatives. Each is computed by taking the limit of the appropriate difference quotient, with all variables but one fixed:

If $f(x_1, x_2, \ldots x_i, \ldots x_n)$ is a function of n variables, then

$$\frac{\partial f}{\partial x_i} = \lim_{\Delta x_i \to 0} \frac{f(x_1, x_2, \ldots, x_i + \Delta x_i, \ldots, x_n) - f(x_1, x_2, \ldots, x_i, \ldots, x_n)}{\Delta x_i}.$$

Calculating partial derivatives

In computing a partial derivative, all but one of the variables are held fixed when taking the limit of difference quotients. For the purposes of partial differentiation, we can treat all but one of the variables as *constants*. This makes it a fairly simple matter to compute partial derivatives of a multi-variable function built up from the standard functions.

☞ **For partial differentiation, consider all the variables except the one of differentiation as constants, and use the usual rules of differentiation for single variable functions (product rule, quotient rule, chain rule, etc.).**

EXAMPLE 1 Calculate $\dfrac{\partial f}{\partial x}$ and $\dfrac{\partial f}{\partial y}$ and evaluate them at the point $(-2, 3)$ if

$$f(x, y) = x^2 + 5xy - y^3.$$

Solution Treating y as a constant, and differentiating with respect to x, we have

$$\frac{\partial f}{\partial x} = 2x + 5y.$$

(Note that if y is considered constant, then y^3 is also a constant.) Treating x as a constant, and differentiating with respect to y, we have

$$\frac{\partial f}{\partial y} = 5x - 3y^2.$$

The values of the partial derivatives at $(-2, 3)$ are:

$$\left.\frac{\partial f}{\partial x}\right|_{(-2,3)} = 2(-2) + 5(3) = 11 \qquad \text{and} \qquad \left.\frac{\partial f}{\partial y}\right|_{(-2,3)} = 5 \cdot (-2) - 3(-3)^2 = -37.$$

∎

EXAMPLE 2 Calculate $\dfrac{\partial s}{\partial u}$ and $\dfrac{\partial s}{\partial v}$ and evaluate them at the $(0, 0)$ if $s = \sin(2u - 3v)$.

Solution

$$\frac{\partial s}{\partial u} = 2\cos(2u - 3v) \qquad \Longrightarrow \qquad \left.\frac{\partial s}{\partial u}\right|_{(0,0)} = 2\cos 0 = 2.$$

$$\frac{\partial s}{\partial v} = -3\cos(2u - 3v) \qquad \Longrightarrow \qquad \left.\frac{\partial s}{\partial v}\right|_{(0,0)} = -3\cos 0 = -3.$$

∎

EXAMPLE 3   Find $f_x, f_y,$ and f_z for the function

$$f(x, y, z) = \frac{x^2 e^{yz}}{y}.$$

Solution

$$f_x = \frac{2x e^{yz}}{y}, \qquad f_y = \frac{zx^2 e^{yz} \cdot y - x^2 e^{yz} \cdot 1}{y^2}, \qquad f_z = \frac{yx^2 e^{yz}}{y} = x^2 e^{yz}.$$

☞   **Note that each partial derivative is itself a function of several variables.**

Geometrical and physical interpretations

In the last chapter, we discussed how freezing the values of all but one variable of a multivariable function f creates a *slice function* that can be graphed like any function of one variable. The partial derivative with respect to this variable gives us the local slope of this slice function. Figures 15.1 and 15.2 illustrate the idea for a function of two variables.

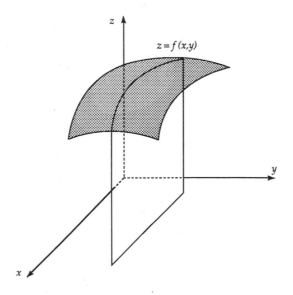

Figure 15.1 $\partial f / \partial x$ gives the local slope of a slice with y fixed.

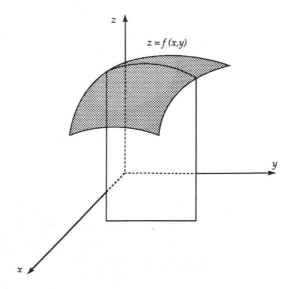

Figure 15.2 $\partial f / \partial y$ gives the local slope of a slice with x fixed.

Geometrically, you can think of a partial derivative as measuring the local slope of a slice of the surface as we move parallel to the coordinate axis represented by the variable of differentiation.

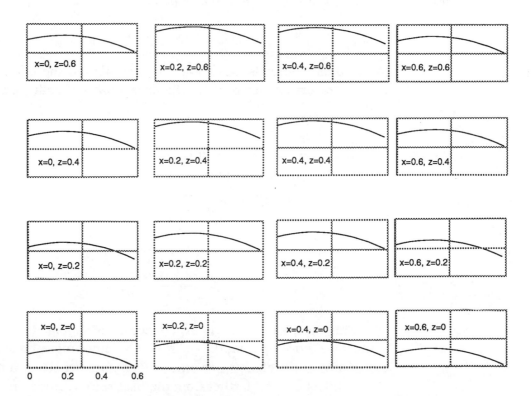

Figure 15.3 Slices of a function graph $w = f(x, y, z)$ with respect to y.

EXAMPLE 4   The y-slices of the function

$$w = f(x, y, z) = 2.62 + 0.6x + 0.4y + z - x^2 - y^2 - z^2$$

over the region $[0, 0.6] \times [0, 0.6] \times [0, 0.6]$ are shown in Figure 15.3 for all possible combinations of $x = 0$, 0.2, 0.4, 0.6 and $z = 0$, 0.2, 0.4, 0.6.

The partial derivative

$$\frac{\partial w}{\partial y} = 0.4 - 2y$$

gives us the local slope at each point y on every one of the graphs shown. In this particular case, note that $\frac{\partial w}{\partial y}$ depends only on y, and not on the values of x and z. This means that all the slices of the graph should be "parallel," since they must all have the same local slope with respect to y. Note that the slices in Figure 15.3 illustrate this phenomenon.   ■

Physically, a partial derivative measures the instantaneous rate of change of output per unit change when we move the input point in a direction parallel to one of the coordinate axes.

EXAMPLE 5   The temperature in ° Kelvin at a point (x, y, z) is given by the function

$$T(x, y, z) = \frac{300}{1 + x^2 + y^2 + z^2}.$$

(A 300° heat source is located at the origin.) Find the instantaneous rate of temperature change per unit when we move vertically up from the point $(1, 2, 3)$.

Solution   A vertical move up is parallel to the z-axis, with z increasing. Therefore, we need the partial derivative with respect to z:

$$\frac{\partial T}{\partial z} = \frac{-600z}{(1 + x^2 + y^2 + z^2)^2}.$$

Evaluating at the point $(1, 2, 3)$ we have

$$\frac{\partial T}{\partial z}\bigg|_{(1,2,3)} = \frac{-1800}{1 + 1 + 4 + 9)^2} = \frac{-1800}{225} = -8.$$

This means that we will experience a $-8°$ change in temperature per unit as we move *up* from the point $(1, 2, 3)$. If we move *down*, then we will experience a $+8°$ change per unit. This makes sense, as a move up from $(1, 2, 3)$ takes us a bit farther from the heat source, but a move down takes us a bit closer to it.   ■

████████████   **EXERCISES**

For each function in exercises 1-10, find

$$\frac{\partial f}{\partial x} \quad and \quad \frac{\partial f}{\partial y}$$

and evaluate each at the indicated point (a, b).

1. $f(x, y) = \dfrac{x + y}{x - y}$ $(a, b) = (1, 2)$.

2. $f(x, y) = xy + x/y$, $(a, b) = (-1, 3)$.

3. $f(x, y) = e^{x^2 + y^2}$, $(a, b) = (1, -2)$.

4. $f(x, y) = 2x - 3y$, $(a, b) = (0, 0)$.

5. $f(x, y) = \sin(x^2 y)$, $(a, b) = (2, 0)$.

6. $f(x, y) = \arctan(y/x)$, $(a, b) = (-3, -3)$.

7. $f(x, y) = \tan(y/x) - \cot(y)$, $(a, b) = (2, \pi/2)$.

8. $f(x, y) = x + \ln(x^2 + y^2 + 1)$, $(a, b) = (-1, -2)$.

9. $f(x, y) = \sqrt{x^2 + y^2}$, $(a, b) = (-3, 4)$.

10. $f(x, y) = x^y$, $(a, b) = (e, 2)$.

For each function in exercises 11-15, find

$$f_x, \qquad f_y, \qquad f_z$$

and evaluate each at the indicated point (a, b, c).

11. $f(x, y, z) = \dfrac{z}{x + y}$, $(a, b, c) = (1, 2, 3)$.

12. $f(x, y, z) = x^3 y^4 z^5$, $(a, b, c) = (1, -2, -1)$.

13. $f(x, y, z) = 5x - 4y + 3z$, $(a, b, c) = (-3, 2, 4)$.

14. $f(x, y, z) = \exp 1 - x^2 - y^2 - z^2$, $(a, b, c) = (0, 1, -1)$.

15. $f(x, y, z) = x \arcsin z - y \arctan z$, $(a, b, c) = (.5, .5, .5)$.

An ant is standing on a hill whose height is given by

$$H(x, y) = x^4 + y^4 - 4x^2 y^2.$$

Take "North" to be the direction of the positive y-axis and "East" to be the direction of the positive x-axis and use partial derivatives to answer exercises 16-19.

16.   If the ant is standing at the point $(2, 3)$ and begins walking North, is it going uphill, downhill, or neither?

17. If the ant is standing at the point $(-1, 2)$ and begins walking South, is it going uphill, downhill, or neither?

18. If the ant is standing at the point $(3, -2)$ and begins walking East, is it going uphill, downhill, or neither?

19. If the ant is standing at the point $(-2, -1)$ and begins walking West, is it going uphill, downhill, or neither?

Suppose now that the height of the hill is given by

$$H(x, y) = x/y.$$

20. If the ant is standing at the point $(2, 3)$ and begins walking North, is it going uphill, downhill, or neither?

21. If the ant is standing at the point $(-1, 2)$ and begins walking South, is it going uphill, downhill, or neither?

22. If the ant is standing at the point $(3, -2)$ and begins walking East, is it going uphill, downhill, or neither?

23. If the ant is standing at the point $(-2, -1)$ and begins walking West, is it going uphill, downhill, or neither?

Two charged objects with charge 2 and -1 units, respectively, located in the plane at locations $(0.2, 0.01)$ and $(0.8, 0.01)$ produce a potential function on the plane given by

$$P(x, y) = \frac{2}{\sqrt{(x - 0.2)^2 + (y - 0.01)^2}} + \frac{-1}{\sqrt{(x - 0.8)^2 + (y - 0.01)^2}}.$$

Use this potential function to answer questions 24-27.

24. If a test particle is located at the origin $(0, 0)$, how fast will the potential change if it starts moving along the positive x-axis?

25. If a test particle is located at the origin $(0, 0)$, how fast will the potential change if it starts moving along the positive y-axis?

26. If a test particle is located at $(0.5, -0.12)$, how fast will the potential change if it starts moving straight towards the y-axis along the line $y = -0.12$?

27. If a test particle is located at $(0.5, -0.12)$, how fast will the potential change if it starts moving straight towards the x-axis along the line $x = 0.5$?

Suppose now that the field-producing objects had charge 2 and 1 instead of 2 and −1. The potential would then be given by

$$P(x,y) = \frac{2}{\sqrt{(x - 0.2)^2 + (y - 0.01)^2}} + \frac{1}{\sqrt{(x - 0.8)^2 + (y - 0.01)^2}}.$$

28. If a test particle is located at the origin $(0,0)$, how fast will the potential change if it starts moving along the positive x-axis?

29. If a test particle is located at the origin $(0,0)$, how fast will the potential change if it starts moving along the positive y-axis?

30. If a test particle is located at $(-0.35, -.42)$, how fast will the potential change if it starts moving straight towards the y-axis along the line $y = -0.42$.

31. If a test particle is located at $(-0.35, -0.42)$, how fast will the potential change if it starts moving straight towards the x-axis along the line $x = -0.35$?

The temperature in ° Kelvin at a point (x, y, z) is given by the function

$$T(x, y, z) = \frac{300}{1 + x^2 + y^2 + z^2}.$$

(A 300° heat source is located at the origin.) Use this information to answer questions 32-35.

32. Find the instantaneous rate of temperature change per unit when we move from the point $(1, 0, 0)$ straight towards the origin.

33. Find the instantaneous rate of temperature change per unit when we move from the point $(0, -3, 0)$ straight towards the origin.

34. Find the instantaneous rate of temperature change per unit when we move from the point $(1, 2, 3)$ straight towards the point $(1, 5, 3)$.

35. At which points in space will the temperature decrease regardless of whether we move straight up or straight down?

15.2   THE TOTAL DERIVATIVE

In the last section, we discussed *partial* derivatives of multivariable functions. In this section we'll talk about the **total derivative** of a multivariable function.

Limits and continuity of multivariable functions

Limits provide a convenient language for talking about continuity and differentiability of multivariable functions, just as they do for single variable functions. A few comments are in order regarding the notion of limit as it applies to multivariable functions. Let's start by reviewing the basic ideas for single variable functions.

For a function of a single variable $y = f(x)$, we write

$$\lim_{x \to a} f(x) = L.$$

This just means that the difference $|f(x) - L|$ can be made as small as we wish, provided the input x is chosen sufficiently close (but not equal) to a.

Recall that the actual function output $f(a)$ need not equal L (nor even be defined) for the limit to exist. However, if we *do* have

$$\lim_{x \to a} f(x) = f(a),$$

then we say f is **continuous at** $x = a$.

We can make a similar definition for the limit of a multivariable function. If $z = f(x, y)$ is a function of two variables, we write

$$\lim_{(x,y) \to (a,b)} f(x, y) = L$$

to mean that the difference $|f(x, y) - L|$ can be made as small as we wish by choosing (x, y) sufficiently close (but not equal) to (a, b).

A function of two variables is said to be **continuous at** (a, b) if

$$\lim_{(x,y) \to (a,b)} f(x, y) = f(a, b).$$

For a function of three variables $w = f(x, y, z)$ we write

$$\lim_{(x,y,z) \to (a,b,c)} f(x, y, z) = L$$

to indicate that $|f(x, y, z) - L| \to 0$ as $(x, y, z) \to (a, b, c)$. The function f is said to be **continuous at** (a, b, c) if

$$\lim_{(x,y,z) \to (a,b,c)} f(x, y, z) = f(a, b, c).$$

☞ **Virtually all of the standard functions (algebraic, trigonometric, inverse trigonometric, exponential, logarithmic) are continuous wherever they are defined. Multivariable functions built up from these through algebraic operations or composition will also be continuous wherever defined.**

EXAMPLE 6   The function

$$z = f(x,y) = \frac{e^{xy}}{\sin(x^2 + y^2)}$$

is continuous at every point $(x,y) \neq (0,0)$. At $(0,0)$ the function is *undefined* since the denominator $\sin(0^2 + 0^2) = 0$. No limit exists at $(0,0)$ since the numerator approaches 1 and the denominator approaches 0 as $(x,y) \to (0,0)$. ∎

We have already noted that the partial derivatives of a multivariable function are themselves multivariable functions. When all these partial derivatives are continuous, the function deserves a special term.

Definition 2 | If all the partial derivatives of a multivariable f are *continuous* at a point, then we say the function f is C^1-**differentiable** at that point.

EXAMPLE 7   The function $z = f(x,y) = \dfrac{e^{xy}}{\sin(x^2 + y^2)}$ is C^1-differentiable at every point except $(0,0)$, since its partial derivatives

$$\frac{\partial z}{\partial x} = \frac{ye^{xy}\sin(x^2+y^2) - 2xe^{xy}\cos(x^2+y^2)}{\sin^2(x^2+y^2)}$$

$$\frac{\partial z}{\partial y} = \frac{xe^{xy}\sin(x^2+y^2) - 2ye^{xy}\cos(x^2+y^2)}{\sin^2(x^2+y^2)}$$

are both continuous at every point $(x,y) \neq (0,0)$. ∎

What is a total derivative?

If a function $y = f(x)$ of one variable is differentiable at a point $x = a$, then we can think of the function as being approximately *locally linear*. The derivative $f'(a)$ gives us a simple multiplication factor (namely, the local slope) for measuring change in output for small changes in input. In other words, if we make a small change in input Δx from $x = a$, then we just multiply by $f'(a)$ to find the approximate change in output:

$$f(a + \Delta x) - f(a) \approx f'(a)\Delta x.$$

We want to use a similar criteria for differentiability of a function of several variables:

If a function of *several* variables is approximately *locally linear*, then for small changes in the inputs, the total change in output must be approximately a simple *linear combination* of the small changes in the various inputs. By this, we mean that we can multiply each input's change by its own multiplication factor and add up the results to approximate the change in output.

The partial derivatives of a multivariable function give us the local slope multiplication factors for each individual input, with all other inputs fixed. If all of the partial derivatives are *continuous* (in other words, f is C^1-differentiable), then this guarantees that the function behaves approximately locally linear. In this case, we can talk about the **total derivative** of f.

Definition 3 | The **total derivative** Df of a C^1-differentiable scalar-valued function f is represented by the single row matrix of partial derivative values.

The total derivative of a function of two variables $z = f(x, y)$ is

$$Df = \begin{bmatrix} \dfrac{\partial f}{\partial x} & \dfrac{\partial f}{\partial y} \end{bmatrix}.$$

For a function of three variables $w = f(x, y, z)$:

$$Df = \begin{bmatrix} \dfrac{\partial f}{\partial x} & \dfrac{\partial f}{\partial y} & \dfrac{\partial f}{\partial z} \end{bmatrix},$$

and in general, for a function f of n variables

$$Df = \begin{bmatrix} \dfrac{\partial f}{\partial x_1} & \dfrac{\partial f}{\partial x_2} & \cdots & \dfrac{\partial f}{\partial x_n} \end{bmatrix}.$$

Calculating total derivatives

If we calculate the total derivative of a function of two variables

$$z = f(x, y)$$

at a point (a, b), we obtain the 1×2 matrix

$$Df(a, b) = \left[\left. \frac{\partial f}{\partial x} \right|_{(a,b)} \quad \left. \frac{\partial f}{\partial y} \right|_{(a,b)} \right].$$

EXAMPLE 8   Find the matrix of the total derivative of the function

$$z = f(x, y) = x^2 + 5xy - y^3$$

at the points $(-2, 3)$ and $(1.2, -2.1)$.

Solution   Since $\dfrac{\partial f}{\partial x} = 2x + 5y$ and $\dfrac{\partial f}{\partial y} = 5x - 3y^2$, we have

$$Df(-2, 3) = \begin{bmatrix} 11 & -37 \end{bmatrix}$$

and

$$Df(1.2, -2.1) = \begin{bmatrix} -8.1 & -7.23 \end{bmatrix}.$$

■

Note that we can get a *different* matrix at each point (a, b).

EXAMPLE 9   Find the matrix of the total derivative of $w = g(x, y, z) = \ln(x^2 + y^2 + z^2)$ at the point $(-2, 1, 5)$.

$$Df = \begin{bmatrix} \dfrac{\partial g}{\partial x} & \dfrac{\partial g}{\partial y} & \dfrac{\partial g}{\partial z} \end{bmatrix} = \begin{bmatrix} \dfrac{2x}{x^2 + y^2 + z^2} & \dfrac{2y}{x^2 + y^2 + z^2} & \dfrac{2z}{x^2 + y^2 + z^2} \end{bmatrix}$$

$$Dg(-2, 1, 5) = \begin{bmatrix} \dfrac{-2}{15} & \dfrac{1}{15} & \dfrac{2}{3} \end{bmatrix}.$$

■

For functions of a single variable, we usually think of the derivative $f'(a)$ as a number (the local slope at $x = a$). Strictly speaking, this number represents the 1×1 matrix of the total derivative.

Interpreting the total derivative

What exactly does the total derivative tell us? In a very real sense, Df captures the "total rate of change" of a multivariable function f. Let's illustrate this statement for a function f of two variables.

If we move away from the point (a, b) a small amount in *any* direction we arrive at a new point

$$(a + \Delta x, b + \Delta y).$$

The total derivative Df at (a, b) can tell us approximately the change in output of our function f by applying the total derivative to the *vector of changes* $\langle \Delta x, \Delta y \rangle$. In other words,

$$\Delta z = f(a + \Delta x, b + \Delta y) - f(a, b) \approx Df(a, b) \begin{bmatrix} \Delta x \\ \Delta y \end{bmatrix},$$

or written in terms of the partial derivatives:

$$\Delta z \approx \left. \frac{\partial f}{\partial x} \right|_{(a,b)} \Delta x + \left. \frac{\partial f}{\partial y} \right|_{(a,b)} \Delta y.$$

This approximation becomes better and better as $\langle \Delta x, \Delta y \rangle$ becomes smaller and smaller.

EXAMPLE 10   Use the total derivative to approximate the change in output

$$\Delta z = f(-2.05, +3.02) - f(-2, 3)$$

for the function $z = f(x, y) = x^2 + 5xy - y^3$.

Solution   We have already computed

$$Df(2, -3) = \begin{bmatrix} 11 & -37 \end{bmatrix}.$$

The vector of changes is

$$\langle \Delta x, \Delta y \rangle = \langle -2.05, 3.02 \rangle - \langle -2, 3 \rangle = \langle -.05, .02 \rangle.$$

Therefore,

$$\Delta z \approx \begin{bmatrix} 11 & -37 \end{bmatrix} \begin{bmatrix} -.05 \\ .02 \end{bmatrix} = 11(-.05) - 37(.02) = -.55 - .74 = -1.29.$$

Comparing this to the actual change in the output

$$\Delta z = f(-2.05, 3.02) - f(-2, 3) = -54.296108 - (-53) = -1.296108,$$

we see that using the total derivative approximates the change in output accurate to within .01 of the actual change in output. ∎

Best linear approximations of multivariable functions

If a function $y = f(x)$ of one variable is differentiable at a point $x = a$, then it has a best linear approximation at that point. Geometrically, the graph of this best linear approximation is the *tangent line* to the graph of $y = f(x)$ at the point $(a, f(a))$. The derivative value $f'(a)$ is the slope of this tangent line. The equation of the tangent line is called the *first-degree Taylor polynomial approximation of f about $x = a$*:

$$f(x) \approx f(a) + f'(a)(x - a).$$

In the last chapter, we discussed how a *linear* function of two variables

$$z = m_1 x + m_2 y + c$$

has a *plane* for a graph. This plane has one slope in the x-direction, computed by the difference quotient $m_1 = \Delta z / \Delta x$ for any two distinct points in the plane with the same y-coordinates. This plane also has a slope in the y-direction, computed by the difference quotient $m_2 = \Delta z / \Delta y$ for any two distinct points in the plane with the same x-coordinates. The x-slope measures "rise/run" as we move parallel to the x-axis, and the y-slope measures "rise/run" as we move parallel to the y-axis.

A function f of two variables has a total derivative at a point (a, b) if both its partial derivatives are continuous at that point. Geometrically, this means that the surface $z = f(x, y)$ has a *tangent plane* at the point $(a, b, f(a, b))$ (see Figure 15.4).

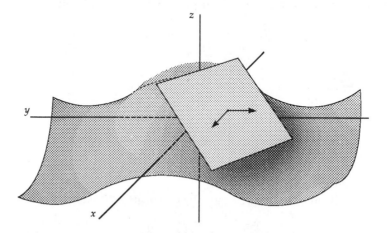

Figure 15.4 Tangent plane to the graph of $z = f(x, y)$.

The x-slope and y-slope of this tangent plane to $z = f(x, y)$ at the point $(a, b, f(a, b))$ are the *partial derivatives* of f at (a, b).

☞   **The equation of the tangent plane to the graph of** $z = f(x, y)$ **at the point** $(a, b, f(a, b))$ **is**

$$z = f(a, b) + f_x(a, b)(x - a) + f_y(a, b)(y - b).$$

Definition 4

> For a C^1-differentiable function of two variables $f(x, y)$, the **first-degree Taylor polynomial approximation about** $(x, y) = (a, b)$ is
>
> $$f(x, y) \approx f(a, b) + f_x(a, b)(x - a) + f_y(a, b)(y - b).$$

This is the best linear approximation of the function f at the point $(a, b, f(a, b))$. Written in terms of the total derivative, we have

$$f(x, y) \approx f(a, b) + Df(a, b) \begin{bmatrix} x - a \\ y - b \end{bmatrix}.$$

Notice the similarity to the the first degree Taylor approximation for a function of one variable:

$$f(x) \approx f(a) + f'(a)(x - a).$$

EXAMPLE 11   Find the best linear approximation to the function

$$z = f(x, y) = x^2 + 5xy - y^3$$

at the point $(-2, 3)$.

Solution   We have $f(-2, 3) = (-2)^2 + 5(-2)3 - 3^3 = 4 - 30 - 27 = -53$. We have already computed

$$Df(-2, 3) = \begin{bmatrix} 11 & -37 \end{bmatrix}.$$

Thus, the best linear approximation for the function is

$$f(x, y) \approx f(-2, 3) + \begin{bmatrix} 11 & -37 \end{bmatrix} \begin{bmatrix} x + 2 \\ y - 3 \end{bmatrix} = -53 + 11(x + 2) - 37(y - 3).$$

■

Definition 5

For a C^1-differentiable function of three variables $f(x, y, z)$, the **first-degree Taylor polynomial approximation of f about** $(x, y, z) = (a, b, c)$ is

$$f(x, y, z) \approx f(a, b, c) + f_x(a, b, c)(x - a) + f_y(a, b, c)(y - b) + f_z(a, b, c)(z - c)$$

$$= f(a, b, c) + Df(a, b, c) \begin{bmatrix} x - a \\ y - b \\ z - c \end{bmatrix}.$$

EXAMPLE 12   Find the best linear approximation to the function $w = g(x, y, z) = \ln(x^2 + y^2 + z^2)$ at the point $(-2, 1, 5)$.

Solution   We have $g(-2, 1, 5) = \ln((-2)^2 + 1^2 + 5^2) = \ln(30)$. We have already computed

$$Dg(-2, 1, 5) = \begin{bmatrix} \dfrac{-2}{15} & \dfrac{1}{15} & \dfrac{2}{3} \end{bmatrix}.$$

Thus, the best linear approximation for the function is

$$w = g(x, y, z) \approx \ln(30) + \begin{bmatrix} \dfrac{-2}{15} & \dfrac{1}{15} & \dfrac{2}{3} \end{bmatrix} \begin{bmatrix} x + 2 \\ y - 1 \\ z - 5 \end{bmatrix}$$

$$= \ln(30) - \frac{2}{15}(x + 2) + \frac{1}{15}(y - 1) + \frac{2}{3}(z - 5).$$

■

EXERCISES

For each function in exercises 1-10, find the total derivative Df, and use it to find the equation of the best linear approximation to f about the indicated point (a, b).

1. $f(x, y) = \dfrac{x + y}{x - y},$ $(a, b) = (1, 2)$.

2. $f(x, y) = xy + x/y,$ $(a, b) = (-1, 3)$.

3. $f(x, y) = e^{x^2 + y^2},$ $(a, b) = (0, 0)$.

4. $f(x, y) = 2x - 3y,$ $(a, b) = (0, 0)$.

5. $f(x, y) = \sin(x^2 y),$ $(a, b) = (2, 0)$.

6. $f(x, y) = \arctan(y/x),$ $(a, b) = (-3, -3)$.

7. $f(x, y) = \tan(y/x) - \cot(y),$ $(a, b) = (2, \pi/2)$.

8. $f(x, y) = x + \ln(x^2 + y^2 + 1),$ $(a, b) = (0, 0)$.

9. $f(x,y) = \sqrt{x^2 + y^2}$, $(a,b) = (-3,4)$.

10. $f(x,y) = x^y$, $(a,b) = (e,2)$.

For each function in exercises 11-15, find the total derivative Df and use it to find the equation of the best linear approximation to f at the indicated point (a,b,c).

11. $f(x,y,z) = \dfrac{z}{x+y}$, $(a,b,c) = (1,2,3)$.

12. $f(x,y,z) = x^3 y^4 z^5$, $(a,b,c) = (1,-2,-1)$.

13. $f(x,y,z) = 5x - 4y + 3z$, $(a,b,c) = (-3,2,4)$.

14. $f(x,y,z) = \exp 1 - x^2 - y^2 - z^2$, $(a,b,c) = (0,1,-1)$.

15. $f(x,y,z) = x \arcsin z - y \arctan z$, $(a,b,c) = (.5,.5,.5)$.

An ant is standing on a hill whose height is given by

$$H(x,y) = x^4 + y^4 - 4x^2 y^2.$$

Take "North" to be the direction of the positive y-axis and "East" to be the direction of the positive x-axis and use this information to answer exercises 16-19.

16. If the ant is standing at the $(2,3)$ and walks 0.1 units North, then 0.1 units East, what is its total change in elevation?

17. What is the total derivative of the height function at the point $(2,3)$? Use this to approximate the change in elevation computed in the previous exercise, and compare your results.

18. If the ant is standing at the $(-1,2)$ and walks 0.2 units South, then 0.3 units West, what is the total change in elevation?

19. What is the total derivative of the height function at the point $(-1,2)$? Use this to approximate the change in elevation computed in the previous exercise, and compare your results.

Suppose now that the height of the hill is given by

$$H(x,y) = x/y.$$

20. If the ant is standing at the $(3,-2)$ and walks 0.2 units East, then 0.1 units South, what is its total change in elevation?

21. What is the total derivative of the height function at the point $(3,-2)$? Use this to approximate the change in elevation computed in the previous exercise, and compare your results.

22. If the ant is standing at the $(-2, -1)$ and walks 0.1 units North, then 0.2 units West, what is its total change in elevation?

23. What is the total derivative of the height function at the point $(-2, -1)$? Use this to approximate the change in elevation computed in the previous exercise, and compare your results.

The temperature in ° Kelvin at a point (x, y, z) is given by the function

$$T(x, y, z) = \frac{300}{1 + x^2 + y^2 + z^2}.$$

(A 300° heat source is located at the origin.) Use this information to answer questions 24 and 25.

24. Find the total temperature change when we move from point $(1, 2, 3)$ to the point $(1.1, 1.9, 3.2)$.

25. Find the total derivative of the temperature function. Use it to approximate the change in temperature computed in the previous exercise, and compare your results.

You'll need 3D-graphing tools for exercises 26-30, which are intended to investigate how well the tangent plane approximates the surface of a graph. For each exercise, find the equation of the tangent plane to the graph of the indicated function at the point $(1, 2, f(1, 2))$. Then, plot the contours of the original function and the line contours of the tangent plane for the inputs (x, y) in $[0, 2] \times [1, 3]$. (Use level curves corresponding to values of z spaced 0.1 units apart.) Then graph the wireframe plot of each over the same set of inputs.

26. $f(x, y) = \dfrac{x + y}{x - y}.$

27. $f(x, y) = xy + x/y.$

28. $f(x, y) = e^{x^2 + y^2}.$

29. $f(x, y) = \sqrt{x^2 + y^2}.$

30. $f(x, y) = x^y.$

15.3 THE GRADIENT AND ITS PROPERTIES—APPLICATIONS

The total derivative Df is represented by a single row matrix of partial derivatives. If we think of this matrix as a *vector*, we call it the **gradient** of the multivariable function f.

Definition 6

> The **gradient** of a function f is the vector of partial derivatives of f. The gradient is denoted by ∇f, and read as "grad f " or "del f."

Hence, the gradient of a function $f(x, y)$ of two variables is

$$\nabla f = \left\langle \frac{\partial f}{\partial x}, \frac{\partial f}{\partial y} \right\rangle;$$

the gradient of a function of three variables, $f(x, y, z)$ is

$$\nabla f = \left\langle \frac{\partial f}{\partial x}, \frac{\partial f}{\partial y}, \frac{\partial f}{\partial z} \right\rangle;$$

and, in general, the gradient of a function of n variables $f(x_1, x_2, \ldots, x_n)$ is

$$\nabla f = \left\langle \frac{\partial f}{\partial x_1}, \frac{\partial f}{\partial x_2}, \ldots, \frac{\partial f}{\partial x_n} \right\rangle.$$

EXAMPLE 13 Find the gradient of the function $z = f(x, y) = y \arctan x$ and evaluate it at $(x, y) = (1, 3)$.

Solution The gradient function is

$$\nabla f = \left\langle \frac{\partial f}{\partial x}, \frac{\partial f}{\partial y} \right\rangle = \left\langle \frac{y}{1 + x^2}, \arctan x \right\rangle.$$

Therefore,

$$\nabla f(1, 3) = \left\langle \frac{3}{1 + 1^2}, \arctan 1 \right\rangle = \left\langle \frac{3}{2}, \frac{\pi}{4} \right\rangle.$$

■

EXAMPLE 14 Find $\nabla f(-3, -2, -1)$ if $f(x, y, z) = e^{xyz}$.

Solution The gradient function is

$$\nabla f = \left\langle \frac{\partial f}{\partial x}, \frac{\partial f}{\partial y}, \frac{\partial f}{\partial z} \right\rangle = \langle yze^{xyz}, xze^{xyz}, xye^{xyz} \rangle = e^{xyz} \langle yz, xz, xy \rangle.$$

Therefore, $\nabla f(-3, -2, -1) = e^{-6} \langle 2, 3, 6 \rangle.$ ∎

Notice that the gradient gives us a *vector value* at each point. For this reason, we can say that the gradient defines a *vector field*. In this section, we want to discuss some important interpretations and uses of the gradient vector.

We already know that the total derivative can be used to approximate the change in output of a C^1-differentiable function f. Written in terms of the gradient, this change is the *dot product of the gradient vector with the vector of changes in the inputs.*

For example, if $z = f(x, y)$ is a C^1-differentiable function of two variables, then a change in input $\langle \Delta x, \Delta y \rangle$ results in a change in output

$$\Delta z \approx \frac{\partial f}{\partial x} \Delta x + \frac{\partial f}{\partial y} \Delta y = \nabla f \cdot \langle \Delta x, \Delta y \rangle.$$

For a C^1-differentiable function of three variables $w = f(x, y, z)$, a change in input of $\langle \Delta x, \Delta y, \Delta z \rangle$ results in the change in output

$$\Delta w \approx \frac{\partial f}{\partial x} \Delta x + \frac{\partial f}{\partial y} \Delta y + \frac{\partial f}{\partial z} \Delta z = \nabla f \cdot \langle \Delta x, \Delta y, \Delta z \rangle.$$

With this in mind, we will be able to use the gradient to measure the rate of change in *any* direction.

Directional derivatives

Let's consider the case of a function of two variables $z = f(x, y)$. Each partial derivative of f measures the instantaneous rate of change of output per unit change in a particular variable. Geometrically, we can think of $\frac{\partial f}{\partial x}$ as measuring the rate of output change of z per unit change in the direction parallel to the x-axis, while $\frac{\partial f}{\partial y}$ measures the rate of output change per unit change in the direction parallel to the y-axis.

These are not the only directions in which we can move in the xy-plane. Indeed, there are infinitely many directions one can move from any point.

Suppose we are at a given input (a, b), and we want to measure the instantaneous rate of change of $z = f(x, y)$ when the input is shifted in the direction

$$\mathbf{u} = \langle u_1, u_2 \rangle,$$

where $\mathbf{u}$ is a *unit vector* ($\|\mathbf{u}\| = \sqrt{u_1^2 + u_2^2} = 1$).

If we move Δt units in this direction, then the change in input is

$$\langle u_1 \Delta t, u_2 \Delta t \rangle,$$

and the corresponding change in output is

$$\Delta z = f(a + u_1 \Delta t, b + u_2 \Delta t) - f(a, b).$$

Thus, the *average rate of change* in the direction $\mathbf{u}$ is

$$\frac{\Delta z}{\Delta t} = \frac{f(a + u_1 \Delta t, b + u_2 \Delta t) - f(a, b)}{\Delta t}.$$

The limit of this difference quotient as $\Delta t \to 0$ gives us the *instantaneous rate of change* per unit in the direction $\mathbf{u}$, and this forms the basis of our definition of a **directional derivative**.

Definition 7

> The **directional derivative** of f at (a, b) in the unit direction $\mathbf{u} = \langle u_1, u_2 \rangle$ is defined as
>
> $$D_{\mathbf{u}}f(a, b) = \lim_{\Delta t \to 0} \frac{f(a + u_1 \Delta t, b + u_2 \Delta t) - f(a, b)}{\Delta t},$$
>
> provided this limit exists.

Now, the gradient gives us a very handy way of computing this directional derivative. If f is a C^1-differentiable function, then the approximate change in output is simply

$$\Delta z \approx \nabla f(a, b) \cdot \langle u_1 \Delta t, u_2 \Delta t \rangle.$$

Dividing by Δt, we have

$$\frac{\Delta z}{\Delta t} \approx \frac{\nabla f(a, b) \cdot \langle u_1 \Delta t, u_2 \Delta t \rangle}{\Delta t} = \nabla f(a, b) \cdot \langle u_1, u_2 \rangle.$$

This approximation becomes better and better as $\Delta t \to 0$. In fact,

$$D_{\mathbf{u}}f(a, b) = \nabla f(a, b) \cdot \mathbf{u}.$$

EXAMPLE 15 Suppose $z = f(x, y) = x^2 y + 2xy - y^3$. Find the instantaneous rate of change of $f(x, y)$ per unit if we move from the input point $(-2, 3)$ straight towards the point $(1, 7)$ in the xy-plane.

Solution First, we need to determine a unit vector describing the direction in which we're moving. The direction vector is given by

$$\langle 1, 7 \rangle - \langle -2, 3 \rangle = \langle 3, 4 \rangle.$$

A unit vector in this same direction is given by

$$\mathbf{u} = \frac{\langle 3, 4 \rangle}{\|\langle 3, 4 \rangle\|} = \frac{\langle 3, 4 \rangle}{\sqrt{3^2 + 4^2}} = \frac{\langle 3, 4 \rangle}{5} = \left\langle \frac{3}{5}, \frac{4}{5} \right\rangle.$$

The gradient of f is

$$\nabla f = \langle 2xy + 2y, \, x^2 + 2x - 3y^2 \rangle.$$

Hence,

$$D_{\mathbf{u}} f(-2, 3) = \nabla f(-2, 3) \cdot \langle 3/5, 4/5 \rangle = \langle -6, -27 \rangle \cdot \langle 3/5, 4/5 \rangle = -25.2.$$

This tells us that as we move from $(-2, 3)$ towards $(1, 7)$, the output $f(x, y)$ is *decreasing* at an instantaneous rate of 25.2 units per unit travelled. ■

The definition of directional derivative is similar for functions of three or more variables.

Definition 8

> The **directional derivative** of f at (a, b, c) with respect to the direction specified by unit vector $\mathbf{u} = \langle u_1, u_2, u_3 \rangle$ is
> $$D_{\mathbf{u}} f(a, b, c) = \lim_{\Delta t \to 0} \frac{f(a + u_1 \Delta t, b + u_2 \Delta t, c + u_3 \Delta t) - f(a, b, c)}{\Delta t}.$$

For C^1-differentiable functions, the same observation regarding the gradient carries through. If f is a C^1-differentiable function of three variables, and $\mathbf{u} = \langle u_1, u_2, u_3 \rangle$ is a unit vector, then

$$D_{\mathbf{u}} f(a, b, c) = \nabla f(a, b, c) \cdot \langle u_1, u_2, u_3 \rangle.$$

☞ **In general, for any C^1-differentiable function f of n variables, and any unit vector $\mathbf{u} \in \mathbb{R}^n$, we have**

$$D_{\mathbf{u}} f = \nabla f \cdot \mathbf{u}.$$

EXAMPLE 16   The temperature in $^\circ$ Kelvin at a point (x, y, z) is given by the function

$$T(x, y, z) = \frac{300}{1 + x^2 + y^2 + z^2}.$$

(A 300° heat source is located at the origin.) Find the instantaneous rate of temperature change per unit when we move from the point $(1, 2, 3)$ in the direction specified by the unit vector $\mathbf{u} = \langle -1/3, -2/3, 2/3 \rangle$.

Solution   First, we calculate the gradient of the temperature function:

$$\nabla T = \left\langle \frac{-600x}{(1 + x^2 + y^2 + z^2)^2}, \frac{-600y}{(1 + x^2 + y^2 + z^2)^2}, \frac{-600z}{(1 + x^2 + y^2 + z^2)^2} \right\rangle.$$

At the point $(1, 2, 3)$, the gradient has the value

$$\nabla T(1, 2, 3) = \left\langle \frac{-600}{225}, \frac{-1200}{225}, \frac{-1800}{225} \right\rangle = \langle -8/3, -16/3, -8 \rangle.$$

Taking the dot product with the given direction, we have

$$D_{\mathbf{u}}T = \nabla T(1, 2, 3) \cdot \mathbf{u} = \left\langle \frac{-8}{3}, \frac{-16}{3}, -8 \right\rangle \cdot \left\langle \frac{-1}{3}, \frac{-2}{3}, \frac{2}{3} \right\rangle = \frac{8}{9} + \frac{32}{9} + \frac{-16}{3} = \frac{-8}{9}.$$

Therefore, we experience a $(-8/9)^\circ$ change in temperature per unit of distance as we move from $(1, 2, 3)$ in the direction $\langle -1/3, -2/3, 2/3 \rangle$.   ■

Direction of fastest change

In what direction should we move in the inputs to realize the *fastest* rate of increase in output of a function f? In terms of the directional derivative, we are asking:

What unit vector u maximizes $D_{\mathbf{u}}f$?

For a C^1-differentiable function, we have just seen that

$$D_{\mathbf{u}}f = \nabla f \cdot \mathbf{u}.$$

If we interpret the dot product geometrically, we have

$$\nabla f \cdot \mathbf{u} = \|\nabla f\| \, \|\mathbf{u}\| \cos \theta = \|\nabla f\| \cos \theta,$$

where θ is the smallest angle between ∇f and $\mathbf{u}$ (provided $\nabla f \neq 0$). Note that we have used the fact that $\|\mathbf{u}\| = 1$ for a unit vector.

Since $-1 \le \cos \theta \le 1$ for all values θ, we can see that the maximum rate of change occurs when $\cos \theta = 1$. But this happens precisely when the two vectors ∇f and $\mathbf{u}$ have exactly the same direction. How remarkable! At every point, the gradient points exactly in the direction we should move to achieve the greatest rate of increase. Furthermore, this maximum rate of change is precisely $\|\nabla f\|$.

When the gradient $\nabla f \neq 0$, it points in the direction of fastest rate of increase of the function f. If u is a unit vector having this direction, (namely $\mathbf{u} = \nabla f / \|\nabla f\|$) then

$$D_{\mathbf{u}} f = \|\nabla f\|.$$

Note that this means that $-\nabla f$ must point in the direction of fastest rate of *decrease* of the function f.

EXAMPLE 17   The temperature in ° Kelvin at a point (x, y, z) is given by the function

$$T(x, y, z) = \frac{300}{1 + x^2 + y^2 + z^2}.$$

(A 300° heat source is located at the origin.) Find a unit vector u that specifies the direction of the fastest increase in temperature change per unit when we move from the point $(1, 2, 3)$. What is the rate of temperature change when we move in this direction?

Solution   First, we have already calculated the gradient of this temperature function:

$$\nabla T = \left\langle \frac{-600x}{(1 + x^2 + y^2 + z^2)^2}, \frac{-600y}{(1 + x^2 + y^2 + z^2)^2}, \frac{-600z}{(1 + x^2 + y^2 + z^2)^2} \right\rangle.$$

At the point $(1, 2, 3)$, the gradient has the value

$$\nabla T(1, 2, 3) = \left\langle \frac{-8}{3}, \frac{-16}{3}, -8 \right\rangle.$$

A unit vector having the same direction is

$$\mathbf{u} = \frac{\nabla T(1, 2, 3)}{\|\nabla T(1, 2, 3)\|} = \frac{\langle -8/3, -16/3, -8 \rangle}{8\sqrt{14}/3} = \frac{1}{\sqrt{14}} \langle -1, -2, -3 \rangle.$$

Notice that this vector points from $(1, 2, 3)$ directly back toward the origin, where the heat source is located.

The maximum instantaneous rate of temperature increase in this direction is

$$\|\nabla T(1, 2, 3)\| = \frac{8\sqrt{14}}{3} \approx 9.98.$$

In other words, if we head straight for the heat source from the point $(1, 2, 3)$, we'll experience an approximate 10° per unit instantaneous rate of temperature increase. ■

Don't forget that a directional derivative $D_{\mathbf{u}}f$ is always with respect to a *unit vector* u.

Rate of change along a path—chain rule

So far, we've been using the gradient to measure rate of change in straight line directions. Now we want to investigate how the gradient can be used to measure rate of change along a curved path. For example, suppose we have a temperature scalar field in space, and we have an object moving through space along some path. How can we measure the rate of change of temperature along this path?

Clearly several things will affect this rate of change:

1) the distribution of temperatures in space represented by the scalar field,

2) the exact positions of the points on the path, and

3) how quickly we traverse the path.

Now, the temperature at each point in space (x, y, z) is given by some scalar field

$$f(x, y, z).$$

If the path is described by the vector position function

$$\mathbf{r}(t) = \langle x(t), y(t), z(t) \rangle,$$

then the temperature at each point along the path is given by composing f with $\mathbf{r}$:

$$f(\mathbf{r}(t)) = f(x(t), y(t), z(t)).$$

This is a scalar-valued function of a single variable t, and the rate of change with respect to t is simply

$$\frac{df(\mathbf{r}(t))}{dt} = \lim_{\Delta t \to 0} \frac{f(\mathbf{r}(t + \Delta t)) - f(\mathbf{r}(t))}{\Delta t}.$$

The gradient again gives us a handy way of computing this derivative. In the numerator of the difference quotient, we have the difference in outputs of f at the two points represented by $\mathbf{r}(t + \Delta t)$ and $\mathbf{r}(t)$.

When Δt is small, the change in input is

$$\mathbf{r}(t + \Delta t) - \mathbf{r}(t) \approx \mathbf{r}'(t)\Delta t,$$

and the corresponding change in output is therefore

$$f(\mathbf{r}(t + \Delta t)) - f(\mathbf{r}(t)) \approx \nabla f(\mathbf{r}(t)) \cdot \mathbf{r}'(t)\Delta t.$$

We can approximate the difference quotient by

$$\frac{f(\mathbf{r}(t + \Delta t)) - f(\mathbf{r}(t))}{\Delta t} \approx \nabla f(\mathbf{r}(t)) \cdot \mathbf{r}'(t),$$

and this approximation becomes better and better as $\Delta t \to 0$.

This final result is known as the gradient form of the **chain rule**:

$$\frac{d}{dt}(f(\mathbf{r}(t)) = \nabla f(\mathbf{r}(t)) \cdot \mathbf{r}'(t).$$

Note that the value of the derivative is written in terms of

1) the gradient of the scalar field f,

2) the position $\mathbf{r}(t)$, and

3) the velocity $\mathbf{r}'(t)$.

EXAMPLE 18 The temperature in ° Kelvin at a point (x, y, z) is given by the function

$$T(x, y, z) = \frac{300}{1 + x^2 + y^2 + z^2}.$$

(A 300° heat source is located at the origin.) A particle is moving along the path

$$\mathbf{r}(t) = \langle t, t^2, t^3.$$

Find the instantaneous rate of temperature change of the particle at time $t = 2$.

Solution We will compute this rate of change in two ways. First, we note that the temperature at any time t (in seconds) along the path is given by

$$T(\mathbf{r}(t)) = \frac{300}{1 + x(t)^2 + y(t)^2 + z(t)^2} = \frac{300}{1 + t^2 + t^4 + t^6}.$$

The derivative of this function is

$$\frac{d}{dt}T(\mathbf{r}(t)) = \frac{-300(2t + 4t^3 + 6t^5)}{(1 + t^2 + t^4 + t^6)^2}.$$

At time $t = 2$, we have

$$\frac{d}{dt}T(\mathbf{r})\bigg|_{t=2} = \frac{-300(4 + 32 + 192)}{(1 + 4 + 16 + 64)^2} = \frac{-68400}{7225} \approx -9.467.$$

Now, let's use the gradient form of the chain rule to obtain the same result. First, the gradient of the temperature field is given by

$$\nabla T = \left\langle \frac{-600x}{(1 + x^2 + y^2 + z^2)^2}, \frac{-600y}{(1 + x^2 + y^2 + z^2)^2}, \frac{-600z}{(1 + x^2 + y^2 + z^2)^2} \right\rangle.$$

The velocity vector of the path is given by

$$\mathbf{r}'(t) = \langle 1, 2t, 3t^2 \rangle.$$

At time $t = 2$, we are at the point $\mathbf{r}(2) = \langle 2, 4, 8 \rangle$, and the temperature gradient there is

$$\nabla T(\mathbf{r}(2)) = \frac{\langle -600(2), -600(4), -600(8) \rangle}{(1 + 2^2 + 4^2 + 8^2)^2} = \frac{\langle -1200, -2400, -4800 \rangle}{7225}.$$

The velocity is $\mathbf{r}'(2) = \langle 1, 4, 12 \rangle$. Hence, the desired rate of change is given by

$$\frac{d}{dt}T(\mathbf{r})\Big|_{t=2} = \nabla T(\mathbf{r}(2)) \cdot \mathbf{r}'(2) = \left\langle \frac{-1200}{7225}, \frac{-2400}{7225}, \frac{-4800}{7225} \right\rangle \cdot \langle 1, 4, 12 \rangle$$

$$= \frac{-1200}{7225} + \frac{-9600}{7225} + \frac{-57600}{7225} = \frac{-68400}{7225} \approx -9.467.$$

We see again that at time $t = 2$, the temperature along the path is decreasing at an instantaneous rate of approximately $-9.467°$ per second. ∎

Directional derivative of f along a path $\mathbf{r}$

The **directional derivative** of f along the path $\mathbf{r}$ at any time t is obtained by computing the directional derivative of f with respect to the *unit tangent vector* $\mathbf{T}(t)$. We write

$$D_{\mathbf{r}}f(t) = \frac{\nabla f(\mathbf{r}(t)) \cdot \mathbf{r}'(t)}{\|\mathbf{r}'(t)\|} = \nabla f(\mathbf{r}(t)) \cdot \mathbf{T}(t).$$

Note that the directional derivative of f along a path $\mathbf{r}$ is simply the rate of change along the path divided by the speed $\|\mathbf{r}'(t)\|$.

EXAMPLE 19   The temperature in ° Kelvin at a point (x, y, z) is given by the function

$$T(x, y, z) = \frac{300}{1 + x^2 + y^2 + z^2}$$

(A 300° heat source is located at the origin.) A particle is moving along the path

$$\mathbf{r}(t) = \langle t, t^2, t^3 \rangle.$$

Find the directional derivative of T along r at $t = 2$.

Solution   In the last example, we computed the rate of temperature change at $t = 2$ to be

$$\nabla T(\mathbf{r}(2)) \cdot \mathbf{r}'(2) \approx -9.467.$$

The speed at $t = 2$ is

$$\|\mathbf{r}'(2)\| = \sqrt{1^2 + 4^2 + 12^2} = \sqrt{161} \approx 12.689.$$

Hence, the directional derivative of T along r at $t = 2$ is

$$D_{\mathbf{r}}f\Big|_{t=2} \approx \frac{-9.467}{12.689} \approx -0.746.$$

So, at time $t = 2$, the temperature along the path is decreasing at an instantaneous rate of approximately $-0.746°$ per unit of distance.   ■

Note that the derivative of $f(\mathbf{r}(t))$ with respect to t is measured in output units per time unit (or whatever other unit t represents), while the directional derivative $D_{\mathbf{r}}(f)$ is measured in output units per unit of distance. This certainly makes sense in terms of dimensional analysis. In our examples using a temperature field T,

$$\frac{d}{dt}f(\mathbf{r}(t)) \text{ is measured in units of } \frac{degrees}{time}.$$

The speed $\|\mathbf{r}'(t)\|$ is measured in units of $distance/time$, so when we divide by the speed, we have

$$D_{\mathbf{r}}f \text{ is measured in units of } \frac{degrees}{distance}.$$

The relationship between the gradient and level curves and surfaces

In the last chapter we noted that given a contour map showing the level curves of a function of two variables, the direction of greatest increase or of greatest decrease at a point is naturally perpendicular to the level curve through that point. We have noted earlier in this section that the gradient gives the direction of greatest rate of increase. That suggests that the gradient vector at a point should be *orthogonal to the level curve through that same point.*

To see why this is reasonable, suppose we have a level curve with equation

$$f(x, y) = c,$$

for some constant c. Now, suppose the differentiable path $\mathbf{r}(t)$ traces out the same curve. If we evaluate the function f along this path, we have

$$f(\mathbf{r}(t)) = c$$

for each t.

Taking the derivative with respect to t:

$$\nabla f(\mathbf{r}(t)) \cdot \mathbf{r}'(t) = 0.$$

The left-hand side is the dot product of the gradient ∇f at the point $\mathbf{r}(t)$ with the velocity vector $\mathbf{r}'(t)$. This tells us that $\nabla f(\mathbf{r}(t))$ is orthogonal to $\mathbf{r}'(t)$, which is *tangent* to the curve. Hence, $\nabla f(\mathbf{r}(t))$ is orthogonal to $\mathbf{r}'(t)$.

The same goes for functions of three variables. Now there could be several differentiable paths $\mathbf{r}(t)$ on a level surface

$$f(x, y, z) = c$$

that pass through a given point on the surface. For each one, $\nabla f(\mathbf{r}(t)) \cdot \mathbf{r}'(t) = 0$. If $\nabla f \neq \mathbf{0}$, then since all the vectors $\mathbf{r}'(t)$ are tangent to the surface, we must have ∇f normal to the tangent plane of the level surface $f(x, y, z) = c$.

Definition 9

A vector normal to the tangent plane of a surface at a point is said to be **normal** to the surface at that same point.

In other words, we have just established that a nonzero gradient ∇f is normal to a level surface $f(x, y, z) = c$.

Using ∇f to find the equation of a tangent plane.

If $f : \mathbb{R}^2 \to \mathbb{R}$, then the tangent plane to the graph of $z = f(x, y)$ at the point $(a, b, f(a, b))$ is given by the equation of the best linear approximation. We can write this in terms of the gradient as follows:

$$z = f(a, b) + \nabla f(a, b) \cdot \langle x - a, y - b \rangle.$$

EXAMPLE 20   Given the function

$$f(x, y) = y^3 e^x,$$

Find the equation of the tangent plane to the graph of $z = f(x, y)$ at $(0, 2, 8)$.

Solution   First, we compute the gradient:

$$\nabla f = \langle \frac{\partial f}{\partial x}, \frac{\partial f}{\partial y} \rangle = \langle y^3 e^x, 3y^2 e^x \rangle$$

and evaluate it at the input point $(a, b) = (0, 2)$:

$$\nabla f(a, b) = \nabla f(0, 2) = (8, 12).$$

The equation of the tangent plane is

$$z = f(a,b) + \nabla f(a,b) \cdot (x - a, y - b)$$

$$= 8 + (8, 12) \cdot (x - 0, y - 2)$$

$$= 8 + (8x) + (12y - 24)$$

$$= 8x + 12y - 16.$$

∎

Recall that if

$$\langle a, b, c \rangle$$

is a vector normal to a plane containing the point (x_0, y_0, z_0), then every point (x, y, z) in the plane satisfies the equation

$$a(x - x_0) + b(y - y_0) + c(z - z_0) = 0.$$

This follows from the observation that the vector

$$\langle x - x_0, y - y_0, z - z_0 \rangle$$

is parallel to the plane, so its dot product with the perpendicular normal vector must be 0:

$$\langle a, b, c \rangle \cdot \langle x - x_0, y - y_0, z - z_0 \rangle = 0.$$

Many surfaces have equations in three variables of the form

$$f(x, y, z) = c.$$

This is simply a level surface for the function f. If (x_0, y_0, z_0) is a point on the surface, then the gradient $\nabla f(x_0, y_0, z_0)$ is normal to the level surface at that point (provided it is not the zero vector). We can use it to find the equation of the tangent plane to the level surface.

EXAMPLE 21   What is the equation of the tangent plane to $\sqrt{x^2 + y^2 + z^2} = 5$ at the point $(0, 3, 4)$?

Solution   The surface described is a level surface for the function

$$f(x, y, z) = \sqrt{x^2 + y^2 + z^2}.$$

The gradient of f is

$$\nabla f = \left\langle \frac{x}{\sqrt{x^2 + y^2 + z^2}}, \frac{y}{\sqrt{x^2 + y^2 + z^2}}, \frac{z}{\sqrt{x^2 + y^2 + z^2}} \right\rangle.$$

At the point $(x_0, y_0, z_0) = (0, 3, 4)$, we have a normal vector given by

$$\nabla f(0, 3, 4) = \left\langle 0, \frac{3}{5}, \frac{4}{5} \right\rangle.$$

Every point in the tangent plane must satisfy

$$\left\langle 0, \frac{3}{5}, \frac{4}{5} \right\rangle \cdot (x - 0, y - 3, z - 4) = \frac{3}{5}(y - 3) + \frac{4}{5}(z - 4) = 0.$$

Therefore, the equation of the tangent plane to the surface at the point $(0, 3, 4)$ is

$$\frac{3}{5}y + \frac{4}{5}z - 5 = 0.$$

∎

EXERCISES

For each function in exercises 1-6, find $D_{\mathbf{u}}f$ at indicated point (a, b) and in the direction of the indicated unit vector $\mathbf{u}$.

1. $f(x, y) = \dfrac{x + y}{x - y}$ $(a, b) = (1, 2)$ $\mathbf{u} = \mathbf{i}/\sqrt{2} + \mathbf{j}/\sqrt{2}$.

2. $f(x, y) = xy + x/y$ $(a, b) = (-1, 3)$ $\mathbf{u} = \langle 3/5, 4/5 \rangle$.

3. $f(x, y) = \sin(x^2 y)$ $(a, b) = (2, 0)$ $\mathbf{u} = \mathbf{j}$.

4. $f(x, y) = \arctan(y/x)$ $(a, b) = (-3, -3)$ $\mathbf{u} = \mathbf{i}$.

5. $f(x, y) = x + \ln(x^2 + y^2 + 1)$ $(a, b) = (0, 0)$ $\mathbf{u} = \langle -0.8, 0.6 \rangle$.

6. $f(x, y) = \sqrt{x^2 + y^2}$. $(a, b) = (-3, 4)$ $\mathbf{u} = \langle 1/\sqrt{2}, -1/\sqrt{2} \rangle$.

For each function in exercises 7-11, find the unit vector $\mathbf{u}$ that gives the direction of the fastest rate of increase from the indicated point.

7. $f(x, y, z) = \dfrac{z}{x + y}$ $(a, b, c) = (1, 2, 3)$.

8. $f(x, y, z) = x^3 y^4 z^5$ $(a, b, c) = (1, -2, -1)$.

9. $f(x, y, z) = 5x - 4y + 3z$ $(a, b, c) = (-3, 2, 4)$.

10. $f(x, y, z) = \exp 1 - x^2 - y^2 - z^2$ $(a, b, c) = (0, 1, -1)$.

11. $f(x, y, z) = x \arcsin z - y \operatorname{arcsec} z$ $(a, b, c) = (.5, .5, .5)$.

The temperature at the point (x, y, z) is given by the scalar field

$$T(x, y, z) = 3x^2 + 2y^2 - 4z,$$

where the temperature is measured in degrees Celsius. Exercises 12-17 refer to this temperature field.

12. Find the gradient function ∇T.

13. What is the rate of change of temperature at the point $(-1, -3, 2)$ if one moves directly toward the origin?

14. What is the rate of change of temperature at the point $(5, -3, -4)$ if one moves in the direction $2\mathbf{i} - 6\mathbf{j} + 3\mathbf{k}$?

15. What unit vector gives the direction of greatest rate of increase of temperature at the point $(4, -1, 0)$?

16. A particle is moving along the path described by

$$\mathbf{r}(t) = \langle \cos t, \sin t, t \rangle,$$

where t is measured in seconds. Find how fast the temperature is changing (in degrees per second) when $t = 0$.

17. Find the directional derivative of the temperature (in degrees per distance unit) along the same path at time $t = 4$.

A scalar field $f : \mathbb{R}^3 \to \mathbb{R}$ is given by

$$f(x, y, z) = \arctan(x^2 + y^2 + z^2).$$

Exercises 18-21 refer to this scalar field.

18. Find ∇f.

19. Verify that

$$D_{\mathbf{i}} f = f_x \qquad D_{\mathbf{j}} f = f_y \qquad D_{\mathbf{k}} f = f_z.$$

20. If $\mathbf{r}(t) = \langle \cos 8t, \sin 6t, e^{4t} \rangle$, find $D_{\mathbf{r}} f$ at $t = 0$.

21. Find the best linear approximation to f at the point $(1, 0, 0)$.

An ant is standing on a hill whose height is given by

$$H(x, y) = x^4 + y^4 - 4x^2y^2.$$

Take "North" to be the direction of the positive y-axis and "East" to be the direction of the positive x-axis and use this information to answer exercises 22-25.

22. If the ant is standing at the point $(2, 3)$ and walks North-East, how fast is the elevation changing? Describe the direction (by a unit vector) that the ant should move to go uphill fastest.

23. If the ant is standing at the point $(-1, 2)$ and walks South-West, how fast is the elevation changing? Describe the direction (by a unit vector) that the ant should move to go downhill fastest.

24. If the ant is standing at the point $(3, -2)$ and walks North-West, how fast is the elevation changing? Describe the direction (by a unit vector) that the ant should move to go uphill fastest.

25. If the ant is standing at the point $(-2, -1)$ and walks South-East, how fast is the elevation changing? Describe the direction (by a unit vector) that the ant should move to go downhill fastest.

Suppose now that the height of the hill is given by

$$H(x, y) = x/y.$$

26. If the ant is standing at the point $(2, 3)$ and walks North-West, how fast is the elevation changing? Describe the direction (by a unit vector) that the ant should move to go downhill fastest.

27. If the ant is standing at the point $(-1, 2)$ and walks South-East, how fast is the elevation changing? Describe the direction (by a unit vector) that the ant should move to go uphill fastest.

28. If the ant is standing at the point $(3, -2)$ and walks North-East, how fast is the elevation changing? Describe the direction (by a unit vector) that the ant should move to go downhill fastest.

29. If the ant is standing at the point $(-2, -1)$ and walks South-West, how fast is the elevation changing? Describe the direction (by a unit vector) that the ant should move to go uphill fastest.

---

30. For each function f in exercises 1-6, find a unit normal vector and the equation of the tangent plane to the surface $z = f(x, y)$ at the indicated point (a, b).

15.4 HIGHER ORDER PARTIAL DERIVATIVES

A partial derivative of a multivariate function f is also a function of several variables. Hence, each partial derivative itself has partial derivatives. We call these the **second-order partial derivatives** of f.

For example, if f is a function of two variables, we can consider

$$\frac{\partial}{\partial x}\left(\frac{\partial f}{\partial x}\right), \qquad \frac{\partial}{\partial y}\left(\frac{\partial f}{\partial x}\right), \qquad \frac{\partial}{\partial x}\left(\frac{\partial f}{\partial y}\right), \qquad \frac{\partial}{\partial y}\left(\frac{\partial f}{\partial y}\right).$$

The shorthand notations for these are

$$\frac{\partial^2 f}{\partial x^2}, \qquad \frac{\partial^2 f}{\partial y\partial x}, \qquad \frac{\partial^2 f}{\partial x\partial y}, \qquad \frac{\partial^2 f}{\partial y^2}.$$

EXAMPLE 22 Suppose $z = f(x,y) = x^2 y + 2xy - y^3$. Find the second-order partial derivatives of f.

Solution The first order partial derivatives of f are:

$$\frac{\partial f}{\partial x} = 2xy + 2y \quad \text{and} \quad \frac{\partial f}{\partial y} = x^2 + 2x - 3y^2.$$

The second-order partial derivatives with respect to x are:

$$\frac{\partial^2 f}{\partial x^2} = 2y \quad \text{and} \quad \frac{\partial^2 f}{\partial x\partial y} = 2x + 2.$$

The second-order partial derivatives with respect to y are:

$$\frac{\partial^2 f}{\partial y\partial x} = 2x + 2 \quad \text{and} \quad \frac{\partial^2 f}{\partial y^2} = -6y.$$

■

Note that the "mixed" second-order partial derivatives in this example are equal. This need not be the case. However, we can state the following:

If either of the mixed derivatives is continuous then so is the other and they must match.

An alternative notation for second-order partial derivatives is

$$f_{xx} = \frac{\partial^2 f}{\partial x^2}, \qquad f_{yx} = \frac{\partial^2 f}{\partial x \partial y}, \qquad f_{xy} = \frac{\partial^2 f}{\partial y \partial x}, \qquad f_{yy} = \frac{\partial^2 f}{\partial y^2}.$$

Note carefully the notation for the mixed partial derivatives, for it is an easy matter to confuse the order. Perhaps the best way to remember the notation is to consider a second-order derivative as the derivative of a derivative. For example,

$$\frac{\partial^2 f}{\partial x \partial y} = \frac{\partial}{\partial x} \left(\frac{\partial f}{\partial y} \right) = (f_y)_x = f_{yx}.$$

Fortunately, the mixed partial derivatives match anyway when they are continuous.

For a function of three variables $f(x, y, z)$, there are a total of *nine* second-order partial derivatives:

$$\frac{\partial^2 f}{\partial x^2} \qquad \frac{\partial^2 f}{\partial x \partial y} \qquad \frac{\partial^2 f}{\partial x \partial z}$$

$$\frac{\partial^2 f}{\partial y \partial x}, \qquad \frac{\partial^2 f}{\partial y^2}, \qquad \frac{\partial^2 f}{\partial y \partial z}$$

$$\frac{\partial^2 f}{\partial z \partial x} \qquad \frac{\partial^2 f}{\partial z \partial y} \qquad \frac{\partial^2 f}{\partial z^2}.$$

For a function of n variables $f(x_1, x_2, \ldots, x_n)$, the second-order partial derivative

$$\frac{\partial^2 f}{\partial x_i \partial x_j} = \frac{\partial}{\partial x_i} \left(\frac{\partial f}{\partial x_j} \right)$$

denotes the partial derivative with respect to x_i of the partial derivative of f with respect to x_j.

EXAMPLE 23   Suppose $f(x, y, z) = e^{xyz}$. Find all the second-order partial derivatives of f.

Solution   The first-order partial derivatives of f are:

$$\frac{\partial f}{\partial x} = yze^{xyz}, \qquad \frac{\partial f}{\partial y} = xze^{xyz}, \qquad \frac{\partial f}{\partial z} = xye^{xyz}.$$

The second-order partial derivatives of f are:

$$\frac{\partial^2 f}{\partial x^2} = y^2z^2e^{xyz}, \qquad \frac{\partial^2 f}{\partial x \partial y} = ze^{xyz} + xyz^2e^{xyz}, \qquad \frac{\partial^2 f}{\partial x \partial z} = ye^{xyz} + xy^2ze^{xyz},$$

$$\frac{\partial^2 f}{\partial y \partial x} = ze^{xyz} + xyz^2e^{xyz}, \qquad \frac{\partial^2 f}{\partial y^2} = x^2z^2e^{xyz}, \qquad \frac{\partial^2 f}{\partial y \partial z} = xe^{xyz} + x^2yze^{xyz},$$

$$\frac{\partial^2 f}{\partial z \partial x} = ye^{xyz} + xy^2ze^{xyz}, \qquad \frac{\partial^2 f}{\partial z \partial y} = xe^{xyz} + x^2yze^{xyz}, \qquad \frac{\partial^2 f}{\partial z^2} = x^2y^2e^{xyz}.$$

Again, note that the mixed partial derivatives involving the same variables of differentiation match. ∎

If all its second-order partial derivatives are continuous, a function f is called C^2-**differentiable**. This is a nice property for a function to have, since it guarantees that the mixed partial derivatives match. For example, knowing that a function $f : \mathbb{R}^3 \to \mathbb{R}$ is C^2-differentiable means only six of the nine second-order partial derivatives need to be calculated.

We can compute third and higher-order partial derivatives. The number of partial derivatives increases exponentially with the order. There are $8 = 2^3$ third-order partial derivatives of a function of two variables:

$$\frac{\partial^3 f}{\partial x^3}, \quad \frac{\partial^3 f}{\partial x^2 \partial y}, \quad \frac{\partial^3 f}{\partial x \partial y \partial x}, \quad \frac{\partial^3 f}{\partial x \partial y^2}, \quad \frac{\partial^3 f}{\partial y \partial x^2}, \quad \frac{\partial^3 f}{\partial y \partial x \partial y}, \quad \frac{\partial^3 f}{\partial y^2 \partial x}, \quad \frac{\partial^3 f}{\partial y^3},$$

and $27 = 3^3$ third-order partial derivatives of a function of three variables!

Functions built up out of the standard algebraic, trigonometric, inverse trigonometric, exponential, and logarithmic functions are continuously differentiable *infinitely many times* at most points at which they are defined. We could say that they are C^∞-differentiable. Most of the functions we'll discuss are certainly at least C^2-differentiable.

Geometrical and physical interpretations

The second derivative $\dfrac{d^2 y}{dx^2}$ gives us information about the concavity of the graph of a function of one variable $y = f(x)$. The second-order partial derivative of a function of two variables gives us information about the shape of the graph of the function $z = f(x, y)$.

If $\dfrac{\partial^2 f}{\partial x^2} > 0$ then the slice of $z = f(x, y)$ with y fixed is concave up.

If $\dfrac{\partial^2 f}{\partial x^2} < 0$ then the slice of $z = f(x, y)$ with y fixed is concave down.

Similarly, the sign of $\dfrac{\partial f}{\partial y^2}$ tells us whether the slice with x fixed is concave up or down.

The mixed partials give us information about how slopes change from slice to slice. For example, if $\dfrac{\partial^2 f}{\partial x \partial y} > 0$, this means that the y-slopes are increasing as we move from slice to slice in the direction of increasing x.

Perhaps the most effective way to incorporate all the information that the second derivatives supply is to use them to find the **best quadratic approximation** to the function at that point. By definition, the best quadratic approximation to a C^2-differentiable function $f(x, y)$ at a point (a, b) is the quadratic function q satisfying

$$q(a, b) = f(a, b), \qquad q_x(a, b) = f_x(a, b), \qquad q_y(a, b) = f_y(a, b),$$

$$q_{xx}(a, b) = f_{xx}(a, b), \qquad q_{xy}(a, b) = f_{xy}(a, b), \qquad q_{yy}(a, b) = f_{yy}(a, b).$$

In other words, q and f and all their corresponding first and second-order derivatives must match at the point (a, b). These six requirements completely determine the quadratic approximation q, as illustrated in the following example.

EXAMPLE 24 Find the best quadratic approximation to

$$f(x, y) = \sin(xy) + e^{2x} + e^y$$

at the point $(0, 0)$.

Solution We need to find a quadratic function

$$q(x, y) = c_{11} x^2 + c_{12} xy + c_{22} y^2 + c_1 x + c_2 y + c_0$$

such that

$$q(0, 0) = f(0, 0),$$

$$\frac{\partial q}{\partial x}\bigg|_{(0,0)} = \frac{\partial f}{\partial x}\bigg|_{(0,0)}, \qquad \frac{\partial q}{\partial y}\bigg|_{(0,0)} = \frac{\partial f}{\partial y}\bigg|_{(0,0)},$$

and

$$\frac{\partial^2 q}{\partial x^2}\bigg|_{(0,0)} = \frac{\partial^2 f}{\partial x^2}\bigg|_{(0,0)}, \qquad \frac{\partial^2 q}{\partial x \partial y}\bigg|_{(0,0)} = \frac{\partial^2 f}{\partial x \partial y}\bigg|_{(0,0)}, \qquad \frac{\partial^2 q}{\partial y^2}\bigg|_{(0,0)} = \frac{\partial^2 f}{\partial y^2}\bigg|_{(0,0)}.$$

First, we gather the needed information regarding the function f:

$$f(x,y) = \sin(xy) + e^{2x} - 2e^{y} \qquad \Longrightarrow \qquad f(0,0) = -1$$

$$f_x(x,y) = y\cos(xy) + 2e^{2x} \qquad \Longrightarrow \qquad f_x(0,0) = 2$$

$$f_y(x,y) = x\cos(xy) - 2e^{y} \qquad \Longrightarrow \qquad f_y(0,0) = -2$$

$$f_{xx}(x,y) = -y^2\sin(xy) + 4e^{2x} \qquad \Longrightarrow \qquad f_{xx}(0,0) = 4$$

$$f_{xy}(x,y) = f_{yx}(x,y) = \cos(xy) - xy\sin(xy) \qquad \Longrightarrow \qquad f_{xy}(0,0) = 1$$

$$f_{yy}(x,y) = -x^2\sin(xy) - 2e^{y} \qquad \Longrightarrow \qquad f_{yy}(0,0) = -2$$

Now, we do the same for the quadratic q and solve for the coefficients that guarantee matching values with f:

$$q(0,0) = c_0 = -1$$

$$q_x(x,y) = 2c_{11}x + c_{12}y + c_1 \qquad \Longrightarrow \qquad q_x(0,0) = c_1 = 2$$

$$q_y(x,y) = c_{12}x + 2c_{22}y + c_2 \qquad \Longrightarrow \qquad q_y(0,0) = c_2 = -2$$

$$q_{xx}(x,y) = 2c_{11} \qquad \Longrightarrow \qquad q_{xx}(0,0) = 2c_{11} = 4 \qquad \Longrightarrow \qquad c_{11} = 2$$

$$q_{xy}(x,y) = q_{yx}(x,y) = c_{12} \qquad \Longrightarrow \qquad c_{12} = 1$$

$$q_{yy}(x,y) = 2c_{22} = -2 \qquad \Longrightarrow \qquad c_{22} = -1$$

Hence the best quadratic approximation to f at $(0,0)$ is

$$2x^2 + xy - y^2 + 2x - 2y - 1.$$

■

The technique used in this example can be applied to any C^2- differentiable function f. The coefficients of the best quadratic approximation to $f(x,y)$ at the point $(0,0)$ are

$$c_0 = f(0,0)$$

$$c_1 = f_x(0,0) \qquad c_2 = f_y(0,0)$$

$$c_{11} = \frac{1}{2}f_{xx}(0,0) \qquad c_{12} = f_{xy}(0,0) \qquad c_{22} = \frac{1}{2}f_{yy}(0,0)$$

Definition 10

> This best quadratic approximation is called the **second-degree Taylor polynomial approximation to** f **at** $(0,0)$. For (x,y) sufficiently close to $(0,0)$ we have
>
> $$f(x,y) \approx f(0,0) + f_x(0,0)x + f_y(0,0)y$$
>
> $$+ \frac{1}{2}\left(f_{xx}(0,0)x^2 + 2f_{xy}(0,0)xy + f_{yy}(0,0)y^2\right).$$

We can find the best quadratic approximation of f at a point other than the origin.

Definition 11

> The **second-degree Taylor polynomial approximation to** f **at** (a,b) is
>
> $$f(x,y) \approx f(a,b) + f_x(a,b)(x-a) + f_y(a,b)(y-b)$$
>
> $$+ \frac{1}{2}(f_{xx}(a,b)(x-a)^2 + 2f_{xy}(a,b)(x-a)(y-b) + f_{yy}(a,b)(y-b)^2).$$

EXAMPLE 25   Find the second-degree Taylor polynomial approximation to $f(x,y) = x^2 \ln y$ at $(-2,1)$.

Solution   First, we compute the first and second order derivatives:

$$f_x = 2x\ln y \qquad f_y = \frac{x^2}{y}$$

$$f_{xx} = 2\ln y \qquad f_{xy} = \frac{2x}{y} \qquad f_{yy} = \frac{-x^2}{y^2}$$

Therefore,

$$f(x, y) \approx f(-2, 1) + f_x(-2, 1)(x + 2) + f_y(-2, 1)(y - 1)$$

$$+ \frac{1}{2} \left(f_{xx}(-2, 1)(x + 2)^2 + 2f_{xy}(-2, 1)(x + 2)(y - 1) + f_{yy}(-2, 1)(y - 1)^2 \right)$$

$$= 0 + 0(x + 2) + 4(y - 1) + \frac{1}{2} \left(0(x + 2)^2 + 2(-4)(x + 2)(y - 1) + (-4)(y - 1)^2 \right)$$

$$= 4(y - 1) - 4(x + 2)(y - 1) - 2(y - 1)^2.$$

∎

The second-degree Taylor polynomial approximation is very useful in helping us analyze extrema of functions of two variables, as we'll see in the next section.

EXERCISES

In exercises 1-8, first find the second-order partial derivatives, and then use these to find the best quadratic approximation to each function at $(0, 0)$. Use the discriminant (see Chapter 14) to determine the shape ("bowl up," "bowl down," "saddle," "trough," or "flat,") of the quadratic approximation function near the input $(0, 0)$.

1. $f(x, y) = x^2 + y^2$.

2. $f(x, y) = x^3 + y^3$.

3. $f(x, y) = x^3 - y^2$.

4. $f(x, y) = x^4 - y^4$.

5. $f(x, y) = xy$.

6. $f(x, y) = x^2 y$.

7. $f(x, y) = xy^3$.

8. $f(x, y) = x^2 y^2$.

---

9. At a point (a, b), a student finds the following for a particular function f:

$$f_x(a, b) > 0 \qquad f_y(a, b) < 0$$

$$f_{xx}(a,b) > 0 \qquad f_{xy}(a,b) < 0 \qquad f_{yy}(a,b) > 0.$$

Describe as completely as you can what this information means in terms of the slices of the graph of $z = f(x,y)$ at the point $(a, b, f(a,b))$.

10. For a function of two variables $z = f(x,y)$, how many fourth-degree partial derivatives are there? How about for a function of three variables?

For each function in exercises 11-15, find all the second-order partial derivatives of the function f.

11. $f(x, y, z) = \dfrac{z}{x+y}$.

12. $f(x, y, z) = x^3 y^4 z^5$.

13. $f(x, y, z) = 5x - 4y + 3z$.

14. $f(x, y, z) = \exp 1 - x^2 - y^2 - z^2$.

15. $f(x, y, z) = x \arcsin z - y \operatorname{arcsec} z$.

For each function in exercises 16-20, find all the second-order partial derivatives, then use them to find the best quadratic approximation to the function at the indicated point (a, b).

16. $f(x, y) = \dfrac{x+y}{x-y}$, \qquad $(a, b) = (1, 2)$.

17. $f(x, y) = xy + x/y$, \qquad $(a, b) = (-1, 3)$.

18. $f(x, y) = e^{x^2 + y^2}$, \qquad $(a, b) = (1, -2)$.

19. $f(x, y) = \sin(x^2 y)$, \qquad $(a, b) = (2, 0)$.

20. $f(x, y) = \arctan(y/x)$, \qquad $(a, b) = (-3, -3)$.

15.5 FINDING EXTREMA OF MULTIVARIABLE FUNCTIONS—OPTIMIZATION

In this section we'll show how the derivatives of multivariable functions can be used to detect the locations of and analyze its extrema (maxima and minima). First, let's review some of the basic terminology regarding extrema of single-variable functions.

An input x_0 is the location of a *relative* or *local maximum* for $y = f(x)$ if there is a neighborhood of x_0 (an open interval centered at x_0) such that $f(x_0) \geq f(x)$ for all x in the neighborhood. The input x_0 is the location of a *relative* or *local minimum* for $y = f(x)$ if $f(x_0) \leq f(x)$ for all x in some neighborhood.

We can make very similar definitions for continuous multivariable functions.

Definition 12

The input (x_0, y_0) for a function f of two variables is the location of a **relative** or **local maximum** if there is some neighborhood (the *interior of a circle* centered at (x_0, y_0)) such that

$$f(x_0, y_0) \geq f(x, y)$$

for all (x, y) in the neighborhood.
The input (x_0, y_0) is the location of a **relative** or **local minimum** if

$$f(x_0, y_0) \leq f(x, y)$$

for all (x, y) in some neighborhood of (x_0, y_0).
For a function of three variables, a point (x_0, y_0, z_0) is the location of either a **relative maximum** or **relative minimum** if there is a neighborhood (the *interior of a sphere* centered at (x_0, y_0, z_0)) such that

$$f(x_0, y_0, z_0) \geq f(x, y, z) \qquad \text{or} \qquad f(x_0, y_0, z_0) \leq f(x, y, z),$$

respectively, for all points (x, y, z) in the neighborhood.

Critical points

The key observation for a continuous function f of a single variable is that its relative extrema can only occur at *critical points*. A critical point x_0 is an input in the domain of f such that

$$f'(x_0) = 0 \qquad \text{or} \qquad f'(x_0) \text{ is undefined.}$$

Now, for a continuous multivariable function, a relative extremum will also appear to be a relative extremum on every *slice* of the function at that point. Geometrically, all this means is that a high or low point on the surface graph must also be a high or low point on every vertical cross-section through that point. If we hold all the variables constant except one and graph the resulting single variable slice function, then a relative extremum must be a *critical point* for this slice. The corresponding partial derivative must be zero or undefined at the extremum. This leads us to make the following definition.

Definition 13

> If f is a continuous multivariable function, then an input point having a neighborhood in the domain of f is called a **critical point** if either
>
> $$\nabla f = 0 \qquad \text{or} \qquad \nabla f \text{ is undefined}$$
>
> at that point.

In other words, a critical point for f is some point where *all the partial derivatives are zero*, or where *at least one partial derivative is undefined.*

Definition 14

> If f is a C^1-differentiable function, then a critical point where $\nabla f = 0$ is called a **stationary point**.

This terminology is well-suited, since the instantaneous rate of change in *any* direction from a stationary point is 0 (why?). For a C^1-function of two variables, a stationary point will be the location of a *horizontal tangent plane $z = c$*, where c is the function value at the stationary point.

EXAMPLE 26   Find the stationary critical points of the function $f(x,y) = x^2 + 5xy - y^3$. What is the equation of the tangent plane at each of these points?

Solution $\nabla f = \langle 2x + 5y, 5x - 3y^2 \rangle = \langle 0,0 \rangle$ only when we simultaneously have

$$2x + 5y = 0$$

$$5x - 3y^2 = 0.$$

Solving the first equation for x gives us $x = -5y/2$. Substituting into the second equation gives us

$$5(-5y/2) - 3y^2 = (y/2)(-25 - 6y) = 0 \implies y = 0 \quad \text{or} \quad y = -25/6.$$

When $y = 0$, $x = -5(0)/2 = 0$. When $y = -25/6$, $x = 125/12$. The two stationary critical points are

$$(0,0) \qquad \text{and} \qquad (125/12, -25/6).$$

The horizontal tangent plane at $(0,0)$ is

$$z = f(0,0) = 0.$$

The horizontal tangent plane at $(125/12, -25/6)$ is

$$z = f(125/12, -25/6) = -15625/432 \approx -36.17.$$

■

EXAMPLE 27   Find the stationary critical points of the function $f(x, y, z) = e^{xyz}$.

Solution $\nabla f = \langle yze^{xyz}, xze^{xyz}, xye^{xyz} \rangle = \langle 0, 0, 0 \rangle$ only when we simultaneously have

$$yze^{xyz} = 0$$

$$xze^{xyz} = 0$$

$$xye^{xyz} = 0.$$

The factor e^{xyz} is never 0. We can see that all three of the partial derivatives are 0 whenever at least two of x, y, and z are 0. The set of stationary critical points of f consists of every point on the three coordinate-axes.   ■

Second derivative test for extrema of functions of two variables

For a function of two variables, three possible shapes the graph of $z = f(x, y)$ could have at a stationary critical point are a local minimum, local maximum, or a saddle point. Figure 15.5 illustrates each of these shapes.

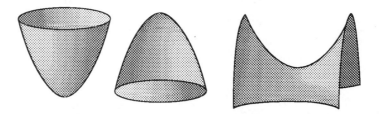

Figure 15.5 Stationary critical points: local minimum, local maximum, saddle point

At a stationary critical point, we know that both of the first-order partial derivatives are zero. Now, we'll show how the second-order partial derivatives can often be used to detect the shape of the graph.

If f is a C^2-differentiable function of two variables, then it has a second-degree Taylor approximation at any point (a, b) which can be written

$$f(x, y) \approx f(a, b) + \nabla f(a, b) \cdot \langle x - a, y - b \rangle$$

$$+ \frac{1}{2}(f_{xx}(a, b)(x - a)^2 + 2f_{xy}(a, b)(x - a)(y - b) + f_{yy}(a, b)(y - b)^2).$$

If (a, b) happens to be a stationary critical point for the function f, then

$$\nabla f(a, b) = 0,$$

and the Taylor polynomial approximation simplifies to

$$f(x, y) \approx f(a, b) + \frac{1}{2}(f_{xx}(a, b)(x - a)^2 + 2f_{xy}(a, b)(x - a)(y - b) + f_{yy}(a, b)(y - b)^2).$$

If we write

$$A = f_{xx}(a, b), \qquad B = f_{xy}(a, b), \qquad C = f_{yy}(a, b),$$

then we have

$$f(x, y) \approx f(a, b) + \frac{1}{2}(A(x - a)^2 + 2B(x - a)(y - b) + C(y - b)^2).$$

In the last chapter, we saw how the *discriminant* of a quadratic polynomial of the form

$$A(x - a)^2 + 2B(x - a)(y - b) + C(y - b)^2$$

is the quantity

$$D = AC - B^2.$$

The sign of this discriminant allows us to analyze the shape of the graph of the quadratic at (a, b).

The graph of our original function $z = f(x, y)$ has a shape very similar to this quadric surface near the point $(a, b, f(a, b))$. For this reason we say the quantity

$$\Delta = AC - B^2,$$

where

$$A = \frac{\partial^2 f}{\partial x^2}, \qquad B = \frac{\partial^2 f}{\partial x \partial y}, \qquad C = \frac{\partial^2 f}{\partial y^2},$$

is the **discriminant** of the function $z = f(x, y)$. We can use this discriminant Δ for a C^2-differentiable function f in essentially the same way we use the discriminant D for quadratic functions, as summarized in the following theorem.

Theorem 15.1

> **Second Derivative Test for Functions of Two Variables.**
> Suppose $z = f(x, y)$ is a C^2-differentiable function, and let $\Delta = AC - B^2$, where
>
> $$A = \frac{\partial^2 f}{\partial x^2}, \qquad B = \frac{\partial^2 f}{\partial x \partial y}, \qquad C = \frac{\partial^2 f}{\partial y^2}.$$
>
> **Hypothesis**: The gradient $\nabla f(a, b) = \langle 0, 0 \rangle$.
> **Case 1**: $\Delta < 0$ at (a, b).
> **Conclusion:** A saddle point occurs at (a, b).
>
> **Case 2**: $\Delta > 0$ and $A > 0$ at (a, b).
> **Conclusion:** A relative minimum occurs at (a, b).
>
> **Case 3**: $\Delta > 0$ and $A < 0$ at (a, b).
> **Conclusion:** A relative maximum occurs at (a, b).
>
> **Case 4**: $\Delta = 0$ at (a, b).
> **Conclusion:** None.

$\square$

A brief comment is in order regarding Case 4 of the theorem. Here, "no conclusion" simply means that when $\Delta = 0$, we do not have enough information to identify the shape of the graph. It is still possible that the critical point is the location of a relative minimum, relative maximum, or a saddle point, but we will need to use some other method to make that identification (such as graphing a wireframe plot or examining a contour map of level curves).

If we form the 2×2 matrix of second-order partial derivatives

$$\begin{bmatrix} \dfrac{\partial^2 f}{\partial x^2} & \dfrac{\partial^2 f}{\partial x \partial y} \\[2ex] \dfrac{\partial^2 f}{\partial y \partial x} & \dfrac{\partial^2 f}{\partial y^2} \end{bmatrix},$$

then note that we can think of the discriminant Δ as the *determinant* of this matrix:

$$\Delta = \begin{vmatrix} \dfrac{\partial^2 f}{\partial x^2} & \dfrac{\partial^2 f}{\partial x \partial y} \\[2ex] \dfrac{\partial^2 f}{\partial y \partial x} & \dfrac{\partial^2 f}{\partial y^2} \end{vmatrix} = \begin{vmatrix} A & B \\ B & C \end{vmatrix} = AC - B^2.$$

Note that we are using the fact that $\dfrac{\partial^2 f}{\partial x \partial y} = \dfrac{\partial^2 f}{\partial y \partial x}$ for a C^2-differentiable function.

EXAMPLE 28   Use the second derivative test to classify each stationary critical point of the function $z = f(x, y) = x^2 + 5xy - y^3$ as a saddle point or relative extremum, if possible.

Solution   We found the stationary critical points to be $(0,0)$ and $(125/12, -25/6)$. The second-order partial derivatives are

$$\frac{\partial^2 f}{\partial x^2} = 2, \qquad \frac{\partial^2 f}{\partial x \partial y} = 5, \qquad \frac{\partial^2 f}{\partial y^2} = -6y.$$

At $(0,0)$, we have

$$A = 2, \qquad B = 5, \qquad C = 0,$$

and so

$$\Delta = AC - B^2 = 0 - 25 = -25 < 0.$$

We conclude that $(0,0)$ is the location of a saddle point.

At $(125/12, -25/6)$, we have

$$A = 2, \qquad B = 5, \qquad C = 25,$$

and so

$$\Delta = AC - B^2 = 50 - 25 = 25 > 0.$$

Since $A = 2 > 0$, we conclude that $(125/12, -25/6)$ is the location of a relative minimum.   ∎

To find the **absolute extrema** of a function of two variables over some domain D in the xy-plane, we must check the output values at

1) the locations of all relative extrema that lie in D.

2) points along the boundary of D.

Along the boundary of the domain, we may be able to write the function in terms of a single variable, and use calculus to locate the relevant extrema along the boundary.

EXAMPLE 29   Find the absolute minimum and maximum of $z = f(x, y) = x^2 + 5xy - y^3$ over the domain $D = [0, 2] \times [0, 2]$.

Solution   The domain in question is a square with corners at $(0, 0)$, $(2, 0)$, $(2, 2)$, and $(0, 2)$ (see Figure 15.6).

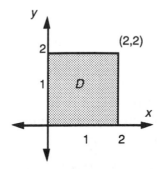

Figure 15.6 Domain D is a square.

Now, in the previous example, we found that the only relative extremum occurs at $(125/12, -25/6)$, which lies outside the domain D. Hence, we need to examine the points along the boundary.

First, points along the bottom edge of the square have coordinates

$$(x, 0) \qquad \text{for } 0 \leq x \leq 2.$$

Substituting into the function, we have

$$z = f(x, 0) = x^2,$$

so the minimum value along the bottom edge is $f(0, 0) = 0$, and the maximum value is $f(2, 0) = 4$.

Along the left edge of the square, the points have coordinates

$$(0, y) \qquad \text{for } 0 \leq y \leq 2.$$

Substituting into the function, we have

$$z = f(0, y) = -y^3,$$

so the minimum value along the left edge is $f(0, 2) = -8$, and the maximum value is $f(0, 0) = 0$.

Along the top edge of the square, the points have coordinates

$$(x, 2) \qquad \text{for } 0 \leq x \leq 2.$$

Substituting into the function, we have

$$z = f(x, 2) = x^2 + 10x - 8.$$

If we treat this as a function of a single variable x, we note that its derivative is $dz/dx = 2x + 10$. The critical point $x = -5$ lies outside the interval of x values, so the extrema must occur at the endpoints $x = 0$ and $x = 2$. The minimum value is $f(0, 2) = -8$, and the maximum value is $f(2, 2) = 16$.

Finally, along the right edge of the square, the points have coordinates

$$(2, y) \qquad \text{for } 0 \le y \le 2.$$

Substituting into the function, we have

$$z = f(2, y) = 4 + 10y - y^3.$$

If we treat this as a function of a single variable y, we note that its derivative is $dz/dy = 10 - 3y^2$. Of the two critical points $y = \pm\sqrt{10/3}$, one *does* lie in the interval of y values, since

$$0 \le \sqrt{10/3} \le 2.$$

Hence, we check $y = 0$, $y = 2$, and $y = \sqrt{10/3}$. The corresponding output values are

$$f(2, 0) = 4, \qquad f(2, \sqrt{10/3}) \approx 16.17, \qquad f(2, 2) = 16.$$

Now, we can see that the absolute minimum value $z = -8$ occurs at the corner $(0, 2)$, while the absolute maximum value $z \approx 16.17$ occurs at $(2, \sqrt{10/3})$ along the right edge of the square.   ∎

EXAMPLE 30   Platypus Corporation must make a rectangular steel box whose three dimensions must have a sum of 120 cm. What is the maximum volume the box can have, and what are its dimensions?

Solution   If we label the dimensions of the box x, y, and z, then we must maximize the volume

$$V = xyz,$$

given that

$$x + y + z = 120.$$

The constraint on the dimensions allows us to write the volume as a function of two variables. Since $z = 120 - x - y$, we have

$$V(x, y) = xy(120 - x - y) = 120xy - x^2y - xy^2.$$

The critical points are where

$$\frac{\partial V}{\partial x} = 120y - 2xy - y^2 = y(120 - 2x - y) = 0,$$

and

$$\frac{\partial V}{\partial y} = 120x - x^2 - 2xy = x(120 - x - 2y) = 0.$$

We observe that both of these equations are satisfied if $(x, y) = (0, 0)$. (These are extremely small dimensions for a box, though, and certainly won't give us a maximum volume.)

Now, if $y \neq 0$, we can solve $\partial V/\partial x = 0$ for x:

$$x = \frac{y^2 - 120y}{-2y} = 60 - \frac{y}{2},$$

and substitute this into the equation $\partial V/\partial y = 0$:

$$(60 - y/2)(120 - (60 - y/2) - 2y) = (60 - y/2)(60 - 3y/2) = 0.$$

This last equation is satisfied when $y = 120$ or when $y = 40$. If $y = 120$, then $x = z = 0$ and the box has zero volume. The value

$$y = 40 \text{ cm}$$

means

$$x = 60 - y/2 = 60 - 40/2 = 60 - 20 = 40 \text{ cm}.$$

If we check the second derivatives of V, we find

$$\frac{\partial^2 V}{\partial x^2} = -2y, \qquad \frac{\partial^2 V}{\partial x \partial y} = 120 - 2x - 2y, \qquad \frac{\partial^2 V}{\partial y^2} = -2x.$$

At $(x, y) = (40, 40)$, we have $A = -2(40) = -80$, and

$$\Delta = AC - B^2 = (-80)(-80) - (120 - 80 - 80)^2 = 6400 - 1600 = 4800 > 0.$$

The second derivative test tells us that a maximum occurs when $x = y = 40$ cm. The third dimension is

$$z = 120 - x - y = 120 - 40 - 40 = 40 \text{ cm},$$

and we conclude that the maximum volume of the box is

$$V = xyz = (40)(40)(40) = 64000 \text{ cm}^3,$$

obtained when the box is in the shape of a cube 40 cm on a side. ■

The method of Lagrange multipliers—extrema under constraints

The second derivative test for functions of two variables is a handy tool. In this section we discuss a technique that can be used in some optimization problems involving functions of two, three, or more variables.

Sometimes we are faced with a problem of finding the maximum or minimum value of a multivariable function under some constraint on the possible inputs. That is, we're given some condition or conditions that the inputs must satisfy, and our task is to find an extremum of a function under those conditions. Here's an example of such a problem:

Problem: Find the maximum temperature that occurs at a point on a given surface in space.

The problem translates to finding the extreme value of a function $f(x, y, z)$ (in this case, temperature) under the constraint that $g(x, y, z) = c$ (the equation of the surface). Let's examine the geometric implications of the problem in terms of the gradients of both f and g.

First, let's assume that of the points on the surface described by $g(x, y, z) = c$, an extreme value of f is attained at (x_0, y_0, z_0), and that

$$\nabla g(x_0, y_0, z_0) \neq \mathbf{0}.$$

From properties of the gradient, we know that

    $\nabla g(x_0, y_0, z_0)$ **is normal to the level surface** $g(x, y, z) = c$.

Now, suppose we have a differentiable path $\mathbf{r}(t)$ whose image curve lies completely on the surface $g(x, y, z) = c$, with

$$\mathbf{r}(t_0) = (x_0, y_0, z_0) \qquad \text{and} \qquad \mathbf{r}'(t_0) \neq \mathbf{0}.$$

In other words, you can imagine that we have drawn a smooth curve on the surface that passes through the point (x_0, y_0, z_0).

Since f attains an extreme value at (x_0, y_0, z_0), we must also have $f(\mathbf{r}(t))$ attaining an extreme value at $t = t_0$. This means that $t = t_0$ must be a *critical point* for $f(\mathbf{r}(t))$, so that

$$\frac{d}{dt} f(\mathbf{r}(t_0)) = 0.$$

The chain rule tells us that

$$\frac{d}{dt} f(\mathbf{r}(t)) = \nabla f(\mathbf{r}(t)) \cdot \mathbf{r}'(t),$$

so we must have

$$\nabla f(\mathbf{r}(t_0)) \cdot \mathbf{r}'(t_0) = 0.$$

Geometrically, this zero dot product tells us that $\nabla f(\mathbf{r}(t_0))$ is orthogonal to $\mathbf{r}'(t_0)$. But $\mathbf{r}'(t_0)$ is a vector that is tangent to the image curve of $\mathbf{r}$, and so it must also be tangent to the surface $g(x, y, z) = c$. Since $\nabla f(x_0, y_0, z_0)$ is perpendicular to every such image curve passing through (x_0, y_0, z_0), we conclude that

☞ $\nabla f(x_0, y_0, z_0)$ **is normal to the level surface** $g(x, y, z) = c$.

The two gradients must point in the same or opposite directions! Put another way, there must be some nonzero scalar λ such that

$$\nabla f(x_0, y_0, z_0) = \lambda \nabla g(x_0, y_0, z_0).$$

Let's summarize the results of our reasoning in the statement of a theorem.

Theorem 15.2

Lagrange Multiplier Theorem (one constraint).
Hypothesis 1: The functions f and g are C^1-differentiable.
Hypothesis 2: The function f has an extremum under the constraint $g(x, y, z) = c$ at the point (x_0, y_0, z_0).
Hypothesis 3: The gradient $\nabla g(x_0, y_0, z_0) \neq 0$.
Conclusion: For some nonzero scalar λ, we must have

$$\nabla f(x_0, y_0, z_0) = \lambda \nabla g(x_0, y_0, z_0).$$

Note: The scalar λ is called a **Lagrange multiplier**, after the Italian-French mathematician Lagrange (1736-1813).

$\square$

Note that the equation

$$\nabla f = \lambda \nabla g$$

actually represents three equations:

$$\frac{\partial f}{\partial x} = \lambda \frac{\partial g}{\partial x}, \qquad \frac{\partial f}{\partial y} = \lambda \frac{\partial g}{\partial y}, \qquad \frac{\partial f}{\partial z} = \lambda \frac{\partial g}{\partial z}.$$

Along with the original constraint $g(x, y, z) = c$, this gives us a total of four equations in four unknowns (x, y, z, and λ). The particular value of λ may not be of much interest, but its introduction as an *auxilliary variable* allows us to solve for $x = x_0$, $y = y_0$, and $z = z_0$, the coordinates of the potential location of an extremum. This strategy is a direct application of the theorem, and is known as the **method of Lagrange multipliers**. We outline the method below:

Problem: Find the extreme values of $f(x, y, z)$ subject to the constraint $g(x, y, z) = c$.

Method of Lagrange Multipliers:

Step 1. Compute the gradient ∇f.

Step 2. Compute the gradient ∇g.

Step 3. Solve the following system of equations for x, y, z, and λ:

$$\frac{\partial f}{\partial x} = \lambda \frac{\partial g}{\partial x}$$

$$\frac{\partial f}{\partial y} = \lambda \frac{\partial g}{\partial y}$$

$$\frac{\partial f}{\partial z} = \lambda \frac{\partial g}{\partial z}$$

$$g(x, y, z) = c.$$

Step 4. Check the values x_0, y_0, z_0 back in the original problem to determine whether $f(x_0, y_0, z_0)$ is an extreme value subject to the constraint.

EXAMPLE 31   Find the point on the plane $x - y + 2z = 5$ that is closest to the origin.

Solution   The distance from any point (x, y, z) to the origin is

$$d = \sqrt{x^2 + y^2 + z^2}.$$

If we can minimize the function

$$f(x, y, z) = x^2 + y^2 + z^2$$

subject to the constraint $g(x, y, z) = x - y + 2z = 5$, we will have our solution.

The gradient of f is

$$\nabla f = \langle 2x, 2y, 2z \rangle,$$

and the gradient of our constraint function g is

$$\nabla f = \langle 1, -1, 2 \rangle.$$

The method of Lagrange multipliers leads us to solve

$$\nabla f = \lambda \nabla g$$

along with the original constraint. That is, we must solve

$$2x = \lambda, \qquad 2y = -\lambda, \qquad 2z = 2\lambda,$$

and

$$x - y + 2z = 5.$$

The three equations involving λ lead us to

$$x = \lambda/2, \qquad y = -\lambda/2, \qquad z = \lambda.$$

Substituting into the original constraint equation, we have

$$(\lambda/2) - (-\lambda/2) + 2\lambda = 5,$$

which yields $3\lambda = 5$ or $\lambda = 5/3$. The coordinates of the point we seek are

$$x = 5/6, \qquad y = -5/6, \qquad z = 5/3.$$

In other words, $(5/6, -5/6, 5/3)$ is the point on the plane $x - y + 2z = 5$ closest to the origin. You can check that this point indeed satisfies the equation of the plane. (Also note that a normal vector to the plane is $\langle 1, -1, 2 \rangle$, while the vector from the origin to the point $(5/6, -5/6, 5/3)$ is

$$\langle 5/6, -5/6, 5/3 \rangle = (5/6)\langle 1, -1, 2 \rangle.$$

In other words, the point we found is on the perpendicular between the origin and the plane, and is therefore the closest point to the origin.) ∎

EXAMPLE 32 Platypus Corporation must make a rectangular steel box whose three dimensions must have a sum of 120 cm. What is the maximum volume the box can have, and what are its dimensions?

Solution We have solved this problem before. This time we wish to illustrate the method of Lagrange multipliers. Our function to maximize is

$$V(x, y, z) = xyz,$$

and our constraint is

$$g(x, y, z) = x + y + z = 120.$$

The gradient equation is

$$\nabla V = \langle yz, xz, xy \rangle = \lambda \langle 1, 1, 1 \rangle = \nabla g.$$

From this, we can see that x, y, and z must all be nonzero, and

$$yz = \lambda \qquad \Longrightarrow \qquad y = \frac{\lambda}{z},$$

$$xz = \lambda \qquad \Longrightarrow \qquad x = \frac{\lambda}{z},$$

$$xy = \lambda \qquad \Longrightarrow \qquad \frac{\lambda^2}{z^2} = \lambda \qquad \Longrightarrow \qquad \lambda = z^2.$$

If we substitute the third result into the first two, we see that

$$x = y = z.$$

The original constraint equation

$$x + y + z = 120$$

tells us that we must have $x = y = z = 40$ cm, and the maximum volume of the box is $V = xyz = 64000$ cm^3. ∎

   The Lagrange Multiplier Theorem also holds for functions of two variables. That is, if f and g are C^1-differentiable functions of two variables and f has an extremum at (x_0, y_0) under the constraint $g(x, y) = c$, then

$$\nabla f(x_0, y_0) = \lambda \nabla g(x_0, y_0),$$

for some Lagrange multiplier λ (provided $\nabla g(x_0, y_0) \neq 0$).

EXERCISES

For each of the following functions of two variables in exercises 1-15, find all the stationary critical points (where $\nabla f = 0$), and then use the second derivative test to classify each point as a relative minimum, relative maximum, or saddle point. If no conclusion is possible, use graphical techniques to decide on the status of the critical point.

1. $f(x, y) = 3x^2 - 2y^2 + 3x - 2y + 5.$ **2.** $f(x, y) = 4xy - x^4 - 2y^2.$

3. $f(x, y) = 3xy - x^3 - y^3.$ **4.** $f(x, y) = 1 - x^2 + y^2.$

5. $f(x, y) = 9 + x^2 - y^2.$ **6.** $f(x, y) = 6x^2 - 2x^3 + 3y^2 + 6xy.$

7. $f(x, y) = x^4 + y^3 - 8x^2 - 3y + 5.$ **8.** $f(x, y) = x^2 - xy + y^2 - 2x + y.$

9. $f(x, y) = 3x^4 - 4x^2y + y^2.$ **10.** $f(x, y) = \dfrac{x^2 + y^2}{e^y}.$

11. $f(x, y) = y^3 - 3x^2y.$ **12.** $f(x, y) = 4xy^3 - 4x^3y.$

13. $f(x, y) = \cos x \sinh y.$ **14.** $f(x, y) = x^4 - y^4.$

15. $f(x, y) = x^2y^2.$

For each of the following functions of two variables in exercises 16-20, find the absolute maximum and minimum values of the function in the square region $D = [-1, 0] \times [-1, 0]$.

16. $f(x, y) = 3x^2 - 2y^2 + 3x - 2y + 5.$

17. $f(x, y) = 4xy - x^4 - 2y^2.$

18. $f(x, y) = 3xy - x^3 - y^3.$

19. $f(x, y) = 1 - x^2 + y^2$.

20. $f(x, y) = 9 + x^2 - y^2$.

For each of the following functions of two variables in exercises 21-25, find the absolute maximum and minimum values of the function in the triangular region bounded by the coordinate axes and the line $y = 1 - x$.

21. $f(x, y) = 6x^2 - 2x^3 + 3y^2 + 6xy$.

22. $f(x, y) = x^4 + y^3 - 8x^2 - 3y + 5$.

23. $f(x, y) = x^2 - xy + y^2 - 2x + y$.

24. $f(x, y) = 3x^4 - 4x^2y + y^2$.

25. $f(x, y) = \dfrac{x^2 + y^2}{e^y}$.

For each of the surfaces in exercises 26-30, use the technique of Lagrange multipliers to find the point on the surface that is closest to the origin.

26. $3x + 2y + z - 6 = 0$.

27. $(x - 1)^2 + (y + 2)^2 - (z - 3)^2 = 4$.

28. $xyz = 8$.

29. $z = x^2 + y^2 - 4$. (Hint: Rewrite the constraint in the form $g(x, y, z) = 0$ first.)

30. $xy + xz + yz = 12$.

---

31. Platypus Corporation is assigned to make a rectangular steel box with a total surface area of 1200 cm^2. What is the maximum volume of the box, and what are its dimensions?

32. Platypus Corporation is assigned to make a rectangular steel box with a fixed volume of 6000 cm^3. What is the minimum surface area of the box, and what are its dimensions?

33. Platypus Corporation is now assigned to make a rectangular steel box *with an open top* and a fixed volume of 6000 cm^3. What is the minimum surface area of the box, and what are its dimensions?

34. Platypus Corporation must now make a rectangular steel box with an open top of fixed volume 6000 cm^3 with a reinforced bottom that costs twice as much per square centimeter as the four sides. What are the dimensions of the box that will cost the least?

35. Platypus Corporation must now make a closed rectangular steel box with a fixed volume of 6000 cm^3 out of material costing 20 cents per square centimeter for the four sides, 30 cents per square centimeter for the top, and

40 cents per square centimeter for the bottom. What are the dimensions of the box that will cost the least?

---

36. It is possible to have more than one constraint and still use the method of Lagrange multipliers. For example, we may be seeking the maximum temperature along a curve in space that is described as the intersection of two surfaces. In this situation, the gradient of the scalar field may be a *linear combination* of the constraint gradients at an extremum.

If we seek an extremum of the function $f(x, y, z)$ under the constraints

$$g(x, y, z) = c \quad \text{and} \quad h(x, y, z) = d,$$

then we look for points (x_0, y_0, z_0) satisfying the constraints as well as the gradient equation

$$\nabla f(x_0, y_0, z_0) = \lambda \nabla g(x_0, y_0, z_0) + \mu \nabla h(x_0, y_0, z_0)$$

for the Lagrange multipliers λ and μ. (However, the method may fail if one of the constraint gradients is a scalar multiple of the other.)

Use the technique of Lagrange multipliers to solve the following problem:

Find the point that lies on both the paraboloid $z = x^2 + y^2$ and the plane $x + y + z = 6$ that is closest to the origin. Also find the point satisfying these constraints that lies farthest from the origin.

16

Integral Calculus of Multivariable Functions

The last chapter discussed the differential calculus of multivariable scalar-valued functions. To understand the principles of integration of functions of several variables, it will be wise for us to review the idea of definite integral for functions of a single variable.

If f is a real-valued function defined over the closed interval $[a, b]$, then we form a *Riemann sum* using the following steps:

Step 1. Form a regular partition of the interval $[a, b]$ by subdividing it into n equal-sized subintervals of length Δx;

Step 2. Choose a single input x_i from each subinterval for $1 \leq i \leq n$;

Step 3. For each x_i, evaluate $f(x_i)$ and multiply by the length of the subinterval Δx;

Step 4. Sum up the results for all the subintervals:

$$\sum_{i=1}^{n} f(x_i) \, \Delta x.$$

Now, if a limiting value exists for these Riemann sums as we take finer and finer partitions (in other words, as $n \to \infty$ and $\Delta x \to 0$), then this limiting value is called the value of the definite integral

$$\int_a^b f(x) \, dx.$$

If $f(x) \geq 0$ for all $x \in [a, b]$, then this definite integral has a geometrical interpretation as the *area* under graph of $y = f(x)$ over this interval (see Figure 16.1).

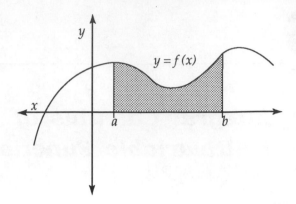

Figure 16.1 Geometric interpretation: $\displaystyle\int_a^b f(x)\,dx$ = area of shaded region.

A definite integral $\displaystyle\int_a^b f(x)\,dx$ can have a variety of *physical* meanings, depending on the interpretation of the function f. For example, we know that if $f(x)$ represents *force* at distance x, then $\displaystyle\int_a^b f(x)\,dx$ represents *work* performed over the interval $[a, b]$.

In this chapter, we'll see how the notion of definite integral can be extended to functions of several variables. These **multiple integrals** also have both geometric and physical interpretations.

16.1 MULTIPLE INTEGRALS

Double integrals—notation and interpretation

Just as a definite integral of a function of one variable has a geometrical interpretation related to *area*, a **double integral** of a function of two variables has a geometrical interpretation related to *volume*. In short, if D is some region in the xy-plane, and if

$$f(x, y) \geq 0$$

for all $(x, y) \in D$, then

$$\iint\limits_D f(x, y)\,dA$$

represents the *volume* between the graph of $z = f(x, y)$ and the xy-plane over the region D (see Figure 16.2).

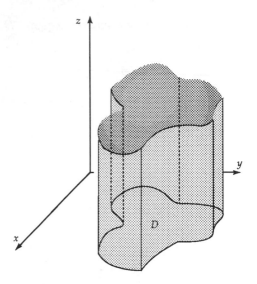

Figure 16.2 Geometric interpretation of $\displaystyle\iint\limits_{D} f(x,y)\,dA$.

Some comments regarding the notation need to be made here. The use of two integral signs $\iint$ reminds us that we're integrating over a 2-dimensional region D. The dA refers to an "area element." (We'll see that the precise form of dA will depend on how we choose to compute the value of the integral.)

Calculating double integrals over rectangular regions

Let's start out by describing the method for the simplest type of region D, that is, a *rectangular region* whose sides lie parallel to the coordinate axes (see Figure 16.3).

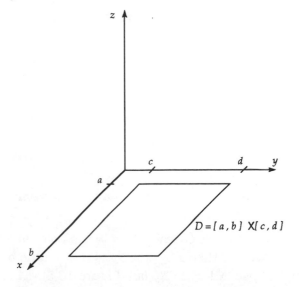

Figure 16.3 Rectangular region $D = [a, b] \times [c, d]$.

In this case, we can represent D as a Cartesian product of closed intervals

$$D = [a, b] \times [c, d]$$

$$= \{(x, y) : a \leq x \leq b \text{ and } c \leq y \leq d\}.$$

Suppose we form a regular partition of each of the intervals $[a, b]$ and $[c, d]$. That is, we subdivide $[a, b]$ into m equal-sized subintervals of length Δx and $[c, d]$ into n equal-sized subintervals of length Δy. (Note that it is *not* necessary to have either $m = n$ or $\Delta x = \Delta y$.) These two partitions create a grid of small rectangles, each of dimension Δx by Δy and having area

$$\Delta A = \Delta x \Delta y,$$

as shown in Figure 16.4.

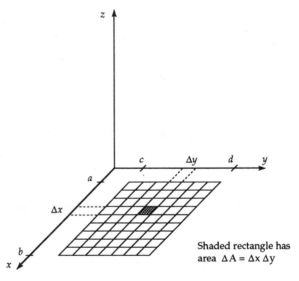

Figure 16.4 Grid partition of $D = [a, b] \times [c, d]$.

We have a total of $m \times n$ rectangles, which we could index with the letters i and j (i indexes a subinterval in the x-direction, so $1 \leq i \leq m$; j represents a subinterval in the y-direction, so $1 \leq j \leq n$). If we choose one point (x_i, y_j) from each rectangle, we can form the *double* Riemann sum

$$\sum_{j=1}^{n} \sum_{i=1}^{m} f(x_i, y_j) \, \Delta x \Delta y,$$

where the double summation signs simply indicate that we sum over all of the rectangles.

Geometrically, if (x, y) is a point in a small rectangle of dimensions Δx by Δy, then $f(x, y) \Delta x \Delta y$ is the volume of a box with the height $f(x, y)$ and base $\Delta A = \Delta x \Delta y$ (see Figure 16.5). This approximates the volume under the graph of $z = f(x, y)$ over this small rectangle.

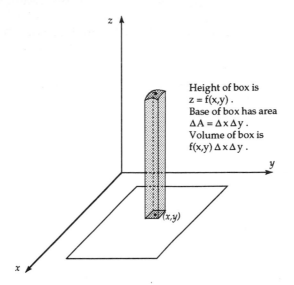

Figure 16.5 $f(x_i, y_j)\Delta x\Delta y$ approximates the volume over a small rectangle.

Hence, the double Riemann sum gives us an approximation to the *total volume* under the graph of $z = f(x,y)$ over the region D. If the values of these double Riemann sum approximations approach a single limit as the grid partition becomes finer and finer, we call this limit the value of the double integral

$$\iint\limits_D f(x,y)\,dA.$$

For a rectangular grid over the region $D = [a,b] \times [c,d]$, we write

$$\int_c^d \int_a^b f(x,y)\,dx\,dy,$$

where the area element

$$dA = dx\,dy$$

indicates that the Riemann sums have been formed with rectangular grid sections.

Computing double integrals iteratively

If f is a continuous function of two variables over the region D, then it's possible to reduce the calculation of the double integral

$$\iint\limits_D f(x,y)\,dA$$

to two single integrals computed *iteratively*.

The idea behind the method is perhaps best explained geometrically. For the time being, let's assume that $f(x, y) \geq 0$ over the entire region $D = [a, b] \times [c, d]$. The graph of $z = f(x, y)$ forms a "cap" over a volume with a rectangular base and vertical sides (see Figure 16.6).

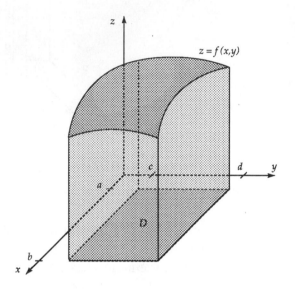

Figure 16.6 Interpreting a double integral as a volume.

Now, imagine that we select a particular $x_0 \in [a, b]$ and slice the volume with the vertical plane $x = x_0$ (see Figure 16.7).

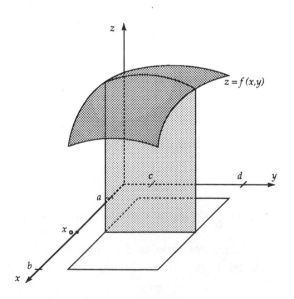

Figure 16.7 $\int_c^d f(x_0, y)\, dy$ is the area sliced by $x = x_0$.

The continuity of f allows us to compute the area of this slice with the definite integral

$$\int_c^d f(x_0, y)\, dy.$$

EXAMPLE 1 Suppose $f(x, y) = x^2 y^3$, and consider the volume bounded by the graph $z = x^2 y^3$ over the rectangular region $D = [-1, 3] \times [1, 4]$. Find the area of the slice of the volume cut out by the vertical plane $x = 2$.

Solution When $x = 2$, we have $f(2, y) = 4y^3$. The area of the slice is

$$\int_1^4 4y^3\, dy = y^4\Big]_{y=1}^{y=4} = 256 - 1 = 255.$$

■

We can perform a similar computation to find the area of the slice for any value $x \in [a, b]$.

EXAMPLE 2 Using the same function as in the previous example, find the area of the slice for *any* value $x \in [-1, 3]$. Use your result to find the area of the slices cut by the end planes $x = -1$ and $x = 3$.

Solution We simply treat x as if it were a particular constant value and find

$$\int_1^4 x^2 y^3\, dy = \frac{x^2 y^4}{4}\Big]_{y=1}^{y=4} = 64x^2 - \frac{x^2}{4} = \frac{255x^2}{4}.$$

For $x = 2$, note that we obtain a cross-sectional area 255 as in the previous example. At the two edges of the volume we have

$$\text{area of slice } = \frac{255}{4} \text{ at } x = -1$$

$$\text{area of slice } = \frac{2295}{4} \text{ at } x = 3.$$

■

The area $\displaystyle\int_c^d f(x, y)\, dy$ naturally depends on the value of x. Indeed, we can think of it as describing a *cross-sectional area function* that outputs the area of the slice corresponding to a value x. As such, if we integrate *this* function over the entire x-interval $[a, b]$, we obtain the *volume*

$$\int_a^b \left(\int_c^d f(x, y)\, dy \right) dx.$$

EXAMPLE 3   Find the volume bounded between the graph of $z = x^2y^3$, the xy-plane, and the vertical planes $x = -1$, $x = 3$, $y = 1$, and $y = 4$.

Solution   This is precisely the volume under the graph of $z = x^2y^3$ over the region

$$D = [-1, 3] \times [1, 4]$$

(since $x^2y^3 \geq 0$ for all $(x, y) \in D$). The volume is given by

$$\int_{-1}^{3} \left(\int_{1}^{4} x^2y^3 \, dy \right) dx = \int_{-1}^{3} \frac{255x^2}{4} \, dx = \frac{255x^3}{12} \Big]_{x=-1}^{x=3} = \frac{2295}{4} + \frac{85}{4} = 595.$$

∎

Changing the order of integration

Of course, we could also take vertical slices parallel to the x-axis (one for each possible value of $y = y_0$) as shown in Figure 16.8.

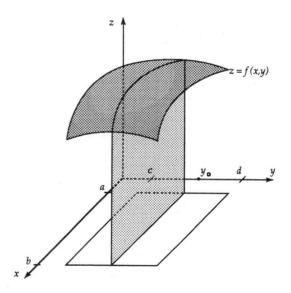

Figure 16.8 $\int_{a}^{b} f(x, y_0) \, dx$ is the area sliced by $y = y_0$.

The volume in this case is given by

$$\int_{c}^{d} \left(\int_{a}^{b} f(x, y) \, dx \right) dy$$

EXAMPLE 4 Find the volume described in the previous example using the opposite order of integration.

Solution

$$\int_1^4 \left(\int_{-1}^3 x^2 y^3 \, dx \right) dy = \int_1^4 \left(\frac{x^3 y^3}{3} \Bigg]_{x=-1}^{x=3} \right) dy$$

$$= \int_1^4 \left(9y^3 + \frac{y^3}{3} \right) dy$$

$$= \int_1^4 \frac{28y^3}{3} \, dy$$

$$= \frac{7y^4}{3} \Bigg]_{y=1}^{y=4}$$

$$= \frac{1792}{3} - \frac{7}{3} = \frac{1785}{3} = 595.$$

■

☞ **As you can see, the order of integration is a matter of choice. This is not always the case, but for** *continuous* **functions** f, **we have**

$$\int_c^d \int_a^b f(x,y) \, dx \, dy = \int_a^b \int_c^d f(x,y) \, dy \, dx.$$

Interpreting the double integral as signed volume

If $f(x,y) < 0$ for $(x,y) \in D$, then the double integral

$$\iint_D f(x,y) \, dA$$

represents the *negative* of the volume between the graph of $z = f(x,y)$ and the xy-plane under the region D. If $f(x,y)$ changes sign over D, then the integral $\iint_D f(x,y) \, dA$ represents the *net volume* above the xy-plane.

EXAMPLE 5 Find $\displaystyle\int_{-2}^{1}\int_{-1}^{3} xy\, dx\, dy.$

Solution The function $f(x, y) = xy$ changes sign over the region

$$D = [-1, 3] \times [-2, 1],$$

and the sign of xy depends on which quadrant (x, y) lies in. Figure 16.9 shows D, the region of integration lying in the xy-plane.

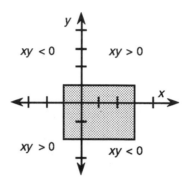

Figure 16.9 The sign of xy depends on the quadrant.

The graph of $z = xy$ lies below the xy-plane in the second and fourth quadrants, and lies above the xy-plane in the first and third. The integral $\displaystyle\int_{-2}^{1}\int_{-1}^{3} xy\, dx\, dy$ represents the *net* volume between the xy-plane and the graph. We have

$$\int_{-2}^{1}\int_{-1}^{3} xy\, dx\, dy = \int_{-2}^{1} \frac{x^2 y}{2}\Bigg]_{x=-1}^{x=3} dy$$

$$= \int_{-2}^{1} \frac{9y}{2} - \frac{y}{2}\, dy$$

$$= \int_{-2}^{1} 4y\, dy$$

$$= 2y^2 \Bigg]_{y=-2}^{y=1}$$

$$= 2 - 8 = -6.$$

The negative value indicates that there are 6 more units of volume *under* the xy-plane than above, between it and the graph of $z = xy$. ∎

EXAMPLE 6   You can check that

$$\int_{-2}^{0} \int_{-1}^{0} xy\, dx\, dy = 1, \qquad \int_{0}^{1} \int_{-1}^{0} xy\, dx\, dy = -\frac{1}{4},$$

$$\int_{-2}^{0} \int_{0}^{3} xy\, dx\, dy = -9, \qquad \int_{0}^{1} \int_{0}^{3} xy\, dx\, dy = \frac{9}{4},$$

and that these sum to the final value of the double integral. ∎

EXAMPLE 7   Find $\displaystyle\int_{0}^{\pi/2} \int_{-1}^{2} y \cos x\, dy\, dx$.

Solution   We have

$$\int_{0}^{\pi/2} \int_{-1}^{2} y \cos x\, dy\, dx = \int_{0}^{\pi/2} (y^2/2) \cos x \Big]_{-1}^{2} dx$$

$$= \int_{0}^{\pi/2} (2 \cos x - (1/2) \cos x)\, dx$$

$$= \int_{0}^{\pi/2} (3/2) \cos x\, dx$$

$$= (3/2) \sin x \Big]_{0}^{\pi/2} = 3/2 - 0 = 3/2.$$

∎

Triple integrals

A triple integral is defined for a function of three variables over a region R in $\mathbb{R}^3$. The notation is

$$\iiint_{R} f(x, y, z)\, dV$$

where dV is the notation for a "volume element," whose form will depend on our method of computation.

The simplest type of region R is a rectangular parallelepiped (box) with sides parallel to the coordinate planes. Such a region can be denoted as a Cartesian product of three closed intervals:

$$R = [a, b] \times [c, d] \times [p, q]$$

$$= \{(x, y.z) : a \leq x \leq b \text{ and } c \leq y \leq d \text{ and } p \leq z \leq q\}$$

(see Figure 16.10).

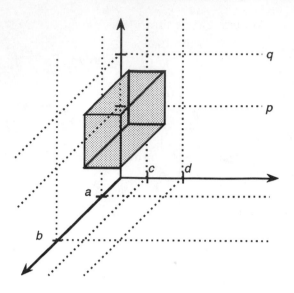

Figure 16.10 The box $[a, b] \times [c, d] \times [p, q]$.

Now, we form a regular partition of each interval:

$[a, b]$ is subdivided into ℓ equal-sized subintervals of length Δx.

$[c, d]$ is subdivided into m equal-sized subintervals of length Δy.

$[p, q]$ is subdivided into n equal-sized subintervals of length Δz.

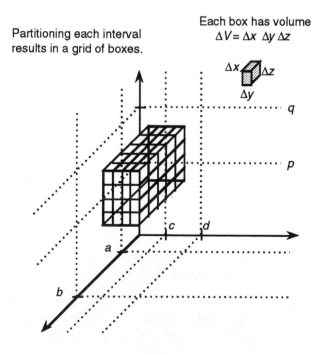

Figure 16.11 Partitioning the box $[a, b] \times [c, d] \times [p, q]$.

Figure 16.11 shows that these partitions create a 3-dimensional grid of rectangular boxes, each of dimension $\Delta x \times \Delta y \times \Delta z$ and having volume:

$$\Delta V = \Delta x \Delta y \Delta z.$$

Now, we choose one point (x_i, y_j, z_k) from each of the $\ell \times m \times n$ boxes ($1 \leq i \leq \ell, 1 \leq j \leq m, 1 \leq k \leq n$), evaluate the function f at each chosen point, and finally form the *triple* Reimann sum

$$\sum_{k=1}^{n} \sum_{j=1}^{m} \sum_{i=1}^{\ell} f(x_i, y_j, z_k) \Delta x \Delta y \Delta z.$$

If, as the partitions are made finer and finer, these sums approach a single limiting value, then that value is denoted by

$$\iiint\limits_{R} f(x, y, z) \, dV$$

or

$$\int_{p}^{q} \int_{c}^{d} \int_{a}^{b} f(x, y, z) dx \, dy \, dz,$$

where $dV = dx \, dy \, dz$ represents the volume element corresponding to a rectangular grid.

Physical interpretation of the triple integral

While we don't have a purely geometric interpretation for this integral, we can provide a *physical* interpretation in terms of density and mass.

If an object is made of a substance of *uniform* density, then one simply multiplies its volume to find its mass. For example, if an object of uniform density $\rho = 12$ g/cm^3 has volume $V = 350$ cm^3, then its mass is

$$\text{total mass} = \rho V = (12 \text{ g/cm}^3)(350 \text{ cm}^3) = 4200 \text{ g}.$$

Now, suppose we have a box R made of material with *variable* density. That is, if we sampled a cubic centimeter of the substance from different parts of the box, we would find that mass would depend on the part of the box we took the sample from. We imagine the density

$$\rho = f(x, y, z)$$

to be a function of the location (x, y, z) of a point in the box.

If f is a continuous function, then dividing the box into an extremely fine grid will determine a collection of tiny boxes, each of approximately constant (uniform) density. Different tiny boxes will have different densities, depending on their location. If each tiny box has dimensions $\Delta x \times \Delta y \times \Delta z$ then the mass of this tiny box is approximately

$$f(x_i, y_j, z_k)\Delta x \Delta y \Delta z$$

where (x_i, y_j, z_k) is a representative point of the box. Summing the masses together (the triple Riemann sum) gives an approximation of the total mass of the box, and this approximation becomes better and better for finer and finer grids. Indeed, we have

$$\text{total mass} = \iiint_R \rho \, dV = \iiint_D f(x, y, z) \, dV.$$

Calculating triple integrals

Triple integrals can be computed iteratively in a manner entirely analogous to double integrals. We'll illustrate with some examples.

EXAMPLE 8   Find $\displaystyle\int_{-1}^{0} \int_{1}^{2} \int_{0}^{1} (z + 2x + 3y) \, dy \, dx \, dz$.

Solution

$$\int_{-1}^{0} \int_{1}^{2} \int_{0}^{1} (z + 2x + 3y) \, dy \, dx \, dz = \int_{-1}^{0} \int_{1}^{2} \left. \left(zy + 2xy + \frac{3y^2}{2} \right) \right]_{y=0}^{y=1} dx \, dz$$

$$= \int_{-1}^{0} \int_{1}^{2} \left(z + 2x + \frac{3}{2} \right) dx \, dz$$

$$= \int_{-1}^{0} \left. zx + x^2 + \frac{3x}{2} \right]_{x=1}^{x=2} dz$$

$$= \int_{-1}^{0} (2z + 4 + 3) - (z + 1 + 3/2) \, dz$$

$$= \int_{-1}^{0} (z + 9/2) \, dz$$

$$= \left. (z^2 + 9z)/2) \right]_{z=-1}^{z=0}$$

$$= (1 - 9)/2) - 0 = -4.$$

■

EXAMPLE 9 Find the mass of the unit cube $[0,1] \times [0,1] \times [0,1]$, if the density ρ at (x,y,z) is e^{-xy} kilograms per cubic unit.

Solution The solution to the problem is given by the triple integral

$$\int_0^1 \int_0^1 \int_0^1 e^{-xy} y \, dx \, dz \, dy = \int_0^1 \int_0^1 -e^{-xy}\Big]_0^1 dz \, dy$$

$$= \int_0^1 \int_0^1 (e^{-y} + 1) dz dy = \int_0^1 z(e^{-y} + 1)\Big]_0^1 dy$$

$$= \int_0^1 (e^{-y} + 1) dy = -e^{-y} + y\Big]_0^1$$

$$= (-e^{-1} + 1) - (-1 + 0) = 2 - e^{-1}.$$

The mass of the cube is $2 - 1/e \approx 1.632$ kilograms. ∎

EXERCISES

Calculate each of the double integrals in exercises 1-10.

1. $\displaystyle\int_1^2 \int_1^4 x + y \, dx \, dy$

2. $\displaystyle\int_0^1 \int_0^1 1 - x - y \, dx \, dy$

3. $\displaystyle\int_0^1 \int_0^1 x^2 + y^2 \, dx \, dy$

4. $\displaystyle\int_1^e \int_1^2 x/y \, dx \, dy$

5. $\displaystyle\int_0^1 \int_1^3 x - 3x^2 y + \sqrt{y} \, dx \, dy$

6. $\displaystyle\int_0^{\pi/2} \int_0^{\pi/2} \sin(x + y) \, dx \, dy$

7. $\displaystyle\int_{-1}^1 \int_0^{\pi/2} x \sin(y) - ye^x \, dy \, dx$

8. $\displaystyle\int_1^4 \int_0^1 y^2/\sqrt{x}, dy \, dx$

9. $\displaystyle\int_0^{\pi/4} \int_0^{\pi/2} \frac{\cos(x)}{\cos^2(y)} \, dx \, dy$

10. $\displaystyle\int_{-2}^{-1} \int_0^1 \frac{y}{1 + x^2} \, dx \, dy$

---

11. Find $\displaystyle\int_0^3 \int_1^2 \int_1^4 x + y + z \, dx \, dy \, dz$. **12.** Find $\displaystyle\int_1^2 \int_{-1}^2 \int_0^2 xy^2 - y/z \, dx \, dy \, dz$.

13. Suppose the cube $[-1,1] \times [-1,1] \times [-1,1]$ is made up of a substance whose density ρ is given by $f(x,y,z) = \dfrac{y^2 e^z}{1 + x^2}$ grams per cubic unit at the point (x,y,z). Find the total mass of the cube.

14. Show that $\displaystyle\int_c^d \int_a^b 1 \, dx \, dy$ gives the area of the rectangle $[a,b] \times [c,d]$.

15. Show that $\displaystyle\int_p^q \int_c^d \int_a^b 1 \, dx \, dy \, dz$ gives the volume of the box $[a,b] \times [c,d] \times [p,q]$.

16.2   DOUBLE INTEGRALS OVER MORE GENERAL REGIONS

In this section we'll discuss how to calculate double integrals when the region D is not a "nice" rectangle (that is, a Cartesian product of closed intervals). First, let's examine some important properties of multiple integrals.

Properties of multiple integrals

Many of the properties of multiple definite integrals are analogous to properties of single definite integrals. For example, if c is a constant, then

$$\iint_D cf(x, y)\, dA = c \iint_D f(x, y)\, dA$$

whenever the integrals on both sides of the equation exist. In other words, we can factor a *constant* through the double integral sign.

Similarly, the additive property extends to multiple integrals:

$$\iint_D (f(x, y) + g(x, y))\, dA = \iint_D f(x, y)\, dA + \iint_D g(x, y)\, dA.$$

In other words, the multiple integral of a sum of two functions is the sum of their integrals.

If the region of integration D is split into two non-overlapping regions D_1 and D_2, then

$$\iint_D f(x, y)\, dA = \iint_{D_1} f(x, y)\, dA + \iint_{D_2} f(x, y)\, dA.$$

And, if $f(x, y) \leq g(x, y)$ for every $(x, y) \in D$, then

$$\iint_D f(x, y)\, dA \leq \iint_D g(x, y)\, dA.$$

All of these properties also hold for triple integrals.

Calculating double integrals by the slicing method

In the last section, we saw how we could think of the double integral

$$\int_a^b \int_c^d f(x,y)\, dy\, dx$$

in terms of *slices*. That is, for each $x \in [a,b]$ we can think of the "inner" integral

$$\int_c^d f(x,y)dy$$

as providing the signed area of the vertical cross-sectional slice of the graph of $z = f(x,y)$ over the y-interval $[c,d]$. We obtain such a value for each $x \in [a,b]$. When we integrate these values over the x-interval $[a,b]$, we obtain the signed volume under the graph of $z = f(x,y)$

$$\int_a^b \left(\int_c^d f(x,y)\, dy \right) dx.$$

If f is continuous, then we can switch the order of integration to

$$\int_c^d \left(\int_a^b f(x,y)\, dx \right) dy.$$

Now, the inner integral represents a slice of the graph of $z = f(x,y)$ over the x-interval $[a,b]$ for each $y \in [c,d]$. We can integrate these values over the y-interval $[c,d]$ to obtain the same final double integral value.

Now, suppose our region of integration D is *not* a rectangle. One possibility is shown in Figure 16.12.

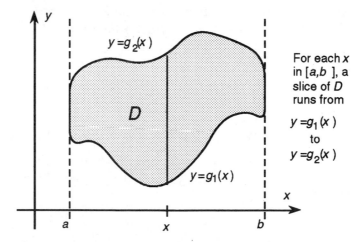

Figure 16.12 Region D is bounded between $y = g_1(x)$ and $y = g_2(x)$.

In this picture, our region D lies entirely between the graphs of $y = g_1(x)$ and $y = g_2(x)$ over the x-interval $[a,b]$.

To integrate $f(x, y)$ over this region, we can still use the method of slicing. For each value $x \in [a, b]$, a slice runs from $y = g_1(x)$ to $y = g_2(x)$. Hence, the inner integral will have the form

$$\int_{g_1(x)}^{g_2(x)} f(x, y) \, dy.$$

Integrating these cross-sections over the interval $[a, b]$ gives us

$$\int_a^b \int_{g_1(x)}^{g_2(x)} f(x, y) \, dy \, dx.$$

Another possiblility is shown in Figure 16.13.

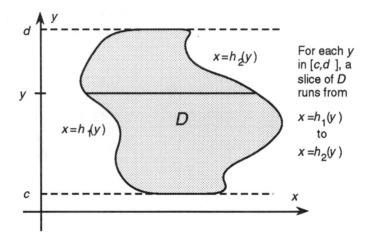

Figure 16.13 Region D is bounded between $x = h_1(y)$ and $x = h_2(y)$.

In this picture, our region D lies entirely between the graphs of $x = h_1(y)$ and $x = h_2(y)$ over the y-interval $[c, d]$. For each $y \in [c, d]$ our slice runs from $x = h_1(y)$ to $x = h_2(y)$. Hence, the inner integral will have the form

$$\int_{h_1(y)}^{h_2(y)} f(x, y) \, dx.$$

Integrating these over the y-interval $[c, d]$ gives us the final result

$$\int_c^d \int_{h_1(y)}^{h_2(y)} f(x, y) \, dx \, dy.$$

EXAMPLE 10 Find $\displaystyle\iint_D (x^2 + xy - y^2)\,dA$ where D is the region bounded by the graph of $y = x^2$ and the lines $y = 0$ and $x = 1$. Use both orders of integration.

Solution The region D is described in two ways in Figure 16.14.

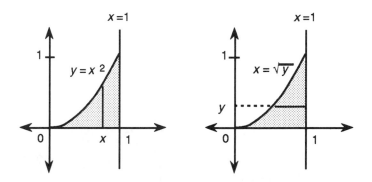

Figure 16.14 The region D bounded by $y = x^2$, $x = 1$, and $y = 0$.

In the picture on the left, the region is considered to lie between $y = x^2$ and $y = 0$ over the x-interval $[0, 1]$. Hence, for each slice corresponding to a value x between 0 and 1, the y values run from 0 to x^2. Therefore, we have

$$\int_0^1 \int_0^{x^2} (x^2 + xy - y^2)\,dy\,dx = \int_0^1 x^2 y + \frac{xy^2}{2} - \frac{y^3}{3} \Big]_{y=0}^{y=x^2} dx$$

$$= \int_0^1 (x^4 + \frac{x^5}{2} - \frac{x^6}{3})\,dx$$

$$= \frac{x^5}{5} + \frac{x^6}{12} - \frac{x^7}{21} \Big]_{x=0}^{x=1}$$

$$= \frac{1}{5} + \frac{1}{12} - \frac{1}{21} \approx 0.2357.$$

In the picture on the right, the region is considered to lie between $x = \sqrt{y}$ and $x = 1$ over the y-interval $[0, 1]$. Hence, for each slice corresponding to a value y between 0 and 1, the x values run from $\sqrt{y}$ to 1. Therefore, we have

$$\int_0^1 \int_{\sqrt{y}}^1 (x^2 + xy - y^2)\,dx\,dy = \int_0^1 \frac{x^3}{3} + \frac{x^2 y}{2} - xy^2 \Big]_{x=\sqrt{y}}^{x=1} dy$$

$$= \int_0^1 (1/3 + y/2 - y^2) - (\frac{y^{3/2}}{3} + \frac{y^2}{2} - y^{5/2})\,dy$$

$$= (y/3 + y^2/4 - y^3/3) - (\frac{2y^{5/2}}{15} + \frac{y^3}{6} - \frac{2y^{7/2}}{7}) \Big]_{y=0}^{y=1}$$

$$= (\frac{1}{3} + \frac{1}{4} - \frac{1}{3}) - (\frac{2}{15} + \frac{1}{6} - \frac{2}{7}) \approx 0.2357.$$

With either order of integration, we obtain the same result.

EXAMPLE 11   Find $\displaystyle\iint_D (x^2 + y^2)\, dA$ where D is the region bounded by the coordinate axes and the line $y = (10 - 3x)/4$.

Solution   The region D is the triangle pictured in Figure 16.15.

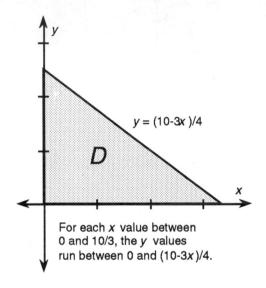

For each x value between 0 and 10/3, the y values run between 0 and (10-3x)/4.

Figure 16.15 The triangular region D.

If we slice the region at each value $x \in [0, 10/3]$, we see that the corresponding y-values run from 0 to $(10 - 3x)/4$. Hence, we can write

$$\iint_D (x^2 + y^2)\, dA = \int_0^{10/3} \int_0^{(10-3x)/4} (x^2 + y^2)\, dy\, dx$$

$$= \int_0^{10/3} x^2 y + \frac{y^3}{3} \Bigg]_{y=0}^{y=(10-3x)/4} dx$$

$$= \int_0^{10/3} \left[\frac{x^2(10 - 3x)}{4} + \frac{((10 - 3x)/4)^3}{3} \right] dx.$$

Now, to calculate the outer integral, we make a substitution:

Let $u = -\dfrac{3x}{4} + \dfrac{5}{2}$ so that $x = -\dfrac{4}{3}\left(u - \dfrac{5}{2}\right).$

Differentiating, we find

$$du = -\frac{3}{4}\, dx \qquad \text{so that} \quad dx = -\frac{4}{3}\, du.$$

As for the limits of integration: when $x = 0$, we have $u = 5/2$; and when $x = 10/3$, we have $u = 0$.

Our integral is transformed by this substitution to the form

$$\int_{5/2}^0 \left[\frac{16u}{9}(u-\frac{5}{2})^2 + \frac{u^3}{3} \right] (-\frac{4}{3}du) = \int_0^{5/2} \frac{4}{3}(\frac{16}{9}u^3 - \frac{80}{9}u^2 + \frac{100}{9}u + \frac{u^3}{3})du.$$

(Note both the change in limits of integration and the corresponding change in sign of the integrand.)

Multiplying out the integrand and combining like terms gives us

$$\int_0^{5/2} (\frac{76}{27}u^3 - \frac{320}{27}u^2 + \frac{400}{27}u)\,du = \frac{1}{27}(76\frac{u^4}{4} - 320\frac{u^3}{3} + 400\frac{u^2}{2})\Big]_{u=0}^{u=5/2}$$

$$= \frac{1}{27}(19 \cdot \frac{5^4}{2^4} - 320 \cdot \frac{5^3}{3 \cdot 2^3} + 200 \cdot \frac{5^2}{2^2})$$

$$\approx 12.0563.$$

∎

EXAMPLE 12   Find $\displaystyle\int_{-1}^0 \int_0^{2\sqrt{1-x^2}} x\,dy\,dx.$

Solution   We have

$$\int_{-1}^0 \int_0^{2\sqrt{1-x^2}} x\,dy\,dx = \int_{-1}^0 xy\Big]_{y=0}^{y=2\sqrt{1-x^2}}\,dx = \int_{-1}^0 2x\sqrt{1-x^2}\,dx.$$

To complete the computation, we make the following substitution:

Let $u = 1 - x^2$. Differentiating, we have $du = -2x\,dx$. Substituting for the limits of integration, we have $u = 1$ when $x = 0$, and we have $u = 0$ when $x = -1$. Thus, our integral is transformed to

$$-\int_0^1 \sqrt{u}\,du = -\frac{2u^{\frac{3}{2}}}{3}\Big]_0^1 = -\frac{2}{3} - 0 = -\frac{2}{3}.$$

∎

Subdividing the region of integration

The next two examples illustrate how it may be desirable to subdivide the region of integration.

EXAMPLE 13   Find $\displaystyle\int_{-2}^{1}\int_{-2|x|}^{|x|} e^{x+y}\, dy\, dx$.

Solution   Notice that we can graph the region of integration by examining the limits and order of integration carefully. The region must be bounded between the graphs of $y = |x|$ and $y = -2|x|$ between $x = -2$ and $x = 1$. Figure 16.16 illustrates this region.

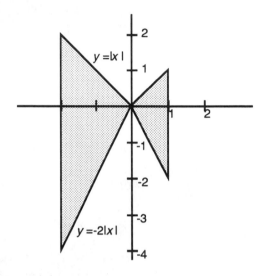

Figure 16.16 The region between $y = |x|$ and $y = -2|x|$ over $[-2, 1]$.

If we use x-slices, then it is convenient to break the region at $x = 0$ (since $|x|$ depends on the *sign* of x):

$$\int_{-2}^{1}\int_{-2|x|}^{|x|} e^{x+y}\, dy\, dx = \int_{-2}^{0}\int_{2x}^{-x} e^{x+y}\, dy\, dx + \int_{0}^{1}\int_{-2x}^{x} e^{x+y}\, dy\, dx$$

$$= \int_{-2}^{0} e^{x+y}\Big]_{y=2x}^{y=-x}\, dx + \int_{0}^{1} e^{x+y}\Big]_{y=-2x}^{y=x}\, dx$$

$$= \int_{-2}^{0} (1 - e^{3x})\, dx + \int_{0}^{1} (e^{2x} - e^{-x})\, dx$$

$$= (x - e^{3x}/3)\Big]_{x=-2}^{x=0} + (e^{2x}/2 + e^{-x})\Big]_{x=0}^{x=1}$$

$$= \left[-\frac{1}{3} - \left(-2 - \frac{e^{-6}}{3}\right)\right] + \left[\frac{e^2}{2} + e^{-1} - \left(\frac{1}{2} + 1\right)\right]$$

$$\approx 4.23.$$

EXAMPLE 14 Find $\iint\limits_{D} e^{x-y}\, dA$ where D is the triangular region with vertices $(0,0)$, $(1,3)$, and $(2,2)$.

Solution The region D is shown in Figure 16.17, where we have subdivided it into two subregions D_1 and D_2.

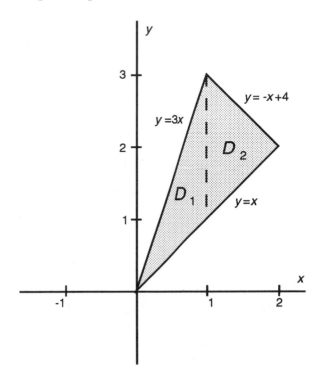

Figure 16.17 Subdividing the region of integration.

Each side of the triangle is part of a straight line. By finding the equation of each of these lines, we can note the following:

In region D_1, x runs from 0 to 1 while y runs from x to $3x$.

In region D_2, x runs from 1 to 2 while y runs from x to $-x + 4$.

Thus, we can write

$$\iint\limits_{D} e^{x-y}\, dy\, dx = \iint\limits_{D_1} e^{x-y}\, dA + \iint\limits_{D_2} e^{x-y}\, dA$$

$$= \int_0^1 \int_x^{3x} e^{x-y}\, dy\, dx + \int_1^2 \int_x^{-x+4} e^{x-y}\, dy\, dx$$

$$= \int_0^1 -e^{x-y}\bigg]_{y=x}^{y=3x} dx + \int_1^2 -e^{x-y}\bigg]_{y=x}^{y=-x+4} dx$$

$$= \int_0^1 (-e^{-2x} + e^0)\, dx + \int_1^2 (-e^{2x-4} + e^0)\, dx$$

$$= \left(\frac{e^{-2x}}{2} + x\right)\bigg]_{x=0}^{x=1} + \left(\frac{e^{2x-4}}{2} + x\right)\bigg]_{x=1}^{x=2}$$

$$= \left[\left(\frac{e^{-2}}{2} + 1\right) - (\frac{1}{2} + 0)\right] + \left[(\frac{1}{2} + 2) - (\frac{e^{-2}}{2} + 1)\right]$$

$$= 2.$$

Changing the order of integration

The choice of order of integration can be important. There are cases when one order of integration may simplify our computations over another, or prevent us from having to subdivide a region.

EXAMPLE 15   Find $\displaystyle\int_{-1}^1 \int_{|x|}^1 (x + y)^2\, dy\, dx$.

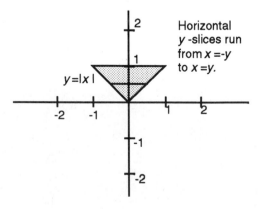

Figure 16.18 Region of integration for $\displaystyle\int_{-1}^1 \int_{|x|}^1 (x + y)^2\, dy\, dx$.

Solution First, let's look at the region under consideration (see Figure 16.18). The given order of integration indicates that the values x run over the interval $[-1, 1]$, and for each x-value in this interval, y runs from $|x|$ to 1.

If we integrate "$dx\,dy$" instead of "$dy\,dx$" we have the values y running over the interval $[0, 1]$, and for each y-value, x runs from $-y$ to y. This gives us the equivalent integral

$$\int_0^1 \int_{-y}^{y} (x+y)^2 \, dx \, dy.$$

We compute this integral as follows:

$$\int_0^1 \int_{-y}^{y} (x+y)^2 \, dx \, dy = \int_0^1 \int_{-y}^{y} (x^2 + 2xy + y^2) \, dx \, dy$$

$$= \int_0^1 \left(\frac{x^3}{3} + x^2 y + xy^2 \right) \Big]_{x=-y}^{x=y} dy$$

$$= \int_0^1 \left[\left(y^3 + y^3 + \frac{y^3}{3} \right) - \left(-y^3 + y^3 - \frac{y^3}{3} \right) \right] dy$$

$$= \int_0^1 \frac{8}{3} y^3 \, dy = \frac{2}{3} y^4 \Big]_{y=0}^{y=1} = \frac{2}{3}.$$

∎

Here is another example of a double integral that is somewhat easier to compute if we change the order of integration.

EXAMPLE 16 Compute $\displaystyle\int_{-3}^{3} \int_{-\sqrt{9-y^2}}^{\sqrt{9-y^2}} x^2 \, dx \, dy.$

Solution Let's first look at the region of integration. Note that the values y run over the interval $[-3, +3]$ and the region is bounded between the two curves $x = -\sqrt{9-y^2}$ and $x = \sqrt{9-y^2}$. Figure 16.19 illustrates that these two curves form the circle $x^2 + y^2 = 9$.

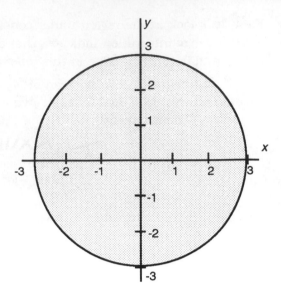

Figure 16.19 A circular region of integration.

Of course, we could look at this same region as being bounded by the two curves $y = -\sqrt{9 - x^2}$ and $y = \sqrt{9 - x^2}$. Using this, we could rewrite the integral:

$$\int_{-3}^{3} \int_{-\sqrt{9-y^2}}^{\sqrt{9-y^2}} x^2 \, dx \, dy = \int_{-3}^{3} \int_{-\sqrt{9-x^2}}^{\sqrt{9-x^2}} x^2 \, dy \, dx$$

$$= \int_{-3}^{3} \left(x^2 y \right]_{y=-\sqrt{9-x^2}}^{y=\sqrt{9-x^2}} \right) dx$$

$$= \int_{-3}^{3} \left[x^2\sqrt{9 - x^2} - x^2(-\sqrt{9 - x^2}) \right] dx$$

$$= \int_{-3}^{3} 2x^2 \sqrt{9 - x^2} \, dx.$$

Using an integral table or a numerical integrator, we find that this final single integral has the value $81\pi/4 \approx 63.617$. ■

Double integrals in polar coordinates

The region of integration D may be more simply described using polar rather than rectangular coordinates. For example, if the boundary of the region is made up of circular arcs, polar coordinates may be a convenient choice.

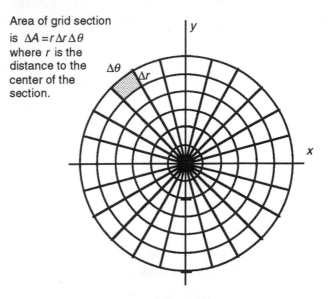

Area of grid section is $\Delta A = r \Delta r \Delta \theta$ where r is the distance to the center of the section.

Figure 16.20 Polar grid of the plane.

If we make a polar grid of a region (see Figure 16.20), one thing that we notice is that the grid sections are not the same size. The further away from the origin, the larger the section. Indeed, the area ΔA of a grid section is

$$\Delta A = r \Delta r \Delta \theta$$

where r is the distance from the origin to the center of the section.

This suggests a procedure for expressing a double integral in terms of polar coordinates.

$$\iint\limits_{D} f(x, y)\, dA$$

in polar coordinates:

Step 1. Describe the region D in terms of θ and r

Step 2. Make the change of variables

$$x = r \cos \theta \qquad y = r \sin \theta$$

Step 3. Use $dA = r\, dr\, d\theta$ as the area element in the integral.

EXAMPLE 17   Find $\iint\limits_{D}(1+x^2+y^2)^{3/2}\,dA$, where D is the interior of the unit circle.

Solution   We'll use a polar coordinate transformation to evaluate the integral. In terms of polar coordinates, D is the region of points (r,θ) satisfying

$$0 \le r \le 1 \quad \text{and} \quad 0 \le \theta \le 2\pi.$$

Substituting $x = r\cos\theta$, $y = r\sin\theta$, and $dA = r\,dr\,d\theta$, we can write

$$\iint\limits_{D}(1+x^2+y^2)^{3/2}\,dA = \int_0^{2\pi}\int_0^1 (1+r^2)^{3/2}r\,dr\,d\theta.$$

Let's make the substitution $u = 1 + r^2$; $du = 2r\,dr$. Our new limits of integration are $u = 1$ (when $r = 0$) and $u = 2$ (when $r = 1$). We have

$$\int_0^{2\pi}\frac{1}{2}\int_1^2 u^{3/2}\,du\,d\theta = \int_0^{2\pi}\frac{1}{5}(u)^{5/2}\Big]_1^2\,d\theta$$

$$= \int_0^{2\pi}(\frac{2^{5/2}}{5}-\frac{1^{5/2}}{5})\,d\theta = (\frac{4\sqrt{2}}{5}-\frac{1}{5})\theta\Big]_0^{2\pi}$$

$$= \frac{8\sqrt{2}\pi}{5}-\frac{2\pi}{5}.$$

∎

EXAMPLE 18   Find $\iint\limits_{D} x^2\,dx\,dy$, where D is the set of points (x,y) satisfying $0 \le x \le y$ and $x^2 + y^2 \le 1$.

Solution   A picture of the region D is shown below in Figure 16.21.

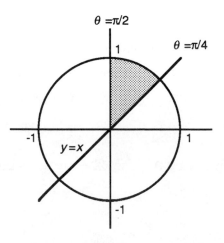

Figure 16.21 Region $D = \{(x,y):\ 0 \le x \le y \text{ and } x^2 + y^2 \le 1\}$.

Using polar coordinates:

$$\int_{\pi/4}^{\frac{\pi}{2}} \int_0^1 r^2 \cos^2 \theta r \, dr \, d\theta = \int_{\frac{\pi}{4}}^{\frac{\pi}{2}} \int_0^1 r^3 \cos^2 \theta \, dr \, d\theta = \int_{\frac{\pi}{4}}^{\frac{\pi}{2}} \frac{r^4}{4} \cos^2 \theta \Big]_0^1 \, d\theta$$

$$= \frac{1}{4} \int_{\frac{\pi}{4}}^{\frac{\pi}{2}} \cos^2 \theta \, d\theta + \frac{1}{4} \int_{\frac{\pi}{4}}^{\frac{\pi}{2}} \frac{1 + \cos 2\theta}{2} \, d\theta$$

$$= \frac{1}{8} \int_{\frac{\pi}{4}}^{\frac{\pi}{2}} d\theta + \frac{1}{8} \int_{\frac{\pi}{4}}^{\frac{\pi}{2}} \cos 2\theta \, d\theta$$

$$= \frac{1}{8} \theta \Big]_{\frac{\pi}{4}}^{\frac{\pi}{2}} + \frac{1}{16} \sin 2\theta \Big]_{\frac{\pi}{4}}^{\frac{\pi}{2}}$$

$$= \left(\frac{\pi}{16} - \frac{\pi}{32} \right) + \left(0 - \frac{1}{16} \right)$$

$$= \frac{\pi}{32} - \frac{1}{16}.$$

☞ **The area of a region D can be found by simply integrating the constant function $f(x, y) = 1$ over the region:**

$$\textbf{area of } D = \iint_D 1 \, dA.$$

EXAMPLE 19 Find the area of the region bounded between $r = 1$ and $r = 1 + \sin \theta$ for $\theta \in [0, \pi]$.

Solution We have

$$\text{area of } D = \iint_D 1 \, dA = \int_0^\pi \int_1^{1+\sin \theta} r \, dr \, d\theta$$

$$= \int_0^\pi r^2/2 \Big]_{r=1}^{r=1+\sin \theta} d\theta$$

$$= \int_0^\pi (1/2)((1 + \sin \theta)^2 - 1) \, d\theta$$

$$= \int_0^\pi \frac{\sin^2 \theta}{2} + \sin \theta \, d\theta$$

$$= \frac{1}{2} \int_0^\pi \sin^2 \theta \, d\theta - \left(\cos \theta \Big]_0^\pi \right)$$

$$= \frac{1}{2} \int_0^\pi \sin^2 \theta \, d\theta + 2.$$

Using an integral table, we find

$$\frac{1}{2}\int_0^\pi \sin^2\theta\, d\theta = \pi/4 \approx 0.785,$$

so the total area of the region is approximately 2.785.   ■

EXERCISES

Find the value of each integral indicated in exercises 1-8.

1. $\displaystyle\int_{-3}^2\int_0^{y^2}(x^2+y)\,dx\,dy.$ **2.** $\displaystyle\int_0^1\int_{y^2}^y(x^2+y^3)\,dx\,dy.$

3. $\displaystyle\int_0^{\frac{\pi}{2}}\int_{\cos x}^0 y\sin x\,dy\,dx.$ **4.** $\displaystyle\int_{-1}^1\int_{|y|}^1(x+y)^2\,dx\,dy$

5. $\displaystyle\int_1^2\int_{-y}^y(x+y)\,dx\,dy$ **6.** $\displaystyle\int_{-1}^1\int_0^3(y-x^2)\,dx\,dy.$

7. $\displaystyle\int_0^1\int_{-\sqrt{y}}^{\sqrt{y}} yx^2\,dx\,dy.$ **8.** $\displaystyle\int_0^1\int_0^2 e^{x-y}\,dy\,dx.$

---

9. Find

$$\iint_D f(x,y)\,dA$$

where $f(x,y) = y^2\sqrt{x}$ and $D = \{(x,y) : x > 0,\ y > x^2,\ y < 10 - x^2\}$.

10. Sketch the region bounded by $y = \ln x$, the x-axis, and the line $x = e$ and integrate the function $f(x,y) = xe^y$ over this region.

11. Set up (but do not integrate) the previous exercise, reversing the order of integration.

12. Find $\displaystyle\iint_D xy\,dA$, where D is the region bounded by $y = x$, $y = 3x$ and $x = 1$.

13. Find $\displaystyle\iint_D (x+y)\,dA$ where D is the region bounded by $y = x$ and $y = x^3$. (Be careful!)

14. Find $\displaystyle\iint_D 2xy\,dx\,dy$, where D is the region in the xy-plane pictured below:

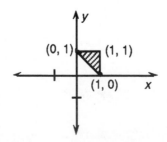

15. Find $\displaystyle\iint_D (x^2 + y^2)\,dx\,dy$, where D is the region in the xy-plane pictured below:

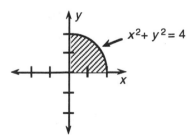

16. For the integral $\displaystyle\int_{-2}^{2}\int_{0}^{\sqrt{4-y^2}} xy\,dx\,dy$, sketch the region of integration, make a change of variables to polar coordinates, then perform the integration.

17. For the integral $\displaystyle\int_{-\sqrt{2}/2}^{\sqrt{2}/2}\int_{-\sqrt{1-x^2}}^{|x|} x^2 y^3\,dy\,dx$, sketch the region of integration, make a change of variables to polar coordinates, then perform the integration.

18. Compute $\displaystyle\iint_D (x^2 + y^2)\,dx\,dy$, where the region D is the unit circle.

Find the area of the region D described in exercises 19-24 by integrating

$$\iint_D 1\,dA.$$

19. D is the region enclosed by the polar graph $r = 1 + \sin(2\theta)$.

20. D is the region enclosed by the polar graph $r = 1 - \cos(2\theta)$.

21. D is the region enclosed by the inner loop of the polar graph $r = 1 + 2\sin(\theta)$.

22. D is the region enclosed by the inner loop of the polar graph $r = 1 - 2\sin(\theta)$.

23. D is the region enclosed by one "leaf" of the polar graph $r = \sin(3\theta)$.

24. D is the region enclosed by one "leaf" of the polar graph $r = \cos(2\theta)$.

16.3 TRIPLE INTEGRALS OVER MORE GENERAL REGIONS

Triple integrals over regions other than nice "boxes" are calculated in a way similar to double integrals.

For example, suppose we can describe the region of integration R as lying between two surfaces $z = g_1(x,y)$ and $z = g_2(x,y)$ over some (2-dimensional) domain D in the xy-plane (see Figure 16.22).

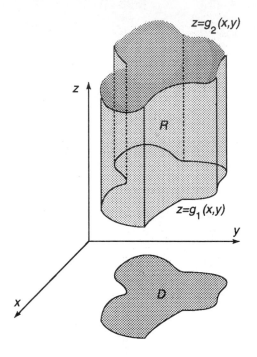

Figure 16.22 Region R is bounded between $z = g_1(x,y)$ and $z = g_2(x,y)$ over D.

The triple integral of a function $f(x,y,z)$ over R can be written in the form

$$\iiint\limits_{R} f(x,y,z)\,dV = \iint\limits_{D} \left(\int_{g_1(x,y)}^{g_2(x,y)} f(x,y,z)\,dz \right) dA.$$

Note that the value of the inner integral will be an expression in terms of x and y alone, so that we have essentially reduced the problem to that of finding a double integral over the 2-dimensional domain D. This problem can then be treated like those of the previous section.

Similarly, if we can express the region of integration R as lying between two surfaces of the form $y = h_1(x,z)$ and $y = h_2(x,z)$ over a domain D in the xz-plane (see Figure 16.23), then we can write the triple integral in the form

$$\iiint_R f(x,y,z)\, dV = \iint_D \left(\int_{h_1(x,z)}^{h_2(x,z)} f(x,y,z)\, dy \right) dA,$$

where the area element dA refers to the xz-plane.

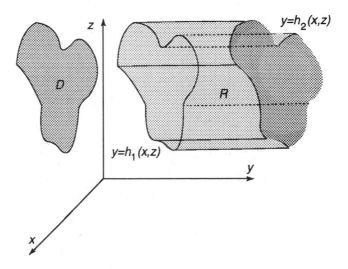

Figure 16.23 Region R is bounded between $y = h_1(x,z)$ and $z = h_2(x,z)$ over D.

A third possibility is that the region R is bounded between the two surfaces $x = k_1(y,z)$ and $x = k_2(y,z)$ over some domain D in the yz-plane. In this case, we could write the triple integral in the form

$$\iiint_R f(x,y,z)\, dV = \iint_D \left(\int_{k_1(y,z)}^{k_2(y,z)} f(x,y,z)\, dx \right) dA,$$

where the area element dA refers to the yz-plane.

EXAMPLE 20   Find $\displaystyle \int_0^1 \int_0^z \int_0^y xy^2 z^3 \, dx\, dy\, dz$.

Solution   We have

$$\int_0^1 \int_0^z \int_0^y xy^2 z^3 \, dx\, dy\, dz = \int_0^1 \int_0^z \frac{x^2 y^2 z^3}{2} \Big]_0^y dy\, dz$$

$$= \int_0^1 \int_0^z \frac{y^4 z^3}{2} \, dy\, dz$$

$$= \int_0^1 \frac{y^5 z^3}{10} \Big]_0^z dz = \int_0^1 \frac{z^8}{10} \, dz = \frac{z^9}{90} \Big]_0^1 = \frac{1}{90}.$$

EXAMPLE 21   Find $\iiint\limits_{R} z\, dV$, where R is the region in space lying between the surfaces $z = \pm\sqrt{xy}$ over the domain in the xy-plane bounded by $y = x^2+1$, $y = 1-x^2$, and $x = 1$.

Solution   The integral can be expressed as

$$\iiint\limits_{R} z\, dV = \int_0^1 \int_{-x^2+1}^{x^2+1} \int_{-\sqrt{xy}}^{\sqrt{xy}} z\, dz\, dy\, dx$$

$$= \int_0^1 \int_{-x^2+1}^{x^2+1} (z^2/2)\Big]_{z=-\sqrt{xy}}^{z=\sqrt{xy}} dy\, dx$$

$$= \int_0^1 \int_{-x^2+1}^{x^2+1} xy\, dy\, dx$$

$$= \int_0^1 (xy^2/2)\Big]_{y=-x^2+1}^{y=x^2+1} dx$$

$$= \int_0^1 \frac{x(x^2+1)^2}{2} - \frac{x(-x^2+1)^2}{2}\, dx$$

$$= \int_0^1 \frac{x(x^2+1)^2}{2} - \frac{x(-x^2+1)^2}{2}\, dx$$

$$= \int_0^1 4x^3\, dx = x^4\Big]_0^1 = 1.$$

∎

☞   **The volume of a region R can be found by simply integrating the constant function $f(x,y,z) = 1$ over the region:**

$$\textbf{volume of } R = \iiint\limits_{R} 1\, dV.$$

EXAMPLE 22   Find the volume of the region R bounded between $z = x^2+y^2$ and $z = -2xy$ over the domain

$$D = \{(x,y) : 0 \le x \le y^2,\ 0 \le y \le 1\}.$$

Solution   We have

$$\iiint\limits_{R} 1\, dV = \int_0^1 \int_0^{y^2} \int_{-2xy}^{x^2+y^2} 1\, dz\, dx\, dy$$

$$= \int_0^1 \int_0^{y^2} z \Big]_{z=-2xy}^{z=x^2+y^2} dx\, dy$$

$$= \int_0^1 \int_0^{y^2} (x^2 + 2xy + y^2)\, dx\, dy$$

$$= \int_0^1 \left(\frac{x^3}{3} + x^2 y + xy^2\right)\Big]_{x=0}^{x=y^2} dy$$

$$= \int_0^1 \left(\frac{y^6}{3} + y^5 + y^4\right) dy$$

$$= \frac{y^7}{21} + \frac{y^6}{6} + \frac{y^5}{5}\Big]_0^1$$

$$= \frac{1}{21} + \frac{1}{6} + \frac{1}{5} = \frac{29}{70}.$$

∎

EXAMPLE 23   Find the volume of the region R bounded by the surface $x = z/y$ and the five planes: $x = 0$, $y = 1$, $y = e$, $z = 2$, and $z = 5$.

Solution   The volume is given by

$$\iiint\limits_{R} 1\, dV = \int_2^5 \int_1^e \int_0^{z/y} 1\, dx\, dy\, dz$$

$$= \int_2^5 \int_1^e x \Big]_{x=0}^{x=z/y} dy\, dz$$

$$= \int_2^5 \int_1^e z/y\, dy\, dz$$

$$= \int_2^5 z \ln y \Big]_{y=1}^{y=e} dz$$

$$= \int_2^5 z\, dz$$

$$= z^2/2 \Big]_2^5 = 25/2 - 4/2 = 21/2.$$

∎

Triple integrals in cylindrical coordinates

Some regions of integration for triple integrals may lend themselves to easier descriptions by the use of *cylindrical* coordinates.

To form a cylindrical partition of a region of space, we first form a polar coordinate grid horizontally (parallel to the xy-plane) of grid dimension Δr by $\Delta\theta$, and then we partition vertically along the z-direction by Δz. The resulting small regions of space carved out by this method look like fragments of a cylindrical shell, each with volume

$$\Delta V \approx r\,\Delta r\,\Delta\theta\,\Delta z.$$

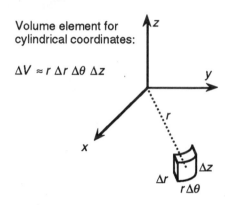

Figure 16.24 Forming a cylindrical partition of space.

Figure 16.24 illustrates a typical cylindrical grid section. If the partition is fine enough, the sections approximate small rectangular boxes of dimensions Δr by $r\Delta\theta$ by Δz.

To express a triple integral

$$\iiint\limits_{R} f(x,y,z)\,dV$$

in terms of cylindrical coordinates:

Step 1. Describe the region R in terms of θ, r, and z.

Step 2. Make the change of variables

$$x = r\cos\theta \qquad y = r\sin\theta \qquad z = z.$$

Step 3. Use $dV = r\,dr\,d\theta\,dz$ as the volume element in the integral.

EXAMPLE 24   A material has density given by $\rho = x^2z + y^2z$ g/cm^3 at any point on or above the xy-plane (units of the three coordinate axes are in centimeters). Find the mass of the cylindrical shell R with inner surface $x^2 + y^2 = 1$, outer surface $x^2 + y^2 = 4$, and lying between the horizontal planes $z = 0$ and $z = 2$.

Solution   The shell is illustrated in Figure 16.25.

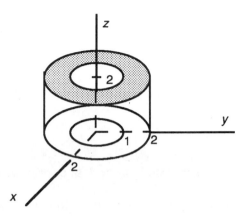

Figure 16.25 Finding the mass of a cylindrical shell.

In terms of cylindrical coordinates, the inner and outer surfaces of the cylindrical shell have the equations $r = 1$ and $r = 2$, respectively. The region R consists of the points

$$\{(r, \theta, z) : 1 \le r \le 2,\ 0 \le \theta \le 2\pi,\ 0 \le z \le 2\}.$$

The density function we are integrating can be written

$$x^2z + y^2z = z(x^2 + y^2) = zr^2$$

in cylindrical coordinates. Hence, we can express the mass integral in cylindrical coordinates as follows:

$$\iiint\limits_{R} (x^2z + y^2z)\, dV = \int_0^2 \int_0^{2\pi} \int_1^2 (zr^2)r\, dr\, d\theta\, dz$$

$$= \int_0^2 \int_0^{2\pi} \int_1^2 zr^3\, dr\, d\theta\, dz.$$

Carrying out the integral computation gives us

$$\int_0^2 \int_0^{2\pi} zr^4/4 \Big]_{r=1}^{r=2} d\theta\, dz = \int_0^2 \int_0^{2\pi} (4z - z/4)\, d\theta\, dz$$

$$= \int_0^2 (15z/4)\theta \Big]_{\theta=0}^{\theta=2\pi} dz$$

$$= \int_0^2 (15\pi z/2)\, dz$$

$$= (15\pi z^2/4) \Big]_{z=0}^{z=2} = 15\pi.$$

The shell has a mass of $15\pi \approx 47.124$ grams.   ■

Triple integrals in spherical coordinates

Similarly, some regions of integration for triple integrals may lend themselves to easier descriptions by the use of *spherical* coordinates (ρ, θ, φ).

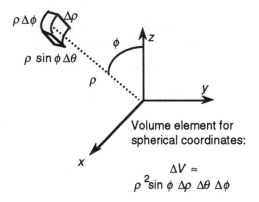

Figure 16.26 Forming a spherical partition of space.

If we partition the distance from the origin by $\Delta\rho$, we obtain a set of concentric spheres. These spheres are then partitioned by $\Delta\theta$ and $\Delta\varphi$ like longitude and latitude lines on the face of the earth. The resulting small regions of space carved out by this method look like fragments of a spherical shell, each with volume

$$\Delta V \approx \rho^2 \sin\varphi\, \Delta\rho\, \Delta\theta\, \Delta\phi.$$

Figure 16.26 illustrates a typical spherical grid section. If the partition is fine enough, the sections approximate small rectangular boxes of dimensions $\Delta\rho$ by $\rho\sin\varphi\Delta\theta$ by $\rho\Delta\varphi$.

To express a triple integral $\iiint\limits_{R} f(x,y,z)\,dV$ in terms of spherical co-ordinates:

Step 1. Describe the region R in terms of ρ, θ, and φ.

Step 2. Make the change of variables

$$x = \rho \sin \varphi \cos \theta \qquad y = \rho \sin \varphi \sin \theta \qquad z = \rho \cos \varphi.$$

Step 3. Use $dV = \rho^2 \sin \varphi \, d\rho \, d\theta \, d\varphi$ as the volume element in the integral.

EXAMPLE 25 Find $\iiint\limits_{R} (x^2 + y^2 + z^2)^{3/2} \, dV$, where R is the region within the unit sphere having all three coordinates $x \geq 0$, $y \geq 0$, $z \geq 0$.

Solution Figure 16.27 illustrates the region in question.

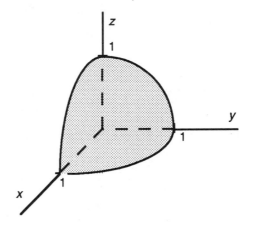

Figure 16.27 Region R is the interior of the unit sphere within the first octant.

In terms of spherical coordinates, R is the set of points satisfying

$$0 \leq \rho \leq 1, \qquad 0 \leq \theta \leq \pi/2, \qquad 0 \leq \varphi \leq \pi/2.$$

Since $x^2 + y^2 + z^2 = \rho^2$ and $dV = \rho^2 \sin \varphi \, d\rho \, d\theta \, d\varphi$, we can express the integral as

$$\iiint\limits_{R} (x^2 + y^2 + z^2)^{3/2} \, dV = \int_0^{\pi/2} \int_0^{\pi/2} \int_0^1 (\rho^3)\rho^2 \sin\varphi \, d\rho \, d\theta \, d\varphi$$

$$= \int_0^{\pi/2} \int_0^{\pi/2} \int_0^1 \rho^5 \sin\varphi \, d\rho \, d\theta \, d\varphi$$

$$= \int_0^{\pi/2} \int_0^{\pi/2} (\rho^6/6) \sin\varphi \Big]_{\rho=0}^{\rho=1} \, d\theta \, d\varphi$$

$$= \int_0^{\pi/2} \int_0^{\pi/2} (1/6) \sin\varphi \, d\theta \, d\varphi$$

$$= \int_0^{\pi/2} (\theta/6) \sin\varphi \Big]_{\theta=0}^{\theta=\pi/2} \, d\varphi$$

$$= \int_0^{\pi/2} (\pi/12) \sin\varphi \, d\varphi$$

$$= (-\pi/12) \cos\varphi \Big]_{\varphi=0}^{\varphi=\pi/2} = 0 - (-\pi/12) = \pi/12.$$

■

EXERCISES

1. Find $\displaystyle\int_0^1 \int_{x^2}^x \int_0^{xy} (x + 2y + 3z) \, dz \, dy \, dx$.

2. Find $\displaystyle\int_{-3}^2 \int_{-y}^y \int_{-x}^x (xy^2 z^3) \, dx \, dz \, dy$.

3. A tetrahedron is bounded by the three coordinate planes and the plane $x + y + z = 1$. Find its volume.

4. Find the volume between the plane $z = 1$ and the cone $z = \sqrt{x^2 + y^2}$.

5. Find the volume between the plane $z = 4$ and the paraboloid $z = x^2 + y^2$.

Consider the region shown in Figure 16.27,

$$R = \{(x, y, z) : x^2 + y^2 + z^2 \le 1, \ x \ge 0, \ y \ge 0, \ z \ge 0\},$$

in exercises 6-10.

6. Express the volume of the region R as a triple integral in rectangular coordinates.

7. Express the volume of the region R as a triple integral in spherical coordinates.

8. The density of the region at any point (x, y, z) is given by $f(x, y, z) = xyz$ kilograms per cubic unit. Find the mass of the region R.

9. If $f(x, y, z)$ is the density at a point in a region R, and M is the total mass of the region, then the **center of mass** of the region is the point

$$(\bar{x},\ \bar{y},\ \bar{z}),$$

whose coordinates are given by the formulas

$$\bar{x} = \frac{1}{M} \iiint\limits_{R} x f(x, y, z)\, dV, \qquad \bar{y} = \frac{1}{M} \iiint\limits_{R} y f(x, y, z)\, dV,$$

and $\qquad \bar{z} = \frac{1}{M} \iiint\limits_{R} z f(x, y, z)\, dV.$

Find the center of mass for the region R of Figure 16.27 if its density at any point is given by $f(x, y, z) = xyz$.

10. The moment of inertia of an object having point density $f(x, y, z)$ about the xy-plane is given by the formula

$$I_{xy} = \iiint\limits_{R} z^2 f(x, y, z)\, dV.$$

Similarly, the moments of inertia of the object about the xz-plane and yz-plane are

$$I_{xz} = \iiint\limits_{R} y^2 f(x, y, z)\, dV \qquad \text{and} \qquad I_{yz} = \iiint\limits_{R} x^2 f(x, y, z)\, dV,$$

respectively. Find the moment of inertia of the region R of Figure 16.27 about each of the coordinate planes if its density at any point is $f(x, y, z) = xyz$.

---

11. If the tetrahedron of exercise 3 has point density $x^2 + y^2 + z^2$, find its mass.

12. If the tetrahedron of exercise 3 has point density $x^2 + y^2 + z^2$, find its center of mass.

13. If the tetrahedron of exercise 3 has point density $x^2 + y^2 + z^2$, find its moment of inertia about each of the coordinate planes.

---

14. Find the volume of the upper hemisphere of the unit sphere using a triple integral.

15. Find the volume of the ellipsoid

$$\frac{x^2}{4} + \frac{y^2}{9} + \frac{z^2}{16} = 1.$$

16.4 NUMERICAL TECHNIQUES FOR MULTIPLE INTEGRALS

In this section, we want to discuss some techniques for numerically approximating the value of a multiple integral. Recall that a single definite integral

$$\int_a^b f(x)dx$$

can be approximated by simply calculating a Reimann sum

$$\sum_{i=1}^n f(x_i)\Delta x$$

for a suitably fine partition of $[a, b]$ and some particular choice of points x_i $(1 \leq i \leq n)$ from the resulting subintervals. The left rectangle, right rectangle, and midpoint rules are all examples of this strategy, corresponding to choosing each point x_i as the left endpoint, right endpoint, or midpoint, respectively, of each subinterval.

A similar strategy can be used to approximate the value of a double integral

$$\iint_D f(x, y)\, dA.$$

If D is a rectangular region, then we can follow these steps:

Step 1. We subdivide the region D using a fine rectangular grid, where each rectangle has dimensions $\Delta x \times \Delta y$.

Step 2. For each rectangle in our grid, we choose a point (x_i, y_j).

Step 3. We calculate $f(x_i, y_j)\Delta x\Delta y$ for each of the rectangles.

Step 4. We sum them up to find

$$\iint_D f(x, y)\, dA \approx \sum_{j=1}^n \sum_{i=1}^m f(x_i, y_j)\, \Delta x\Delta y.$$

We'll illustrate the method by using it to approximate the volume between the paraboloid $z = x^2 + y^2$ and the xy-plane over the unit square $[0, 1] \times [0, 1]$.

First, we note that in this case, we can calculate the volume exactly:

$$\int_0^1 \int_0^1 x^2 + y^2\, dx\, dy = \int_0^1 \frac{x^3}{3} + xy^2 \bigg]_{x=0}^{x=1} dy$$

$$= \int_0^1 \frac{1}{3} + y^2\, dy$$

$$= \frac{y + y^3}{3} \bigg]_{y=0}^{y=1} = 2/3.$$

We have chosen a simple integral for which we know the exact value for purposes of comparing our approximations. The importance of numerical techniques is more evident when we need to approximate the value of a definite integral that is difficult or impossible to calculate exactly.

Suppose we partition $[0, 1] \times [0, 1]$ into 25 small squares by partitioning $[0, 1]$ into 5 equal subintervals along both the x and y axes. In other words, we have

$$\Delta x = 0.2 = \Delta y,$$

so that the area of each small square is

$$\Delta A = 0.04 = \Delta x \Delta y.$$

To use this grid to approximate the value of the integral

$$\int_0^1 \int_0^1 x^2 + y^2 \, dx \, dy$$

we need to

1) pick one point (x, y) from each small square,

2) evaluate $x^2 + y^2$ at each chosen point,

3) add all these values together and multiply by 0.04.

The choice of points to make from each rectangle in our grid is purely arbitrary. However, if we wanted to automate the procedure, we might make some systematic choices, as we do for the left rectangle or right rectangle numerical methods of integration. Figure 16.28 shows three possible systematic choices of points we could make for a rectangular grid.

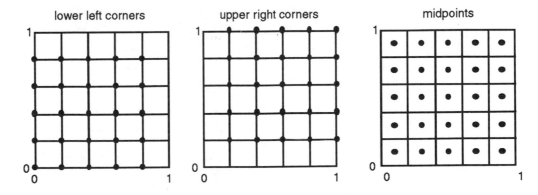

Figure 16.28 Three possible choices of points for a rectangular grid.

EXAMPLE 26   Approximate $\int_0^1 \int_0^1 x^2 + y^2 \, dx \, dy$ using each of the choices of points indicated in Figure 16.28.

Solution   Below, we have assembled the data we need to calculate each approximation, depending on the choice of points from the grid sections.

lower left corners		upper right corners		midpoints	
(x, y)	$x^2 + y^2$	(x, y)	$x^2 + y^2$	(x, y)	$x^2 + y^2$
$(0.0, 0.0)$	0.00	$(0.2, 0.2)$	0.08	$(0.1, 0.1)$	0.02
$(0.2, 0.0)$	0.04	$(0.4, 0.2)$	0.20	$(0.1, 0.3)$	0.10
$(0.4, 0.0)$	0.16	$(0.6, 0.2)$	0.40	$(0.1, 0.5)$	0.26
$(0.6, 0.0)$	0.36	$(0.8, 0.2)$	0.68	$(0.1, 0.7)$	0.50
$(0.8, 0.0)$	0.64	$(1.0, 0.2)$	1.04	$(0.1, 0.9)$	0.82
$(0.0, 0.2)$	0.04	$(0.2, 0.4)$	0.20	$(0.3, 0.1)$	0.10
$(0.2, 0.2)$	0.08	$(0.4, 0.4)$	0.32	$(0.3, 0.3)$	0.18
$(0.4, 0.2)$	0.20	$(0.6, 0.4)$	0.52	$(0.3, 0.5)$	0.34
$(0.6, 0.2)$	0.40	$(0.8, 0.4)$	0.80	$(0.3, 0.7)$	0.58
$(0.8, 0.2)$	0.68	$(1.0, 0.4)$	1.16	$(0.3, 0.9)$	0.90
$(0.0, 0.4)$	0.16	$(0.2, 0.6)$	0.40	$(0.5, 0.1)$	0.26
$(0.2, 0.4)$	0.20	$(0.4, 0.6)$	0.52	$(0.5, 0.3)$	0.34
$(0.4, 0.4)$	0.32	$(0.6, 0.6)$	0.72	$(0.5, 0.5)$	0.50
$(0.6, 0.4)$	0.52	$(0.8, 0.6)$	1.00	$(0.5, 0.7)$	0.74
$(0.8, 0.4)$	0.80	$(1.0, 0.6)$	1.36	$(0.5, 0.9)$	1.06
$(0.0, 0.6)$	0.36	$(0.2, 0.8)$	0.68	$(0.7, 0.1)$	0.50
$(0.2, 0.6)$	0.40	$(0.4, 0.8)$	0.80	$(0.7, 0.3)$	0.58
$(0.4, 0.6)$	0.52	$(0.6, 0.8)$	1.00	$(0.7, 0.5)$	0.74
$(0.6, 0.6)$	0.72	$(0.8, 0.8)$	1.28	$(0.7, 0.7)$	0.98
$(0.8, 0.6)$	1.00	$(1.0, 0.8)$	1.64	$(0.7, 0.9)$	1.30
$(0.0, 0.8)$	0.64	$(0.2, 1.0)$	1.04	$(0.9, 0.1)$	0.82
$(0.2, 0.8)$	0.68	$(0.4, 1.0)$	1.16	$(0.9, 0.3)$	0.90
$(0.4, 0.8)$	0.80	$(0.6, 1.0)$	1.36	$(0.9, 0.5)$	1.06
$(0.6, 0.8)$	1.00	$(0.8, 1.0)$	1.64	$(0.9, 0.7)$	1.30
$(0.8, 0.8)$	1.28	$(1.0, 1.0)$	2.00	$(0.9, 0.9)$	1.62

If we total the values $x^2 + y^2$ for each choice, and multiply by $\Delta x \Delta y = 0.04$, we obtain the corresponding approximations:

Using lower left corners: $\int_0^1 \int_0^1 x^2 + y^2 \, dx \, dy \approx (12)(0.04) = 0.48.$

Using upper right corners: $\int_0^1 \int_0^1 x^2 + y^2 \, dx \, dy \approx (21)(0.04) = 0.84.$

Using midpoints: $\int_0^1 \int_0^1 x^2 + y^2 \, dx \, dy \approx (16.5)(0.04) = 0.66.$

■

Numerical techniques for general regions of integration

For a more general region D, we have an additional complication, since not all of the rectangles will fit entirely within the region. What do we do with those that overlap the boundary of the region? One idea might be to count only those within the region. Another idea would be to count all those rectangles that include any part of the region. Figure 16.29 shows which rectangles from our grid partition of $[0,1] \times [0,1]$ we would include if our region of integration had been the quarter of the unit circle in this square.

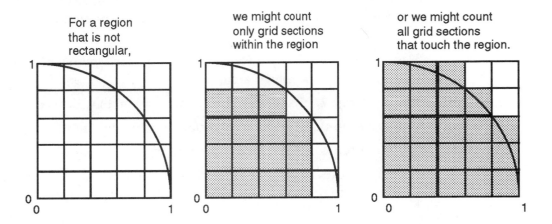

Figure 16.29 Choosing grid sections for a non-rectangular region.

If we choose only those rectangles completely within the region, we stand a good chance of underestimating the value of the integral. On the other hand, if we count *all* the rectangles that include some part of the region, then we stand a good chance of overestimating the area of the region. For a very irregularly shaped region, the ratio of the number of rectangles completely inside the region to the number overlapping any part of the region may be quite small, but for finer and finer grid partitions, we expect this ratio to be nearly 1. (There are some pathologically bizarre regions for which this is not true, however.)

Another strategy for approximating a double integral

$$\iint\limits_{D} f(x, y) \, dy \, dx$$

is outlined below:

Step 1: Partition region D in strips of width Δx in the x-direction;

Step 2: Choose a slice of the region D from each strip;

Step 3: Approximate the area of each slice like any single integral, perhaps using trapezoidal rule or Simpson's Rule;

Step 4: Multiply the area of each slice by Δx and add the results.

(If the order of integration is $dx\,dy$ instead, we can partition D into strips of width Δy in the y-direction in Step 1, and then multiply the area of the chosen slices by Δy in Step 4.)

A technique similar to this one is used by many machine integrators to numerically approximate double integrals.

Monte Carlo method of approximating integrals

A **Monte Carlo** method is one making use of randomly generated numbers. Knowledge of the laws of probability can then be used to our advantage. In Volume I, we discussed how random numbers could be used to estimate the area of a region:

Step 1. Completely enclose the region D with a rectangle R (or some other region of known area).

Step 2. Randomly choose several points within R.

Step 3. Find the proportion of points falling within region D.

Step 4. Multiply this proportion by the area of R.

Figure 16.30 illustrates this Monte Carlo method as applied to approximate the area of one quarter of the unit circle.

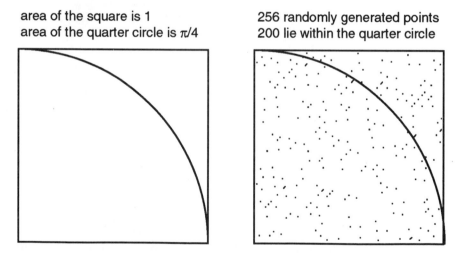

area of the square is 1
area of the quarter circle is $\pi/4$

256 randomly generated points
200 lie within the quarter circle

Figure 16.30 Using randomly selected points to approximate area.

Of the 256 randomly generated points in the unit square R (of area 1), 200 of the points fell within the quarter circle D. This leads to the estimate:

$$\text{area of quarter circle} \approx (200/256) \cdot 1 = 0.78125.$$

In comparison, the exact area of the quarter circle is

$$\pi/4 \approx 0.7854.$$

A similar Monte Carlo method can be used to approximate the value of a double integral over an irregularly shaped region. To approximate

$$\iint_D f(x, y)\, dA$$

we add two more steps:

Step 5. Evaluate $f(x, y)$ at each point that falls in the region D, and find the average of these function values.

Step 6. Multiply this average function value by the estimated area found in Step 4.

This is the approximate value of the integral.

EXAMPLE 27 The average value of $x^2 + y^2$ for the 200 points within the quarter circle D of Figure 16.30 was found to be 0.506. Use this to approximate

$$\iint_D x^2 + y^2\, dA,$$

and compare it to the actual integral value.

Solution The estimated area of the quarter circle was 0.78125. Therefore, the estimated value of the integral is

$$\iint_D x^2 + y^2\, dA \approx (0.78125)(0.506) = 0.3953125.$$

In comparison, we can use polar coordinates to find the actual integral value:

$$\iint_D x^2 + y^2\, dA = \int_0^{\pi/2} \int_0^1 (r^2) r\, dr\, d\theta$$

$$= \int_0^{\pi/2} \int_0^1 r^3\, dr\, d\theta$$

$$= \int_0^{\pi/2} r^4/4 \Big]_{r=0}^{r=1} d\theta$$

$$= \int_0^{\pi/2} 1/4\, d\theta$$

$$= \int \theta/4 \Big]_0^{\pi/2} = \pi/8 \approx 0.3927.$$

We can see that the Monte Carlo estimate compares very favorably with the actual value. ■

With suitable adjustments, these Riemann sum, slicing, and Monte Carlo methods can also be used to approximate triple integrals.

▬▬▬ EXERCISES

1. Using the same grid as in Example 26, approximate the value of the integral

$$\int_0^1 \int_0^1 e^{x^2+y^2} \, dx \, dy$$

using the lower left corners of each grid section.

2. Using the same grid as in Example 26, approximate the value of the integral

$$\int_0^1 \int_0^1 e^{x^2+y^2} \, dx \, dy$$

using the upper right corners of each grid section.

3. Using the same grid as in Example 26, approximate the value of the integral

$$\int_0^1 \int_0^1 e^{x^2+y^2} \, dx \, dy$$

using the midpoints of each grid section.

---

4. Approximate $\iint_D x^2 + y^2 \, dA$, where D is the interior of the quarter circle of Figure 16.29 by using only those grid sections completely inside D and choosing the lower left corner of each section.

5. Approximate $\iint_D x^2 + y^2 \, dA$, where D is the interior of the quarter circle of Figure 16.29 by using all those grid sections having any overlap with D and choosing the upper right corner of each section.

6. Average the results of the previous two exercises and compare with the actual value.

---

7. Approximate $\int_{0.2}^{0.6} \int_{0.4}^{1.0} x^2 + y^2 \, dA$ using midpoints of appropriate grid sections.

8. Find the exact value of

$$\iint_D e^{x^2+y^2} \, dA$$

where D is the same quarter circle. (Hint: change to polar coordinates!)

9.   Generate 100 ordered pairs of numbers (x, y) such that $0 \leq x \leq 1$ and $0 \leq y \leq 1$. Keep only those ordered pairs satisfying $x^2 + y^2 \leq 1$. For each of these ordered pairs, evaluate $e^{x^2 + y^2}$ and average the results. Finally, multiply this average by the percentage of the points that fell within 1 unit of the origin, and compare with the answer from the previous exercise.

---

10.   A lake is roped off within a 100 meter by 200 meter rectangle. Fifty points are chosen at random within the rectangle, and 36 of these points fall within the lake. A depth reading is taken at each of the 36 points, and the average depth reading is 2.6 meters. What is the estimated volume of water in the lake in kiloliters?

17

Vector Analysis

In this chapter, we examine how differential and integral calculus can be used to analyze *vector fields*. We'll examine the different types of derivatives that can be formed on vector fields, and then turn to some generalizations of integral calculus to curves and surfaces in space. We'll see that each of these generalizations in turn gives rise to what could be called the *Fundamental Theorems of Vector Calculus*.

What is a vector field?

A vector-valued multivariable function is called a **vector field**. We can think of a vector-valued function as a vector of scalar-valued functions. Most often, we'll think of the inputs to a vector field as the coordinates specifying the location of an object, either in a plane (2 inputs) or in space (3 inputs).

If $\mathbf{F} : \mathbb{R}^2 \longrightarrow \mathbb{R}^2$ is a vector field having points in the plane as inputs and 2-dimensional vectors as outputs, we can write

$$\mathbf{F}(x, y) = \langle P(x, y), \ Q(x, y) \rangle$$

$$= P(x, y)\mathbf{i} + Q(x, y)\mathbf{j},$$

where the components P and Q are scalar-valued functions of two variables.

If $\mathbf{F} : \mathbb{R}^3 \longrightarrow \mathbb{R}^3$ is a vector field having points in space as inputs and 3-dimensional vectors as outputs, we can write

$$\mathbf{F}(x, y, z) = \langle P(x, y, z), \ Q(x, y, z), \ R(x, y, z) \rangle$$

$$= P(x, y, z)\mathbf{i} + Q(x, y, z)\mathbf{j} + R(x, y, z)\mathbf{k},$$

where the components P, Q, and R are scalar-valued functions of three variables.

Some authors restrict their attention solely to vector fields in space ($\mathbb{R}^3 \longrightarrow \mathbb{R}^3$). There really is no loss, for a vector field

$$\mathbf{F}(x,y) = \langle P(x,y),\ Q(x,y) \rangle$$

can also be thought of as representing

$$\mathbf{F}(x,y,z) = \langle P(x,y),\ Q(x,y),\ 0 \rangle,$$

where our attention is restricted to the xy-plane.

A vector field $\mathbf{F}$ is described as being *continuous, differentiable, C^1-differentiable,* or *C^2-differentiable*, provided all of its scalar-valued components have that property.

Force fields and velocity fields

Much of vector calculus was invented as a tool for working with physics, and it is here that we see the most natural examples of *vector fields*. For motivation and interpretations, we'll appeal to two kinds of vector fields that arise often in physical situations—*force fields* and *velocity fields*.

Many *forces* may act on an object (gravitational, electrostatic, magnetic, etc.). The direction and magnitude of these force vectors may well depend on the precise location of the object relative to another body or an electric charge. Hence, each force can be thought of as a vector-valued function of the three variables that specify its position. If an object is immersed in a fluid that is flowing, then the *velocity* vector of the fluid at each location gives us the instantaneous direction and speed of the fluid flow at that point.

EXAMPLE 1 Consider a satellite in orbit around the earth. The magnitude of the gravitational force exerted on the satellite by the earth is proportional to the product of the masses of the earth and the satellite, and is inversely proportional to the square of the distance between the satellite and the center of the earth. Precisely, the magnitude of the gravitational force $\mathbf{F}$ is

$$\|\mathbf{F}\| = \frac{GMm}{d^2},$$

where M is the (constant) mass of the earth, m is the (constant) mass of the satellite, d is the distance from the satellite to the center of the earth, and G is a gravitational constant.

If we designate the center of the earth as the origin of a rectangular coordinate system, then the position of the satellite can be indicated by some ordered triple of coordinates (x,y,z). The distance is given by

$$d = \sqrt{x^2 + y^2 + z^2}.$$

The direction of the gravitational force **F** is toward the center of the earth (see Figure 17.1).

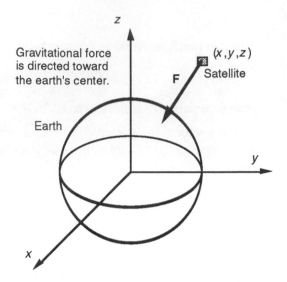

Figure 17.1 Gravitational forces form a vector field.

Since the position vector of the satellite is $\langle x, y, z \rangle$, a *unit vector* pointing back toward the center of the earth is

$$\frac{-\langle x, y, z \rangle}{\sqrt{x^2 + y^2 + z^2}}.$$

Now we can write

$$\mathbf{F}(x, y, z) = \frac{-GMm}{(x^2 + y^2 + z^2)^{3/2}} \langle x, y, z \rangle,$$

so **F** is a vector-valued function of the satellite coordinates x, y, and z. ∎

Visualizing vector fields

We can obtain a visual picture of a vector field by selecting several input points and drawing the output vector of appropriate length and direction emanating from each point. In the plane, this gives us a picture that is similar to a *slope* or *direction field*, except now our line segments have arrowheads, and may have different lengths.

EXAMPLE 2   Suppose a thin sheet of liquid swirling about a central point has a velocity field given by

$$\mathbf{F}(x, y) = \langle -y, x \rangle,$$

where we consider the central point as the origin $(0, 0)$. Here's a table of vectors at a selection of points:

point (x, y)	vector $\mathbf{F}(x, y)$
$(0, 0)$	$\langle 0, 0 \rangle$
$(1, 0)$	$\langle 0, 1 \rangle$
$(0, 1)$	$\langle -1, 0 \rangle$
$(-1, 0)$	$\langle 0, -1 \rangle$
$(0, -1)$	$\langle 1, 0 \rangle$
$(2, 0)$	$\langle 0, 2 \rangle$
$(0, 2)$	$\langle -2, 0 \rangle$
$(-2, 0)$	$\langle 0, -2 \rangle$
$(0, -2)$	$\langle 2, 0 \rangle$

point (x, y)	vector $\mathbf{F}(x, y)$
$(3, 0)$	$\langle 0, 3 \rangle$
$(0, 3)$	$\langle -3, 0 \rangle$
$(-3, 0)$	$\langle 0, -3 \rangle$
$(0, -3)$	$\langle 3, 0 \rangle$
$(\sqrt{2}, \sqrt{2})$	$\langle -\sqrt{2}, \sqrt{2} \rangle$
$(-\sqrt{2}, \sqrt{2})$	$\langle -\sqrt{2}, -\sqrt{2} \rangle$
$(-\sqrt{2}, -\sqrt{2})$	$\langle \sqrt{2}, -\sqrt{2} \rangle$
$(\sqrt{2}, -\sqrt{2})$	$\langle \sqrt{2}, \sqrt{2} \rangle$

Figure 17.2 illustrates these velocity vectors along with some *flow lines* (shown in dashes).

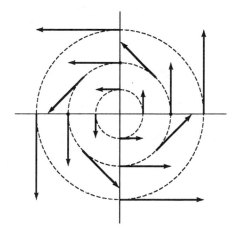

Figure 17.2 Illustration of the vector field $\mathbf{F}(x, y) = \langle -y, x \rangle$.

We can see that the flow lines are circles, and the direction of the flow is counterclockwise. Also, the further out the point from the center, the greater the velocity vector's magnitude. ∎

17.1 DERIVATIVES OF VECTOR FIELDS–DIVERGENCE AND CURL

We have seen that a *scalar field* f has a *vector* derivative, as given by its gradient.

$$\mathbf{grad}\ f = \nabla f,$$

For example, if we have a scalar function of three variables,

$$f(x, y, z),$$

then $\mathbf{grad}\ f$ is given by

$$\nabla f = \left\langle \frac{\partial f}{\partial x}, \frac{\partial f}{\partial y}, \frac{\partial f}{\partial z} \right\rangle.$$

In this section we'll examine derivatives of vector fields. Because of their many physical interpretations, we'll find it useful to form different "types" of derivatives for different purposes.

Divergence of a vector field

One type of derivative of a vector field is known as the *divergence*, and it can be computed whenever all the components of the vector field are differentiable.

Definition 1

The **divergence** of a vector field $\mathbf{F} : \mathbb{R}^2 \longrightarrow \mathbb{R}^2$, where

$$\mathbf{F}(x, y) = \langle P(x, y), Q(x, y) \rangle,$$

is defined as

$$\mathrm{div}\ \mathbf{F} = \frac{\partial P}{\partial x} + \frac{\partial Q}{\partial y}.$$

If $\mathbf{F} : \mathbb{R}^3 \longrightarrow \mathbb{R}^3$, where

$$\mathbf{F}(x, y, z) = P(x, y, z)\mathbf{i} + Q(x, y, z)\mathbf{j} + R(x, y, z)\mathbf{k}$$

then

$$\mathrm{div}\ \mathbf{F} = \frac{\partial P}{\partial x} + \frac{\partial Q}{\partial y} + \frac{\partial R}{\partial z}.$$

Note that the divergence of a vector field is always a *scalar field.*

In general, the divergence can be defined whenever $\mathbf{F} : \mathbb{R}^n \longrightarrow \mathbb{R}^n$, where

$$\mathbf{F} = \langle f_1, f_2, f_3, \ldots, f_n \rangle,$$

and each of the component functions f_i is a differentiable scalar field of the n variables $x_1, x_2, x_3, \ldots, x_n$. In this case, we have

$$\text{div } \mathbf{F} = \frac{\partial f_1}{\partial x_1} + \frac{\partial f_2}{\partial x_2} + \frac{\partial f_3}{\partial x_3} + \cdots + \frac{\partial f_n}{\partial x_n}.$$

EXAMPLE 3 Find the divergence of each of the following vector fields:

$$\mathbf{F}(x, y, z) = \langle x, y, z \rangle$$

$$\mathbf{G}(x, y, z) = \langle y^2 + z^2, x^2 + z^2, x^2 + y^2 \rangle$$

$$\mathbf{H}(x, y, z) = \langle 3x^2, 4y^2, 5z^2 \rangle.$$

Solution

$$\text{div } \mathbf{F} = 1 + 1 + 1 = 3$$

$$\text{div } \mathbf{G} = 0 + 0 + 0 = 0$$

$$\text{div } \mathbf{H} = 6x + 8y + 10z.$$

∎

EXAMPLE 4 A thin sheet of fluid has a velocity field given by

$$\mathbf{F}(x, y) = \left\langle \frac{y}{x^2 + y^2}, \frac{-x}{x^2 + y^2} \right\rangle.$$

Find div $\mathbf{F}$.

Solution We have

$$\text{div } \mathbf{F} = \frac{0(x^2 + y^2) - 2xy}{(x^2 + y^2)^2} + \frac{0(x^2 + y^2) - 2y(-x)}{(x^2 + y^2)^2} = \frac{-2xy + 2xy}{(x^2 + y^2)^2} = 0.$$

∎

Physical interpretation of divergence

What exactly does div **F** tell us about the vector field **F**? Let's think in terms of **F** representing the velocity field of a gas (a gas is a fluid that can compress or expand). At any point, imagine a small cubical volume V of the gas centered at the point. As time passes, the different particles of gas in this cube will move different distances, and the volume may change in both shape and size.

The instantaneous rate of change of the volume with respect to time is dV/dt. A negative rate indicates compression, and a positive rate indicates expansion of the gas. If we divide this rate by the original volume V of the gas, then

$$\frac{dV/dt}{V}$$

gives us the *instantaneous rate of change of volume per unit of volume.*

As the cube of gas V is taken smaller and smaller, the limiting value of this quantity approaches the divergence of **F**:

$$\text{div } \mathbf{F} = \lim_{V \to 0} \frac{dV/dt}{V}.$$

In other words, div **F** gives us a measure of the instantaneous rate of volume change *per unit of volume*. Roughly speaking, if the arrows of a vector field "converge," we'd expect the divergence of **F** to be negative, and if the arrows "diverge," we'd expect the divergence to be positive (see Figure 17.3).

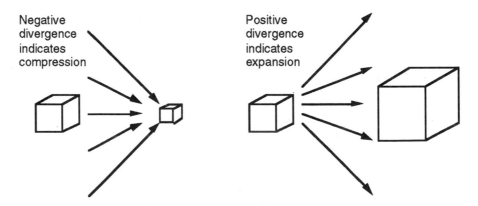

Figure 17.3 Negative and positive divergence.

However, even if the arrows neither converge nor diverge, a vector field can have nonzero divergence as shown in the following example.

EXAMPLE 5 Suppose

$$\mathbf{F}(x, y) = \left(\frac{1}{1 + y}\right)\mathbf{j}$$

gives us the velocity field of a gas for all points (x, y, z) in space (except those in the plane $y = -1$). Since $\mathbf{F}$ does not depend on z, we can picture the flow of gas by looking at the vectors in the xy-plane. Figure 17.4 illustrates some of the vectors in this vector field.

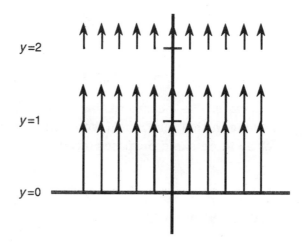

Figure 17.4 The vector field $\mathbf{F}(x, y) = \left(\dfrac{1}{1 + y}\right)\mathbf{j}$.

Looking at this picture, we would not describe the arrows as converging or diverging. However if we calculate the divergence, we have

$$\operatorname{div} \mathbf{F}(x, y, z) = 0 + \frac{-1}{(1 + y)^2} + 0 = \frac{-1}{(1 + y)^2}$$

which is negative at any point where $y \neq -1$. This indicates that a small cube of gas would compress in this field. (Can you see why from Figure 17.4?) ∎

Definition 2

A vector field $\mathbf{F}$ is called **incompressible** at a point (x, y, z) if

$$\operatorname{div} \mathbf{F}(x, y, z) = 0.$$

EXAMPLE 6   The velocity field of a draining bathtub is approximated by

$$\mathbf{F}(x, y, z) = \frac{y}{x^2 + y^2}\mathbf{i} - \frac{x}{x^2 + y^2}\mathbf{j} - \mathbf{k}.$$

Is $\mathbf{F}$ an incompressible vector field?

Solution   We have

$$\operatorname{div} \mathbf{F} = \frac{\partial}{\partial x}\left[\frac{y}{x^2 + y^2}\right] + \frac{\partial}{\partial y}\left[\frac{-x}{x^2 + y^2}\right] + \frac{\partial}{\partial z}(-1)$$

$$= \frac{0 \cdot (x^2 + y^2) - 2xy}{(x^2 + y^2)^2} + \frac{0(x^2 + y^2) - 2y(-x)}{(x^2 + y^2)^2} + 0$$

$$= \frac{-2xy + 2xy}{(x^2 + y^2)^2} = 0.$$

Therefore, $\mathbf{F}$ is incompressible. Physically, this simply means that a fluid moving in this vector field would neither compress nor expand. Water is for all intents and purposes incompressible, so this makes sense.   ■

Curl of a vector field

The *divergence* of a vector field of any dimension produces a new *scalar field*. In contrast, the *curl* of a vector field is defined only for $\mathbb{R}^3$, and produces a new *vector field*, also of dimension three.

Definition 3

> Suppose $\mathbf{F} : \mathbb{R}^3 \longrightarrow \mathbb{R}^3$ is a vector field, with
>
> $$\mathbf{F}(x, y, z) = P(x, y, z)\mathbf{i} + Q(x, y, z)\mathbf{j} + R(x, y, z)\mathbf{k}$$
>
> for some scalar components P, Q, and R. If $\mathbf{F}$ is differentiable then we define the **curl** of $\mathbf{F}$ as
>
> $$\operatorname{curl} \mathbf{F} = \left(\frac{\partial R}{\partial y} - \frac{\partial Q}{\partial z}\right)\mathbf{i} + \left(\frac{\partial P}{\partial z} - \frac{\partial R}{\partial x}\right)\mathbf{j} + \left(\frac{\partial Q}{\partial x} - \frac{\partial P}{\partial y}\right)\mathbf{k}.$$
>
> Note that $\operatorname{curl} \mathbf{F}$ is also a vector field $\mathbb{R}^3 \longrightarrow \mathbb{R}^3$.

The formula looks complicated, but fortunately there is a simple way to remember it. If we form a 3×3 "matrix" with the standard basis vectors $\mathbf{i}$, $\mathbf{j}$, and $\mathbf{k}$ in the first row, the partial derivative operators $\frac{\partial}{\partial x}$, $\frac{\partial}{\partial y}$, and $\frac{\partial}{\partial z}$ in the second row, and the three components of $\mathbf{F}$ (P, Q, and R) in the third row, then $\operatorname{curl} \mathbf{F}$ is simply the "determinant:"

$$\operatorname{curl} \mathbf{F} = \begin{vmatrix} \mathbf{i} & \mathbf{j} & \mathbf{k} \\ \partial/\partial x & \partial/\partial y & \partial/\partial z \\ P & Q & R \end{vmatrix}.$$

Recall that we used a similar device for remembering the *cross product* of two three-dimensional vectors.

EXAMPLE 7   Find curl $\mathbf{F}(0, 1, 0)$ if $\mathbf{F}(x, y, z) = \langle x, 3xy, z \rangle$.

Solution

$$\text{curl } \mathbf{F} = \begin{vmatrix} \mathbf{i} & \mathbf{j} & \mathbf{k} \\ \partial/\partial x & \partial/\partial y & \partial/\partial z \\ x & 3xy & z \end{vmatrix}$$

$$= 0\mathbf{i} - 0\mathbf{j} + (3y - 0)\mathbf{k} = \langle 0, 0, 3y \rangle.$$

Therefore, curl $\mathbf{F}(0, 1, 0) = \langle 0, 0, 3 \rangle$.   ∎

EXAMPLE 8   Find curl $\mathbf{F}(1, 1, 1)$ if $\mathbf{F}(x, y, z) = \langle y, z, x \rangle$.

Solution   We have

$$\text{curl } \mathbf{F} = \begin{vmatrix} \mathbf{i} & \mathbf{j} & \mathbf{k} \\ \partial/\partial x & \partial/\partial y & \partial/\partial z \\ y & z & x \end{vmatrix}$$

$$= 1\mathbf{i} - 1\mathbf{j} - 1\mathbf{k} = \langle 1, -1, 1 \rangle.$$

Therefore, curl $\mathbf{F}(1, 1, 1) = \langle 1, -1, 1 \rangle$.   ∎

EXAMPLE 9   Find curl $\mathbf{F}(2, 0, 1)$ if

$$\mathbf{F}(x, y, z) = (x + y)^3 \mathbf{i} + (\sin xy)\mathbf{j} + (\cos xyz)\mathbf{k}.$$

Solution   We have

$$\text{curl } \mathbf{F} = \begin{vmatrix} \mathbf{i} & \mathbf{j} & \mathbf{k} \\ \partial/\partial x & \partial/\partial y & \partial/\partial z \\ (x + y)^3 & \sin xy & \cos xyz \end{vmatrix}$$

$$= (-xz \sin xyz)\mathbf{i} - (-yz \sin xyz)\mathbf{j} + (y \cos xy - 3(x + y)^2)\mathbf{k}.$$

Therefore, curl $\mathbf{F}(2, 0, 1) = \langle 0, 0, -12 \rangle$.   ∎

Physical interpretation of curl

What exactly does curl **F** tell us about **F**? Let's imagine that **F** describes the velocity field of a fluid. Now suppose we drop a small paddle wheel like the one shown in Figure 17.5 into the fluid.

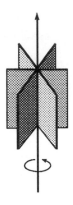

Figure 17.5 A paddle wheel.

Three questions we could ask in this situation are:

1) In what direction should the axis of the paddle wheel point if we are to achieve the fastest rate of rotation of the paddle wheel?

2) If the axis of the paddle wheel is aligned for maximum rotational velocity, just how fast does the paddle wheel rotate?

3) In what direction does the paddle wheel rotate?

The vector curl **F** provides us with answers to all of these questions!

1) At any point in the fluid, the paddle wheel will rotate the fastest if its axis is aligned with the direction of curl **F**.

2) If the axis of the paddle wheel is aligned in this way, then the magnitude of curl **F** is precisely *twice* the maximum angular velocity of the paddle wheel. The Greek letter omega (ω) is often used to indicate angular velocity. We can write

$$\|\text{curl } \mathbf{F}\| = 2\omega.$$

3) If the axis of the paddle wheel is imagined to have *right-hand screw threads*, then the paddle wheel rotates in the direction that results in the "screw" moving in the direction of curl **F**. Figure 17.5 illustrates that the direction of rotation is counterclockwise (looking down on the paddle wheel) if the curl vector points straight up.

Definition 4	A vector field $\mathbf{F}$ is called **irrotational** at a point (x, y, z) if

$$\mathbf{curl}\ \mathbf{F}(x, y, z) = 0 = \langle 0, 0, 0 \rangle.$$

Del Notation

We've already used the "del" notation for the gradient of a scalar field f:

$$\mathbf{grad}\ f = \nabla f.$$

The ∇ notation is also convenient for expressing the divergence and curl of a vector field $\mathbf{F}$. If we think of ∇ as the vector of partial derivative operators, then for functions of the three rectangular coordinate variables x, y, and z, we have

$$\nabla = \left\langle \frac{\partial}{\partial x}, \frac{\partial}{\partial y}, \frac{\partial}{\partial z} \right\rangle.$$

Given a vector field

$$\mathbf{F}(x, y, z) = P(x, y, z)\mathbf{i} + Q(x, y, z)\mathbf{j} + R(x, y, z)\mathbf{k}$$

we can write its divergence as the *dot product* of ∇ and $\mathbf{F}$:

$$\mathbf{div}\ \mathbf{F} = \nabla \cdot \mathbf{F} = \frac{\partial P}{\partial x} + \frac{\partial Q}{\partial y} + \frac{\partial R}{\partial z}.$$

Similarly, we can write the curl $\mathbf{F}$ as the *cross product* of ∇ and $\mathbf{F}$:

$$\mathbf{curl}\ \mathbf{F} = \nabla \times \mathbf{F} = \begin{vmatrix} \mathbf{i} & \mathbf{j} & \mathbf{k} \\ \partial/\partial x & \partial/\partial y & \partial/\partial z \\ P & Q & R \end{vmatrix}$$

$$= \left(\frac{\partial R}{\partial y} - \frac{\partial Q}{\partial z} \right)\mathbf{i} + \left(\frac{\partial P}{\partial z} - \frac{\partial R}{\partial x} \right)\mathbf{j} + \left(\frac{\partial Q}{\partial x} - \frac{\partial R}{\partial y} \right)\mathbf{k}.$$

EXAMPLE 10　Find the divergence and curl of each of the following vector fields:

$$\mathbf{F}(x, y, z) = \langle 2x, 3y, 4z \rangle$$

$$\mathbf{G}(x, y, z) = \langle x^2, y^2, z^2 \rangle$$

$$\mathbf{H}(x, y, z) = \langle x + y, y + z, z + x \rangle.$$

Identify the points at which each of the fields are irrotational or incompressible.

Solution　$\nabla \cdot \mathbf{F} = 2 + 3 + 4 = 9$ at every point, so $\mathbf{F}$ is incompressible nowhere.

$$\nabla \times \mathbf{F} = \begin{vmatrix} \mathbf{i} & \mathbf{j} & \mathbf{k} \\ \partial/\partial x & \partial/\partial y & \partial/\partial z \\ 2x & 3y & 4z \end{vmatrix} = 0\mathbf{i} - 0\mathbf{j} + 0\mathbf{k} = \langle 0, 0, 0 \rangle,$$

so $\mathbf{F}$ is irrotational everywhere.

$\nabla \cdot \mathbf{G} = 2x + 2y + 2z$, so $\mathbf{G}$ is incompressible at every point in the plane $x + y + z = 0$.

$$\nabla \times \mathbf{G} = \begin{vmatrix} \mathbf{i} & \mathbf{j} & \mathbf{k} \\ \partial/\partial x & \partial/\partial y & \partial/\partial z \\ x^2 & y^2 & z^2 \end{vmatrix} = 0\mathbf{i} - 0\mathbf{j} + 0\mathbf{k} = \langle 0, 0, 0 \rangle,$$

so $\mathbf{G}$ is irrotational everywhere.

$\nabla \cdot \mathbf{H} = 1 + 1 + 1 = 3$ at every point, so $\mathbf{H}$ is incompressible nowhere.

$$\nabla \times \mathbf{H} = \begin{vmatrix} \mathbf{i} & \mathbf{j} & \mathbf{k} \\ \partial/\partial x & \partial/\partial y & \partial/\partial z \\ x + y & y + z & z + x \end{vmatrix} = (0 - 1)\mathbf{i} - (1 - 0)\mathbf{j} + (0 - 1)\mathbf{k} = \langle -1, -1, -1 \rangle,$$

so $\mathbf{H}$ is irrotational nowhere.　　　　　　　　　　　　　　■

The laplacian

The del operator can be used as a building block to define other operators. For example, starting with a scalar field f, we can find its *gradient* ∇f. The *divergence* of this new vector field can be written

$$\mathbf{div}\,(\mathbf{grad}\,f) = \nabla \cdot (\nabla f) = \frac{\partial^2 f}{\partial x^2} + \frac{\partial^2 f}{\partial y^2} + \frac{\partial^2 f}{\partial z^2}.$$

This is called the **laplacian.** A shorthand notation for the laplacian of f is

$$\nabla^2 f.$$

EXAMPLE 11   Compute the laplacian of the scalar field $f(x,y,z) = x^2 + y^2 + z^2$.

Solution   The gradient of f is grad $f = \nabla f = \langle 2x, 2y, 2z \rangle$. The laplacian is

$$\text{div (grad } f) = \nabla \cdot (\nabla f) = \nabla^2 f = 2 + 2 + 2 = 6.$$

∎

The laplacian $\nabla^2 f(x,y,z)$ can be thought of roughly as providing a measure of the difference between the *average value* of f at points near (x,y,z), and the *actual value* $f(x,y,z)$. Hence, if f is a temperature field, and $\nabla^2 f(x,y,z)$ is negative, then the temperature at (x,y,z) is lower than the average temperature of the immediate surroundings.

The operator ∇^2 applied to a *vector field* **F** "distributes" over the components. For example, if

$$\mathbf{F}(x,y) = \langle P(x,y), \ Q(x,y) \rangle,$$

then

$$\nabla^2 \mathbf{F} = \left(\frac{\partial^2 P}{\partial x^2} + \frac{\partial^2 P}{\partial y^2} \right) \mathbf{i} \ + \ \left(\frac{\partial^2 Q}{\partial x^2} + \frac{\partial^2 Q}{\partial y^2} \right) \mathbf{j}.$$

Many properties relating div, grad, curl, and the laplacian can be written down very conveniently using the del notation.

EXAMPLE 12   Suppose that

$$\mathbf{F}(x,y,z) = \langle P(x,y,z), Q(x,y,z), R(x,y,z) \rangle$$

and that P, Q, and R are all C^2-differentiable. (In other words, the second order partial derivatives are all continuous.) Show that

$$\text{div (curl } \mathbf{F}) = \nabla \cdot (\nabla \times \mathbf{F}) = 0.$$

Solution   First, we calculate curl **F**:

$$\nabla \times \mathbf{F} = \left\langle \frac{\partial R}{\partial y} - \frac{\partial Q}{\partial z}, \ \frac{\partial P}{\partial z} - \frac{\partial R}{\partial x}, \ \frac{\partial Q}{\partial x} - \frac{\partial P}{\partial y} \right\rangle.$$

Now,

$$\text{div (curl } \mathbf{F}) = \frac{\partial}{\partial x}\left(\frac{\partial R}{\partial y} - \frac{\partial Q}{\partial z} \right) + \frac{\partial}{\partial y}\left(\frac{\partial P}{\partial z} - \frac{\partial R}{\partial x} \right) + \frac{\partial}{\partial z}\left(\frac{\partial Q}{\partial x} - \frac{\partial P}{\partial y} \right)$$

$$= \left(\frac{\partial^2 R}{\partial x \partial y} - \frac{\partial^2 Q}{\partial x \partial z} \right) + \left(\frac{\partial^2 P}{\partial y \partial z} - \frac{\partial^2 R}{\partial y \partial x} \right) + \left(\frac{\partial^2 Q}{\partial z \partial x} - \frac{\partial^2 P}{\partial z \partial y} \right).$$

Since P, Q, and R are C^2-differentiable, the mixed partial derivatives of each must match. If we write div(curl **F**), rearranging the terms, we have

$$\text{div} (\text{curl } \mathbf{F}) = \left(\frac{\partial^2 P}{\partial y \partial z} - \frac{\partial^2 P}{\partial z \partial y} \right) + \left(\frac{\partial^2 Q}{\partial z \partial x} - \frac{\partial^2 Q}{\partial x \partial z} \right) + \left(\frac{\partial^2 R}{\partial x \partial y} - \frac{\partial^2 R}{\partial y \partial x} \right) = 0.$$

∎

EXAMPLE 13   Suppose that f is a C^2-differentiable scalar field. Show that $\nabla \times (\nabla f) = \mathbf{0}$.

Solution

$$\nabla \times (\nabla f) = \begin{vmatrix} \mathbf{i} & \mathbf{j} & \mathbf{k} \\ \partial/\partial x & \partial/\partial y & \partial/\partial z \\ \frac{\partial f}{\partial x} & \frac{\partial f}{\partial y} & \frac{\partial f}{\partial z} \end{vmatrix}$$

$$= \left(\frac{\partial^2 f}{\partial y \partial z} - \frac{\partial^2 f}{\partial z \partial y} \right) \mathbf{i} - \left(\frac{\partial^2 f}{\partial x \partial z} - \frac{\partial^2 f}{\partial z \partial x} \right) \mathbf{j} + \left(\frac{\partial^2 f}{\partial x \partial y} - \frac{\partial^2 f}{\partial y \partial x} \right) \mathbf{k}$$

$$= \langle 0, 0, 0 \rangle,$$

since the mixed partial derivatives of a C^2-differentiable function match. Hence, the curl of a gradient vector field f is always the zero vector, provided f is C^2-differentiable. ∎

EXERCISES

Exercises 1-20 refer to the following scalar and vector fields:

$$f(x, y, z) = x^2 y^3 z^4 \qquad \mathbf{F}(x, y, z) = \langle y^2 + z^2, x^2 + z^2, x^2 + y^2 \rangle$$

$$g(x, y, z) = \ln(xyz) \qquad \mathbf{G}(x, y, z) = \langle xyz, xy^2 z^3, x^3 y^2 z \rangle$$

$$h(x, y, z) = e^{xyz} \qquad \mathbf{H}(x, y, z) = \langle xyz, x^2 y^2 z^2, x^2 + y^2 + z^2 \rangle$$

$$p(x, y, z) = x + y - z \qquad \mathbf{P}(x, y, z) = \langle x + y + z, x + y^2 + z^3, x^3 - y^2 - z \rangle$$

Compute the indicated scalar or vector field.

1.	grad $f = \nabla f$	**2.**	grad $g = \nabla g$
3.	grad $h = \nabla h$	**4.**	grad $p = \nabla p$
5.	div **F** $= \nabla \cdot$ **F**	**6.**	div **G** $= \nabla \cdot$ **G**
7.	div **H** $= \nabla \cdot$ **H**	**8.**	div **P** $= \nabla \cdot$ **P**
9.	curl **F** $= \nabla \times$ **F**	**10.**	curl **G** $= \nabla \times$ **G**
11.	curl **H** $= \nabla \times$ **H**	**12.**	curl **P** $= \nabla \times$ **P**
13.	$\nabla^2 f$	**14.**	$\nabla^2 g$
15.	$\nabla^2 h$	**16.**	$\nabla^2 p$
17.	∇^2 **F**	**18.**	∇^2 **G**

19. $\nabla^2 \mathbf{H}$ **20.** $\nabla^2 \mathbf{P}$

Suppose $f(x, y, z)$ and $g(x, y, z)$ are any two C^2-differentiable scalar fields and $\mathbf{F}(x, y, z)$ and $\mathbf{G}(x, y, z)$ are any two C^2-differentiable vector fields. Verify each of the statements in exercises 21-25.

21. $\nabla \cdot (\nabla f \times \nabla g) = 0$.

22. $\nabla \cdot (\mathbf{F} \times \mathbf{G}) = \mathbf{G} \cdot (\nabla \times \mathbf{F}) - \mathbf{F} \cdot (\nabla \times \mathbf{G})$.

23. $\mathrm{curl}\,(f\mathbf{F}) = f\,\mathrm{curl}\,\mathbf{F} + \nabla f \times \mathbf{F}$.

24. $\nabla \times (\nabla \times \mathbf{G}) = \nabla(\nabla \cdot \mathbf{G}) - (\nabla^2 \mathbf{G})$.

25. $\mathrm{div}\,(f\mathbf{F}) = f\,\mathrm{div}\,\mathbf{F} + \mathbf{F} \cdot \nabla f$.

---

26. A scalar field f that satisfies *Laplace's Equation*:

$$\nabla^2 f = 0,$$

is called **harmonic**. Show that

$$f(x, y, z) = \sin x \sinh y + \cos x \cosh z$$

is harmonic.

27. Which of the scalar fields f, g, h, and p of exercises 1-20 are harmonic?

28. Which of the vector fields $\mathbf{F}$, $\mathbf{G}$, $\mathbf{H}$, and $\mathbf{P}$ of exercises 1-20 are incompressible everywhere?

29. Which of the vector fields $\mathbf{F}$, $\mathbf{G}$, $\mathbf{H}$, and $\mathbf{P}$ of exercises 1-20 are irrotational everywhere?

30. A student says that the curl of any vector field is incompressible and that the gradient of any scalar field is irrotational. Are these statements accurate?

---

31. Some people describe the curl as measuring the "swirl" of a vector field. However, that's a bit misleading, as shown in this exercise. Suppose a fluid has a velocity field given by

$$\mathbf{F}(x, y, z) = \frac{y}{4}\mathbf{i}.$$

Sketch some of the vectors in the xy-plane and note that they do not "swirl." Then find curl $\mathbf{F}$ and note that it is not zero. Does this make sense in terms of the paddle wheel interpretation? Explain why.

17.2 TOTAL DERIVATIVE OF A VECTOR FIELD—JACOBIANS

So far, we've mentioned the divergence and curl as special types of derivatives for vector fields. In this section, we'll turn to the notion of a *total derivative* of a vector field.

The Jacobian

The **total derivative** of a C^1-differentiable vector field $\mathbf{F}$ is represented by the matrix of partial derivatives of its components. This matrix is known as the **Jacobian matrix.**

Definition 5

The **Jacobian matrix** of a vector field $\mathbf{F} : \mathbb{R}^2 \longrightarrow \mathbb{R}^2$, where

$$\mathbf{F}(x,y) = \langle P(x,y),\ Q(x,y) \rangle$$

is defined as

$$D\mathbf{F} = \begin{bmatrix} \partial P/\partial x & \partial P/\partial y \\ \partial Q/\partial x & \partial Q/\partial y \end{bmatrix}.$$

If $\mathbf{F} : \mathbb{R}^3 \longrightarrow \mathbb{R}^3$, where

$$\mathbf{F}(x,y,z) = P(x,y,z)\mathbf{i} + Q(x,y,z)\mathbf{j} + R(x,y,z)\mathbf{k}$$

then

$$D\mathbf{F} = \begin{bmatrix} \partial P/\partial x & \partial P/\partial y & \partial P/\partial z \\ \partial Q/\partial x & \partial Q/\partial y & \partial Q/\partial z \\ \partial R/\partial x & \partial R/\partial y & \partial R/\partial z \end{bmatrix}.$$

Let's make some observations regarding the Jacobian matrix.

Observation 1. Each row of the Jacobian matrix $D\mathbf{F}$ can be viewed as the *gradient* of a scalar component of $\mathbf{F}$. For example, if

$$\mathbf{F}(x,y,z) = \langle P(x,y,z), Q(x,y,z), R(x,y,z) \rangle,$$

then

$$D\mathbf{F} = \begin{bmatrix} \nabla P \\ \nabla Q \\ \nabla R \end{bmatrix} = \begin{bmatrix} \partial P/\partial x & \partial P/\partial y & \partial P/\partial z \\ \partial Q/\partial x & \partial Q/\partial y & \partial Q/\partial z \\ \partial R/\partial x & \partial R/\partial y & \partial R/\partial z \end{bmatrix}.$$

Observation 2. The sum of the diagonal entries of DF (this is called the **trace** of a matrix) is just the divergence of $\mathbf{F}$:

$$\text{div } \mathbf{F} = \frac{\partial P}{\partial x} + \frac{\partial Q}{\partial y} + \frac{\partial R}{\partial z}.$$

Observation 3. If $\mathbf{F} : \mathbb{R}^3 \longrightarrow \mathbb{R}^3$, the determinant of the Jacobian matrix DF is the triple scalar product of the gradients of its components:

$$|DF| = \begin{vmatrix} \partial P/\partial x & \partial P/\partial y & \partial P/\partial z \\ \partial Q/\partial x & \partial Q/\partial y & \partial Q/\partial z \\ \partial R/\partial x & \partial R/\partial y & \partial R/\partial z \end{vmatrix} = \nabla P \cdot (\nabla Q \times \nabla R).$$

The Jacobian matrix gives us a way of measuring the instantaneous rate of change in the vector output of a vector field with respect to change in the inputs. For example, suppose

$$\mathbf{F}(x, y, z) = \langle P(x, y, z), Q(x, y, z), R(x, y, z) \rangle,$$

and we want to measure the change in vector output when we move from the input point (x_0, y_0, z_0) to the point $(x_0 + \Delta x, y_0 + \Delta y, z_0 + \Delta z)$. The resulting changes in each component of $\mathbf{F}$ are

$$\Delta P = P(x_0 + \Delta x, y_0 + \Delta y, z_0 + \Delta z) - P(x_0, y_0, z_0)$$

$$\approx \nabla P(x_0, y_0, z_0) \cdot \langle \Delta x, \Delta y, \Delta z \rangle,$$

$$\Delta Q = Q(x_0 + \Delta x, y_0 + \Delta y, z_0 + \Delta z) - Q(x_0, y_0, z_0)$$

$$\approx \nabla Q(x_0, y_0, z_0) \cdot \langle \Delta x, \Delta y, \Delta z \rangle,$$

$$\Delta R = R(x_0 + \Delta x, y_0 + \Delta y, z_0 + \Delta z) - R(x_0, y_0, z_0)$$

$$\approx \nabla R(x_0, y_0, z_0) \cdot \langle \Delta x, \Delta y, \Delta z \rangle.$$

These three component changes can be conveniently expressed in terms of the Jacobian matrix. We can write

$$\Delta \mathbf{F} = \mathbf{F}(x_0 + \Delta x, y_0 + \Delta y, z_0 + \Delta z) - \mathbf{F}(x_0, y_0, z_0)$$

$$\approx \begin{bmatrix} \nabla P(x_0, y_0, z_0) \cdot \langle \Delta x, \Delta y, \Delta z \rangle \\ \nabla Q(x_0, y_0, z_0) \cdot \langle \Delta x, \Delta y, \Delta z \rangle \\ \nabla R(x_0, y_0, z_0) \cdot \langle \Delta x, \Delta y, \Delta z \rangle \end{bmatrix}$$

$$= DF(x_0, y_0, z_0) \begin{bmatrix} \Delta x \\ \Delta y \\ \Delta z \end{bmatrix}.$$

Similarly, if

$$\mathbf{F}(x, y) = \langle P(x, y), \ Q(x, y) \rangle,$$

then

$$\Delta \mathbf{F} = \mathbf{F}(x_0 + \Delta x, y_0 + \Delta y) - \mathbf{F}(x_0, y_0)$$

$$\approx \begin{bmatrix} \nabla P(x_0, y_0) \cdot \langle \Delta x, \Delta y \rangle \\ \nabla Q(x_0, y_0) \cdot \langle \Delta x, \Delta y \rangle \end{bmatrix}$$

$$= D\mathbf{F}(x_0, y_0) \begin{bmatrix} \Delta x \\ \Delta y \end{bmatrix}.$$

Jacobian determinant as scale factor

There is a very nice geometric interpretation of the determinant of the Jacobian matrix. Let's consider the 2-dimensional case first. Think of the points (x, y) and $(x + \Delta x, y + \Delta y)$ as the opposite corners of a rectangle of dimensions Δx by Δy. If $\mathbf{F} : \mathbb{R}^2 \to \mathbb{R}^2$ is a C^1-differentiable vector field, then for small Δx and Δy, the image of this rectangle under $\mathbf{F}$ will be approximately a parallelogram (see Figure 17.6).

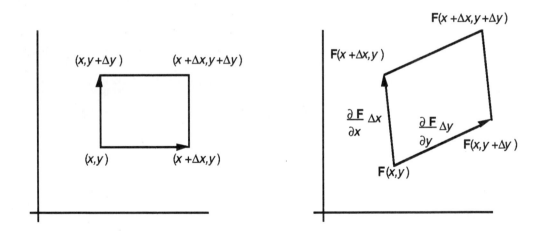

Figure 17.6 A rectangle and its image under $\mathbf{F}$.

The sides of this parallelogram are the vectors

$$\frac{\partial \mathbf{F}}{\partial x}\Delta x = \left\langle \frac{\partial P}{\partial x}\Delta x, \ \frac{\partial Q}{\partial x}\Delta x \right\rangle \quad \text{and} \quad \frac{\partial \mathbf{F}}{\partial y}\Delta y = \left\langle \frac{\partial P}{\partial y}\Delta y, \ \frac{\partial Q}{\partial y}\Delta y \right\rangle,$$

and the area of the parallelogram is given by

$$\left| \frac{\partial P}{\partial x}\Delta x \frac{\partial Q}{\partial y}\Delta y \ - \ \frac{\partial P}{\partial y}\Delta y \frac{\partial Q}{\partial x}\Delta x \right| = \left| \frac{\partial P}{\partial x}\frac{\partial Q}{\partial y} \ - \ \frac{\partial P}{\partial y}\frac{\partial Q}{\partial x} \right| \Delta x \Delta y.$$

The quantity in the absolute value signs on the right-hand side of this equation is the determinant of the Jacobian matrix:

$$\begin{vmatrix} \partial P/\partial x & \partial P/\partial y \\ \partial Q/\partial x & \partial Q/\partial y \end{vmatrix}.$$

If we denote this determinant by $J(x, y)$, we have the

$$\text{area of the parallelogram} = |J(x, y)|\Delta x \Delta y.$$

In other words, the absolute value of the determinant of the Jacobian matrix is the scale factor relating the area of the original rectangle to its image parallelogram under **F**.

In three dimensions, if we think of (x, y, z) and $(x + \Delta x, y + \Delta y, z + \Delta z)$ as the opposite corners of a small rectangular box, then Δx, Δy and Δz are the dimensions, and the

$$\text{volume of the box} = \Delta x \Delta y \Delta z.$$

For very small dimensions, the image of this box under a differentiable vector field

$$\mathbf{F}(x, y, z) = \langle P(x, y, z), Q(x, y, z), R(x, y, z)\rangle,$$

is approximately a parallelepiped, and the

$$\text{volume of the parallelepiped} = |J(x, y, z)|\Delta x \Delta y \Delta z,$$

where

$$J(x, y, z) = \begin{vmatrix} \partial P/\partial x & \partial P/\partial y & \partial P/\partial z \\ \partial Q/\partial x & \partial Q/\partial y & \partial Q/\partial z \\ \partial R/\partial x & \partial R/\partial y & \partial R/\partial z \end{vmatrix}$$

is again the determinant of the Jacobian matrix. In this case, its absolute value provides the scale factor relating the volume of the original box to its image under **F**.

Application to change of variables in double integration

Given a double integral

$$\iint\limits_{D} f(x, y) \, dx \, dy$$

over a domain D in the xy-plane, we can rewrite the integral in the form

$$\iint\limits_{D} f(r\cos\theta, r\sin\theta) r \, dr \, d\theta$$

where the region D is now described in polar coordinates, and the factor r arises in expressing the area element dA in polar coordinates.

This change of variables from (x, y) to (r, θ) can actually be thought of as describing a function from $\mathbb{R}^2$ to $\mathbb{R}^2$, where

$$x = P(r, \theta) = r\cos\theta \quad \text{and} \quad y = Q(r, \theta) = r\sin\theta.$$

In fact, if we label the coordinate axes with r and θ, then we can think of the region D as the image of some region D^* in the $r\theta$-plane.

EXAMPLE 14 The region

$$D = \{(x,y) : 0 \le x,\ 0 \le y,\ x^2 + y^2 \le 4\}$$

can be described as the image of a rectangle

$$D^* = \{(r,\theta) : 0 \le r \le 2,\ 0 \le \theta \le \pi/2\}$$

under the mapping

$$(r,\theta) \longmapsto (r\cos\theta, r\sin\theta) = (x,y).$$

(See Figure 17.7.)   ∎

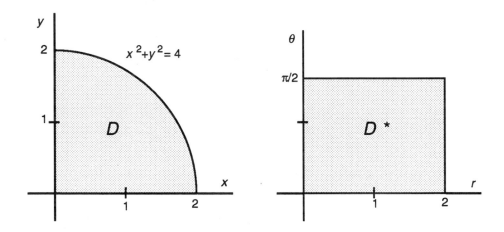

Figure 17.7 The region D is the image of D^* when $x = r\cos\theta$ and $y = r\sin\theta$.

What is the Jacobian scale factor of this particular mapping? If we compute the determinant of the Jacobian matrix, we find

$$J(r,\theta) = \begin{vmatrix} \partial x/\partial r & \partial x/\partial \theta \\ \partial y/\partial r & \partial y/\partial \theta \end{vmatrix} = \begin{vmatrix} \cos\theta & -r\sin\theta \\ \sin\theta & r\cos\theta \end{vmatrix}$$

$$= r\cos^2\theta + r\sin^2\theta = r(\cos^2\theta + \sin^2\theta) = r.$$

Since $r \ge 0$, we have $|J(r,\theta)| = r$, which is precisely the scale factor we should expect in changing to polar coordinates in double integration.

A Jacobian scale factor arises when we make a change of variables in integration. In general, if we wish to compute

$$\iint_D f(x,y)\,dx\,dy$$

by changing variables from (x, y) to (u, v) via the mapping

$$x = P(u, v) \quad \text{and} \quad y = Q(u, v),$$

we must take the following steps:

1) Compute the Jacobian determinant

$$J(u, v) = \begin{vmatrix} \partial P/\partial u & \partial P/\partial v \\ \partial Q/\partial u & \partial Q/\partial v \end{vmatrix} = \frac{\partial P}{\partial u}\frac{\partial Q}{\partial v} - \frac{\partial P}{\partial v}\frac{\partial Q}{\partial u}.$$

2) Find the region D^* in the uv-plane whose image is D under the mapping. You may find it useful to solve

$$x = P(u, v) \quad \text{and} \quad y = Q(u, v)$$

for u and v in terms of x and y. Rewriting the description of the original region D in terms of u and v provides a description of D^* in the uv-plane.

3) Calculate

$$\iint\limits_{D^*} f(P(u, v), Q(u, v)) |J(u, v)| \, du \, dv.$$

We'll illustrate the method with an example.

EXAMPLE 15   Calculate

$$\int_0^2 \int_0^x (x + y) \, dy \, dx$$

both directly, and by making the change of variables:

$$x = u + v \quad \text{and} \quad y = u - v.$$

Solution   We first calculate the double integral in the original variables:

$$\int_0^2 \int_0^x (x + y) \, dy \, dx = \int_0^2 \left(xy + \frac{y^2}{2} \right) \Big]_0^x dx$$

$$= \int_0^2 \left(x^2 + \frac{x^2}{2} \right) dx$$

$$= \int_0^2 \frac{3x^2}{2} dx$$

$$= \frac{x^3}{2} \Big]_0^2 = \frac{8}{2} - 0 = 4.$$

Now, we'll calculate the integral again, using the change of variables indicated.

1) The Jacobian determinant is

$$J(u,v) = \begin{vmatrix} \dfrac{\partial x}{\partial u} & \dfrac{\partial x}{\partial v} \\ \dfrac{\partial y}{\partial u} & \dfrac{\partial y}{\partial v} \end{vmatrix} = \begin{vmatrix} 1 & 1 \\ 1 & -1 \end{vmatrix} = -1 - 1 = -2.$$

Our conversion scale factor will be $|J(u,v)| = 2$.

2) The original region of integration D is defined by $0 \le y \le x$ and $0 \le x \le 2$. If we solve $x = u + v$, $y = u - v$ for u and v, we find:

$$x + y = 2u \qquad x - y = 2v$$

so

$$u = \frac{x+y}{2} \qquad v = \frac{x-y}{2}.$$

The borderlines of the original region D lie on the lines $y = 0$, $x = 2$, and $y = x$. Therefore, the borderlines of the region D^* lie on the lines $u - v = 0$, $u + v = 2$, and $v = 0$. Figure 17.8 illustrates both the original region D and the new region of integration D^*.

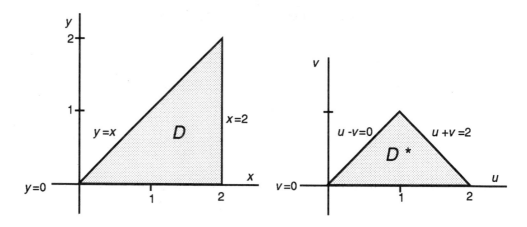

Figure 17.8 The region D is the image of D^* when $x = u + v$ and $y = u - v$.

3) We now calculate the double integral in terms of the new variables

u and v:

$$\iint_D (x+y)\,dy\,dy = \int_{D^*} ((u+v)+(u-v))2\,du\,dv$$

$$= \int_{D^*} 4u\,du\,dv$$

$$= 4\int_0^1 \int_v^{2-v} u\,du\,dv$$

$$= 4\int_0^1 \frac{u^2}{2}\Big]_v^{2-v} dv$$

$$= 4\int_0^1 \left(\frac{(2-v)^2}{2} - \frac{v^2}{2}\right) dv$$

$$= 4\int_0^1 \left(\frac{4}{2} - \frac{4v}{2} + \frac{v^2}{2} - \frac{v^2}{2}\right) dv$$

$$= 4\int_0^1 (2-2v)\,dv$$

$$= 4(2v - v^2)\Big]_0^1$$

$$= (8v - 4v^2)_0^1 = (8-4)-(0-0) = 4.$$

We can see that the two calculations match exactly. ∎

The chain rule for total derivatives

We have talked about vector fields having the same number of components as inputs, but we can also talk about functions

$$\mathbf{f} : \mathbb{R}^m \to \mathbb{R}^n,$$

where $m \neq n$, and

$$\mathbf{f}(x_1, x_2, \ldots, x_m) = \langle f_1(x_1, x_2, \ldots, x_m), \ldots, f_n(x_1, x_2, \ldots, x_m)\rangle,$$

and each component f_i $(1 \le i \le n)$ is a scalar-valued function $\mathbb{R}^m \to \mathbb{R}$.

Provided all the functions f_i are C^1-differentiable, the total derivative of $\mathbf{f}$ is the $n \times m$ matrix of partial derivatives:

$$D\mathbf{f} = \begin{bmatrix} \frac{\partial f_1}{\partial x_1} & \frac{\partial f_1}{\partial x_2} & \frac{\partial f_1}{\partial x_3} & \cdots & \frac{\partial f_1}{\partial x_m} \\ \frac{\partial f_2}{\partial x_1} & \frac{\partial f_2}{\partial x_2} & \frac{\partial f_1}{\partial x_3} & \cdots & \frac{\partial f_2}{\partial x_m} \\ \vdots & \vdots & \vdots & \ddots & \vdots \\ \frac{\partial f_n}{\partial x_1} & \frac{\partial f_n}{\partial x_2} & \frac{\partial f_n}{\partial x_3} & \cdots & \frac{\partial f_n}{\partial x_m} \end{bmatrix}$$

In other words, the ith row of Df consists of the components of the gradient of the ith component function f_i.

EXAMPLE 16 Find the total derivative of $\mathbf{h} : \mathbb{R}^3 \to \mathbb{R}^2$, where

$$\mathbf{h}(x, y, z) = \langle xyz, \sqrt{x^2 + y^2 + z^2} \rangle.$$

Evaluate $D\mathbf{h}(0, 3, 4)$.

Solution $D\mathbf{h}$ is represented by a 2×3 matrix. If

$$h_1(x, y, z) = xyz \qquad \text{and} \qquad h_2(x, y, z) = \sqrt{x^2 + y^2 + z^2},$$

then

$$D\mathbf{h} = \begin{bmatrix} \frac{\partial h_1}{\partial x} & \frac{\partial h_1}{\partial y} & \frac{\partial h_1}{\partial z} \\ \frac{\partial h_2}{\partial x} & \frac{\partial h_2}{\partial y} & \frac{\partial h_2}{\partial z} \end{bmatrix} = \begin{bmatrix} yz & xz & xy \\ \frac{x}{\sqrt{x^2+y^2+z^2}} & \frac{y}{\sqrt{x^2+y^2+z^2}} & \frac{z}{\sqrt{x^2+y^2+z^2}} \end{bmatrix}.$$

So, we have

$$D\mathbf{h}(0, 3, 4) = \begin{bmatrix} 12 & 0 & 0 \\ 0 & \frac{3}{5} & \frac{4}{5} \end{bmatrix}.$$

∎

Properties of the total derivative

1. If $\mathbf{f} : \mathbb{R}^m \to \mathbb{R}^n$, then for any scalar c

$$D(c\mathbf{f}) = cD\mathbf{f}.$$

2. If $\mathbf{f}$ and $\mathbf{g}$ are functions $\mathbb{R}^m \to \mathbb{R}^n$, then

$$D(\mathbf{f} + \mathbf{g}) = D\mathbf{f} + D\mathbf{g}.$$

3. In the special case that $f : \mathbb{R}^n \to \mathbb{R}$, then Df is a $1 \times n$ matrix

$$Df = \begin{bmatrix} \frac{\partial f}{\partial x_1} & \frac{\partial f}{\partial x_2} & \frac{\partial f}{\partial x_3} & \cdots & \frac{\partial f}{\partial x_n} \end{bmatrix},$$

and we can think of Df as the *gradient*:

$$\nabla f = \left\langle \frac{\partial f}{\partial x_1}, \frac{\partial f}{\partial x_2}, \cdots, \frac{\partial f}{\partial x_n} \right\rangle.$$

If f and g are functions $\mathbb{R}^n \to \mathbb{R}$, then

$$D(fg) = gDf + fDg = g\nabla f + f\nabla g,$$

and

$$D\left(\frac{f}{g}\right) = \frac{gDf - fDg}{g^2} = \frac{g\nabla f + f\nabla g}{g^2}$$

whenever g is $\neq 0$.

Note: This product rule and quotient rule only hold for the special case of functions $\mathbb{R}^n \to \mathbb{R}$.

4. If $\mathbf{f} : \mathbb{R}^m \to \mathbb{R}^n$ and $\mathbf{g} : \mathbb{R}^n \to \mathbb{R}^p$ then $\mathbf{g} \circ \mathbf{f} : \mathbb{R}^m \to \mathbb{R}^p$, and at any point $\mathbf{x} = (x_1, x_2, \ldots, x_n)$ in $\mathbb{R}^n$, we have

$$D\left(\mathbf{g} \circ \mathbf{f}\right)(\mathbf{x}) = D\mathbf{g}\left(\mathbf{f}(\mathbf{x})\right) D\mathbf{f}(\mathbf{x}).$$

This is the **chain rule**. Note how it appears virtually identical to the chain rule for single variable calculus, except that the product indicated on the right-hand side of the equation is a *matrix product*.

EXAMPLE 17   Suppose

$$f(u, v) = \frac{u^2 + v^2}{u^2 - v^2}$$

and

$$u(x, y) = e^{-x-y}, \qquad v(x, y) = e^{xy}.$$

Find $D(f \circ \mathbf{g})(x, y)$, where $\mathbf{g}(x, y) = \langle u(x, y), v(x, y) \rangle$.

Solution   We'll compute $D(f \circ \mathbf{g})$ using the chain rule:

$$Df(u, v) = \left[\frac{2u(u^2 - v^2) - 2u(u^2 + v^2)}{(u^2 - v^2)^2} \quad \frac{2v(u^2 - v^2) - (-2v(u^2 + v^2))}{(u^2 - v^2)^2} \right]$$

$$= \left[\frac{-4uv^2}{(u^2 - v^2)^2} \quad \frac{4u^2v}{(u^2 - v^2)^2} \right]$$

and

$$D\mathbf{g}(x, y) = \begin{bmatrix} -e^{-x-y} & -e^{-x-y} \\ ye^{xy} & xe^{xy} \end{bmatrix}.$$

The chain rule says

$$D(f \circ \mathbf{g})(x, y) = Df(\mathbf{g}(x, y))D\mathbf{g}(x, y)$$

$$= \left[\frac{-4e^{-x-y}e^{2xy}}{(e^{2(-x-y)} - e^{2xy})^2} \quad \frac{4e^{2(-x-y)}e^{xy}}{(e^{2(-x-y)} - e^{2xy})^2} \right] \begin{bmatrix} -e^{-x-y} & -e^{-x-y} \\ ye^{xy} & xe^{xy} \end{bmatrix}$$

$$= \left[\frac{4e^{2(-x-y)}e^{2xy} + 4ye^{2(-x-y)}e^{2xy}}{(e^{2(-x-y)} - e^{2xy})^2} \quad \frac{4e^{2(-x-y)}e^{2xy} + 4xe^{2(-x-y)}e^{2xy}}{(e^{2(-x-y)} - e^{2xy})^2} \right]$$

Now, if we substituted for u and v in terms of x and y in the composition $f \circ g$ we have a function of x and y:

$$h(x,y) = f \circ g(x,y) = \frac{e^{2(-x-y)} + e^{2xy}}{e^{2(-x-y)} - e^{2xy}}.$$

You can check directly that the matrix $Dh = [\partial h/\partial x \quad \partial h/\partial y]$ is precisely the one we found above. ∎

EXAMPLE 18   Suppose $\mathbf{f}(x,y) = \langle e^x, x+y \rangle$ and $\mathbf{g}(u,v) = \langle u, \cos v, v+u \rangle$. Find $D(\mathbf{g} \circ \mathbf{f})(0,0)$ in two ways: by direct substitution and differentiation, and by the chain rule.

Solution   By direct substitution, we have

$$\mathbf{g} \circ \mathbf{f}(x,y) = \mathbf{g}(\mathbf{f}(x,y)) = \mathbf{g}(e^x, x+y) = \langle e^x, \cos(x+y), x+y+e^x \rangle.$$

So,

$$D(\mathbf{g} \circ \mathbf{f}(x,y)) = \begin{bmatrix} e^x & 0 \\ -\sin(x+y) & -\sin(x+y) \\ 1+e^x & 1 \end{bmatrix}.$$

At $(x,y)=(0,0)$ we have

$$D(\mathbf{g} \circ \mathbf{f})(0,0) = \begin{bmatrix} 1 & 0 \\ 0 & 0 \\ 2 & 1 \end{bmatrix}.$$

Now, by the chain rule, we have

$$D(\mathbf{g} \circ \mathbf{f})(0,0) = D\mathbf{g}(\mathbf{f}(0,0))D\mathbf{f}(0,0).$$

We compute

$$D\mathbf{g}(u,v) = \begin{bmatrix} 1 & 0 \\ 0 & -\sin v \\ 1 & 1 \end{bmatrix} \quad \text{and} \quad D\mathbf{f}(x,y) = \begin{bmatrix} e^x & 0 \\ 1 & 1 \end{bmatrix}.$$

Since $\mathbf{f}(0,0) = \langle 1, 0 \rangle$, we have

$$D(\mathbf{g} \circ \mathbf{f})(0,0) = D\mathbf{g}(1,0)D\mathbf{f}(0,0)$$

$$= \begin{bmatrix} 1 & 0 \\ 0 & 0 \\ 1 & 1 \end{bmatrix} \begin{bmatrix} 1 & 0 \\ 1 & 1 \end{bmatrix}$$

$$= \begin{bmatrix} 1 & 0 \\ 0 & 0 \\ 2 & 1 \end{bmatrix},$$

which matches exactly the first computation. ∎

EXAMPLE 19 Using the same functions f and g in the previous example, note that the composition f ∘ g does not make sense, since the output of g has *three* components, but f requires *two* inputs. ∎

████████ **EXERCISES**

Exercises 1-16 refer to the following functions:

$$\mathbf{F}(u,v) = \langle 2u - 3v, 3u + v\rangle \qquad \mathbf{G}(u,v) = \langle uv, u/v, v/u\rangle$$

$$\mathbf{H}(x,y,z) = \langle xyz, xy^2z^3\rangle \qquad \mathbf{P}(x,y,z) = \langle x + y + z, x + y^2 + z^3, x^3 - y^2 - z\rangle.$$

Compute the indicated total derivative. If an indicated composition makes sense, then verify the chain rule by direct substitution and differentiation. If an indicated composition does not make sense, explain why.

1. $DF(u,v)$ **2.** $DG(u,v)$

3. $DH(x,y,z)$ **4.** $DP(x,y,z)$

5. $D(\mathbf{F} \circ \mathbf{G})(u,v)$ **6.** $D(\mathbf{G} \circ \mathbf{F})(u,v)$

7. $D(\mathbf{H} \circ \mathbf{P})(x,y,z)$ **8.** $D(\mathbf{P} \circ \mathbf{H})(x,y,z)$

9. $D(\mathbf{F} \circ \mathbf{P})(x,y,z)$ **10.** $D(\mathbf{F} \circ \mathbf{H})(x,y,z)$

11. $D(\mathbf{H} \circ \mathbf{G})(u,v)$ **12.** $D(\mathbf{H} \circ \mathbf{F})(u,v)$

13. $D(\mathbf{G} \circ \mathbf{P})(x,y,z)$ **14.** $D(\mathbf{G} \circ \mathbf{H})(x,y,z)$

15. $D(\mathbf{P} \circ \mathbf{G})(u,v)$ **16.** $D(\mathbf{P} \circ \mathbf{F})(u,v)$

---

17. Substitute $u = xy$, $v = y/x$ to find the area of the first quadrant region bounded by the lines $y = x$, $y = 2x$, and the hyperbolas $xy = 1$, $xy = 2$ (see picture below).

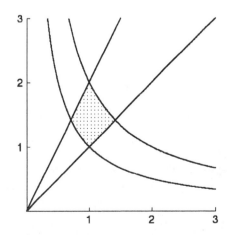

18. Elliptical coordinates are given by $x = ar\cos\theta, y = br\sin\theta$, where a and b are constants. What is the Jacobian scale factor for this change of variables?

A Jacobian scale factor also arises in triple integration. To change variables in the integral

$$\iiint_R f(x, y, z)\, dx\, dy\, dz$$

to u, v, and w via

$$x = P(u, v, w), \qquad y = Q(u, v, w), \qquad z = R(u, v, w),$$

one must

1) calculate the Jacobian determinant

$$J(u, v, w) = \begin{vmatrix} \dfrac{\partial x}{\partial u} & \dfrac{\partial x}{\partial v} & \dfrac{\partial x}{\partial w} \\[2mm] \dfrac{\partial y}{\partial u} & \dfrac{\partial y}{\partial v} & \dfrac{\partial y}{\partial w} \\[2mm] \dfrac{\partial z}{\partial u} & \dfrac{\partial z}{\partial v} & \dfrac{\partial z}{\partial w} \end{vmatrix};$$

2) find the region R^* in uvw-space whose image is R under the given transformation;

3) compute

$$\iiint_{R^*} f(P(u, v, w), Q(u, v, w), R(u, v, w))|J(u, v, w)|\, du\, dv\, dw.$$

19. The transformation from Cartesian to cylindrical coordinates is via

$$x = r \cos \theta, \qquad y = r \sin \theta, \qquad z = z.$$

Verify that the Jacobian determinant

$$J(r, \theta, z) = r.$$

20. The transformation from Cartesian to spherical coordinates is via

$$x = \rho \cos \theta \sin \varphi, \qquad y = \rho \sin \theta \sin \varphi, \qquad z = \rho \cos \varphi.$$

Verify that the Jacobian determinant

$$J(\rho, \theta, \varphi) = -\rho^2 \sin \varphi.$$

17.3 LINE INTEGRALS

Now we turn our attention to integration of both scalar and vector fields. Let's recall the situation for single variable calculus again. A definite integral

$$\int_a^b f(x)dx$$

measures the net effect of the real-valued function f over the interval $[a, b]$. We can interpret this value in a variety of ways, depending on the physical meaning of the function f. For example, if $f(x)$ represents the *force* acting on an object at distance x, then $\int_a^b f(x)dx$ represents the *work* performed as the object moves over the distance interval $[a, b]$.

In this example, we can consider the interval as a simple straight line path of the object. Our interest in this section turns to integrals over much more general paths. The traditional name for such an integral is *line integral*, though perhaps a better name (used occasionally) would be a *path integral*. You can think of a line integral as measuring the net effect of a scalar field or vector field on an object as it moves along some path. In this section we'll define and illustrate how line integrals are calculated and discuss some of their physical properties and interpretations.

Smooth and piecewise smooth curves—arc length

A vector-valued function $\mathbf{r}(t)$ of one real variable t describes a *path*. The image of a continuous path is a *curve* in the plane ($\mathbb{R}^2$) if

$$\mathbf{r}(t) = \langle x(t), y(t) \rangle$$

or in space ($\mathbb{R}^3$), if

$$\mathbf{r}(t) = \langle x(t), y(t), z(t) \rangle.$$

In either case, we say $\mathbf{r}$ parametrizes the curve. A single curve can have infinitely many different parametrizations.

A continuous path $\mathbf{r}$ is **smooth** over the interval $a \leq t \leq b$ if $\mathbf{r}'(t)$ is *continuous* and *nonzero* for $a < t < b$. If we can break the interval $[a, b]$ into a finite number of subintervals, and if $\mathbf{r}$ is smooth over each subinterval then we say $\mathbf{r}$ is **piecewise smooth** over $[a, b]$. Figure 17.9 illustrates the image curves of a smooth and a piecewise smooth path.

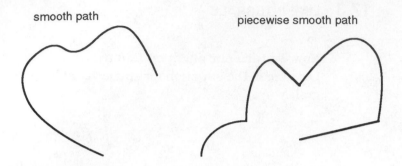

Figure 17.9 A smooth and piecewise smooth path.

We can find the arc length of a smooth curve by integrating the speed $\|\mathbf{r}'(t)\|$ (magnitude of the velocity) over the interval $[a, b]$:

$$\text{arc length } = \int_a^b \|\mathbf{r}'(t)\| \, dt.$$

The arc length of a piecewise smooth curve can be found by finding the lengths of each of its smooth pieces and adding the results.

We can use this to form the arc length function for the path $\mathbf{r}$:

$$s(t) = \int_a^t \|\mathbf{r}'(u)\| \, du.$$

The function s simply gives us the length of the path $\mathbf{r}$ over the interval $[a, t]$, and we have

$$\frac{ds}{dt} = \|\mathbf{r}'(t)\|.$$

In other words, the derivative of the arc length function with respect to t is just the *speed*. We call

$$ds = \|\mathbf{r}'(t)\| \, dt$$

the arc length element, so that

$$\int_a^b ds$$

gives the arc length of a curve with parametrization $\mathbf{r}$ over the interval $a \leq t \leq b$.

A **closed** curve C is one whose initial and terminal points are the same. If $\mathbf{r}$ is a parametrization for a closed curve C defined on $[a, b]$, then

$$\mathbf{r}(a) = \mathbf{r}(b).$$

A **simple** curve has a parametrization $\mathbf{r}$ such that $\mathbf{r}(t_1) \neq \mathbf{r}(t_2)$ whenever $t_1 \neq t_2$, with the only possible exception allowed being the endpoints (see Figure 17.10).

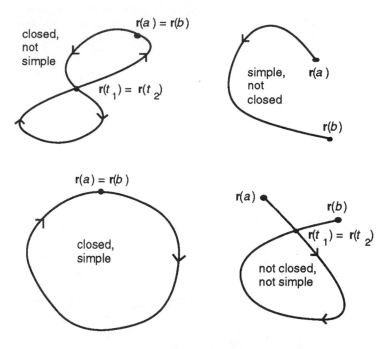

Figure 17.10 Types of curves.

A parametrized curve comes with an *orientation* (direction) determined by the position function **r**. In Figure 17.10, the orientation of each curve is indicated by small arrowheads showing which direction the velocity vector **r**′ points. Notice that for an intersection point of a non-simple curve (where $\mathbf{r}(t_1) = \mathbf{r}(t_2)$, but $t_1 \neq t_2$), the idea of a tangent vector becomes ambiguous, for the direction of the velocity vector depends on the time t we cross that point.

If $\mathbf{r}(t)$ is a parametrization defined over $[a, b]$, we can easily obtain another parametrization that reverses the orientation of the curve:

$$\mathbf{q}(t) = \mathbf{r}(b + a - t), \qquad \text{for } a \leq t \leq b.$$

Note that $\mathbf{q}(a) = \mathbf{r}(b)$, $\mathbf{q}(b) = \mathbf{r}(a)$, and

$$\mathbf{q}'(t) = -\mathbf{r}'(b + a - t).$$

Line integral of a scalar field f with respect to arc length

If f is a scalar field defined in a region of the plane $\mathbb{R}^2$ or space $\mathbb{R}^3$, and **r** is a path whose image curve C lies in the same region, we can define a line integral of f with respect to arc length.

Definition 6

The **line integral** of f with respect to arc length along C is

$$\oint_C f\,ds = \int_a^b f(\mathbf{r}(t))\left(\frac{ds}{dt}\right)dt = \int_a^b f(\mathbf{r}(t))\|\mathbf{r}'(t)\|\,dt.$$

 The small circle in the symbol $\oint$ indicates that we are calculating a line integral, but not necessarily over a *closed* curve.

If we think of C as a *wire* in space and $f(x, y, z)$ measures the mass per unit length (density) at each point (x, y, z), then

$$\oint_C f\,ds \quad = \quad \text{mass of the wire,}$$

while the line integral

$$\oint_C 1\,ds \quad = \quad \text{length of the wire.}$$

EXAMPLE 20   A straight wire has endpoints $(1, 2, 3)$ and $(-2, 3, -1)$ (units measured in cm). Its density at any point is given by $f(x, y, z) = y^2$ grams per cm. Find the mass of the wire.

Solution   We can parametrize the line segment represented by the wire as

$$\mathbf{r}(t) = \langle 1, 2, 3 \rangle + t(\langle -2, 3, -1 \rangle - \langle 1, 2, 3 \rangle)$$

$$= \langle 1, 2, 3 \rangle + t\langle -3, 1, -4 \rangle$$

$$= \langle 1 - 3t,\ 2 + t,\ 3 - 4t \rangle.$$

Note that $\mathbf{r}(0) = \langle 1, 2, 3 \rangle$ and $\mathbf{r}(1) = \langle -2, 3, -1 \rangle$. For $0 \leq t \leq 1$, $\mathbf{r}(t)$ traces out each point along the length of the wire.

The velocity is simply the constant vector $\mathbf{r}'(t) = \langle -3, 1, -4 \rangle$, so the speed is

$$\|\mathbf{r}'(t)\| = \sqrt{(-3)^2 + 1^2 + (-4)^2} = \sqrt{26},$$

and the mass is given by

$$\oint_C y^2\,ds = \int_0^1 \sqrt{26}(2 + t)^2\,dt = \frac{\sqrt{26}}{3}(2 + t)^3\Big]_0^1 = 9\sqrt{26} \approx 45.89.$$

The wire has a mass of approximately 45.89 grams.   ∎

EXAMPLE 21   Find $\displaystyle\oint_C f\,ds$, where $f(x,y,z) = x+y+yz$ and C is the helix parametrized by

$$\mathbf{r}(t) = \langle \sin t, \cos t, t \rangle \qquad \text{for } 0 \leq t \leq 2\pi.$$

Solution   The velocity and speed are

$$\mathbf{r}'(t) = \langle \cos t, -\sin t, 1 \rangle \qquad \text{and} \qquad \|\mathbf{r}'(t)\| = \sqrt{\cos^2 t + \sin^2 t + 1} = \sqrt{2}.$$

Since $x = \sin t$, $y = \cos t$, and $z = t$, we have

$$\oint_C f\,ds = \int_0^{2\pi} f(\sin t, \cos t, t)\sqrt{2}\,dt$$

$$= \int_0^{2\pi} (\sin t + \cos t + t\cos t)\sqrt{2}\,dt$$

$$= \sqrt{2}\left[-\cos t + \sin t + (t\sin t - \cos t)\right]\Big|_0^{2\pi}$$

$$= \sqrt{2}(-1 + 0 + (0-1)) - \sqrt{2}(-1 + 0 + (0-1))$$

$$= -2\sqrt{2} + 2\sqrt{2} = 0.$$

■

EXAMPLE 22   Find $\displaystyle\oint_C f\,ds$, where $f(x,y,z) = x+\cos^2 z$ over the same curve as the previous example.

Solution   The velocity and speed are the same as in the previous example, so

$$\oint_C f\,ds = \int_0^{2\pi} (\sin t + \cos^2 t)\sqrt{2}\,dt$$

$$= \sqrt{2}\left(-\cos t + \frac{t}{2} + \frac{\sin 2t}{4}\right)\Big|_0^{2\pi} = \sqrt{2}\cdot\frac{2\pi}{4} = \frac{\sqrt{2}\pi}{2}.$$

■

EXAMPLE 23   Find $\oint_C f\, ds$, where $f(x, y, z) = x + y + z$ and C is parametrized by

$$\mathbf{r}(t) = \langle t, t^2, \frac{2}{3}t^3 \rangle \qquad \text{for } 0 \leq t \leq 1.$$

Solution   The velocity and speed are

$$\mathbf{r}'(t) = \langle 1, 2t, 2t^2 \rangle \qquad \text{and} \qquad \|\mathbf{r}'(t)\| = \sqrt{1 + 4t^2 + 4t^4} = \sqrt{(1 + 2t^2)^2} = 1 + 2t^2.$$

Therefore,

$$\oint_C f\, ds = \int_0^1 (t + t^2 + \frac{2}{3}t^3)(1 + 2t^2)\, dt$$

$$= \int_0^1 (t + t^2 + \frac{2}{3}t^3 + 2t^3 + 2t^4 + \frac{4}{3}t^5)\, dt$$

$$= \left(\frac{t^2}{2} + \frac{t^3}{3} + \frac{1}{6}t^4 + \frac{1}{2}t^4 + \frac{2}{5}t^5 + \frac{2}{9}t^6 \right)\Bigg]_0^1$$

$$= \left(\frac{1}{2} + \frac{1}{3} + \frac{1}{6} + \frac{1}{2} + \frac{2}{5} + \frac{2}{9} \right) - (0)$$

$$= \frac{163}{90}.$$

■

Line integral of a vector field F

If $\mathbf{F}$ is a vector field defined in some region of the plane $\mathbb{R}^2$ or space $\mathbb{R}^3$, then we can define a line integral of $\mathbf{F}$ as follows.

Definition 7   If C is a curve parametrized by a smooth path $\mathbf{r}$ defined over the interval $[a, b]$, and if $\mathbf{F}(\mathbf{r}(t))$ is defined for $a \leq t \leq b$, then

$$\oint_C \mathbf{F} \cdot d\mathbf{s} = \int_a^b \mathbf{F}(\mathbf{r}(t)) \cdot \mathbf{r}'(t)\, dt$$

is the **line integral** of $\mathbf{F}$ along $\mathbf{r}$. If $\mathbf{r}$ is piecewise smooth, then the line integral of $\mathbf{F}$ along C is obtained by adding the line integrals over the smooth pieces of $\mathbf{r}$.

The notation $\oint_C$ reminds us again that this is a line integral over a curve C. The notation $\mathbf{F} \cdot d\mathbf{s}$ is meant to remind us that a dot product $\mathbf{F}(\mathbf{r}(t)) \cdot \mathbf{r}'(t)$ is involved in the calculation of the line integral.

If **F** represents a *vector force field* and **r** is the path of an object moving through the force field, then

$$\oint_C \mathbf{F} \cdot d\mathbf{s}$$

measures the work performed on the object over the path. Let's see why this makes sense. At any time t, the vector $\mathbf{F}(\mathbf{r}(t))$ gives us the force vector at a point on our path **r**. The velocity vector $\mathbf{r}'(t)$ points in the tangent direction to the curve. Therefore, the dot product

$$\mathbf{F}(\mathbf{r}(t)) \cdot \mathbf{T}(t)$$

gives us exactly the component of the force that points in the same direction as the instantaneous motion of the object at time t (see Figure 17.11).

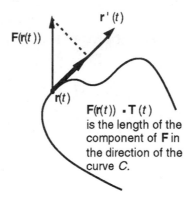

Figure 17.11 The component of a force field along a curve.

Integrating this force component over the interval $[a, b]$ measures the work.

If **F** is the velocity vector field for a fluid and **r** describes a curve C through this fluid, then at any point $\mathbf{r}(t)$ on the curve we can take the dot product of **F** with the unit tangent vector **T**, and this gives us the tangential component of the fluid flow along the curve. In other words, at any time t,

$$\mathbf{F} \cdot \mathbf{T} = \mathbf{F}(\mathbf{r}(t)) \cdot \frac{\mathbf{r}'(t)}{\|\mathbf{r}'(t)\|}$$

measures the flow in the same direction as the curve at that point.

Note that $\mathbf{F} \cdot \mathbf{T}$ is a scalar (it's a dot product). If we take the line integral of this scalar over the curve C with respect to arc length, we'll have a measure of the total fluid flow or *circulation* along the curve:

$$\oint_C \mathbf{F} \cdot \mathbf{T} \, ds = \int_a^b \mathbf{F}(\mathbf{r}(t)) \cdot \frac{\mathbf{r}'(t)}{\|\mathbf{r}'(t)\|} \|\mathbf{r}'(t)\| \, dt.$$

Notice that the speed factor $\|\mathbf{r}'(t)\|$ cancels out, so that

$$\oint_C \mathbf{F} \cdot \mathbf{T} \, ds = \int_a^b \mathbf{F}(\mathbf{r}(t)) \cdot \mathbf{r}'(t) \, dt = \oint_C \mathbf{F} \cdot d\mathbf{s}.$$

This is simply our definition of the line integral of the original vector field over the curve.

EXAMPLE 24 Find $\oint_C \mathbf{F} \cdot \mathbf{T} \, ds$ where $\mathbf{F}(x, y, z) = \langle y, 2x, y \rangle$ and the curve C is parametrized by

$$\mathbf{r}(t) = \langle t, t^2, t^3 \rangle \qquad \text{for } 0 \le t \le 1.$$

Solution We compute $\mathbf{r}'(t) = \langle 1, 2t, 3t^2 \rangle$, so

$$\oint_C \mathbf{F} \cdot \mathbf{T} \, ds = \int_0^1 \mathbf{F}(\mathbf{r}(t)) \cdot \mathbf{r}'(t) \, dt$$

$$= \int_0^1 \langle t^2, 2t, t^2 \rangle \cdot \langle 1, 2t, 3t^2 \rangle \, dt = \int_0^1 (t^2 + 4t + 3t^4) \, dt$$

$$= \frac{t^3}{3} + \frac{4t^3}{3} + \frac{3t^5}{5} \Big]_0^1 = \frac{1}{3} + \frac{3}{2} + \frac{3}{5} = \frac{34}{15}.$$

■

Other notations for line integrals

If $\mathbf{r}$ is a path $[a, b] \longrightarrow \mathbb{R}^2$, then we have

$$\mathbf{r}(t) = \langle x(t), y(t) \rangle$$

and

$$\mathbf{r}'(t) = \left\langle \frac{dx}{dt}, \frac{dy}{dt} \right\rangle.$$

Now, if our vector field $\mathbf{F}$ has the form

$$\mathbf{F}(x, y) = \langle P(x, y), Q(x, y) \rangle,$$

then the line integral of $\mathbf{F}$ along the image curve C can be written

$$\oint \mathbf{F} \cdot d\mathbf{s} = \int_a^b \langle P(x, y), Q(x, y) \rangle \cdot \left\langle \frac{dx}{dt}, \frac{dy}{dt} \right\rangle dt$$

$$= \int_a^b \left[P(x, y) \frac{dx}{dt} + Q(x, y) \frac{dy}{dt} \right] dt.$$

A shorthand notation for this form is simply

$$\oint_C P\,dx + Q\,dy.$$

Similarly, if $\mathbf{r}(t) = \langle x(t), y(t), z(t)\rangle$, and

$$\mathbf{F}(x,y,z) = \langle P(x,y,z), \ Q(x,y,z), \ R(x,y,z)\rangle,$$

then

$$\oint \mathbf{F}\cdot d\mathbf{s} = \oint_C P\,dx + Q\,dy + R\,dz.$$

EXAMPLE 25   Find $\displaystyle\oint_C x\,dy - y\,dx$, where C is parametrized by

$$\mathbf{r}(t) = \langle \cos t, \sin t\rangle \qquad \text{for } 0 \le t \le 2\pi.$$

Solution   We have

$$x = \cos t, \qquad \frac{dx}{dt} = -\sin t, \qquad y = \sin t, \qquad \frac{dy}{dt} = \cos t.$$

Therefore,

$$\oint_C x\,dy - y\,dx = \int_0^{2\pi} [\cos^2 t + \sin^2 t]\,dt = \int_0^{2\pi} 1\,dt = t\Big]_0^{2\pi} = 2\pi.$$

■

EXAMPLE 26   Find $\displaystyle\oint_C yz\,dx + xz\,dy + xy\,dz$, where C consists of straight line segments joining $(1,0,0)$ to $(0,1,0)$ to $(0,0,1)$ as shown in Figure 17.12.

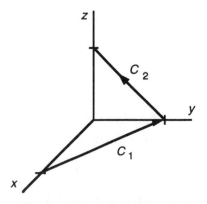

Figure 17.12 The smooth pieces of C are C_1 and C_2.

Solution   We will consider C in two parts. We can parametrize the line segment C_1 by

$$\mathbf{r}_1 : [0,1] \longrightarrow \mathbb{R}^3 \qquad \text{where } \mathbf{r}_1(t) = \langle 1 - t, t, 0 \rangle.$$

Note that $\mathbf{r}_1(0) = \langle 1, 0, 0 \rangle$, $\mathbf{r}_1(1) = \langle 0, 1, 0 \rangle$, and $\mathbf{r}_1'(t) = \langle -1, 1, 0 \rangle$.

We can parametrize the line segment C_2 by

$$\mathbf{r}_2 : [0,1] \longrightarrow \mathbb{R}^3 \qquad \text{where } \mathbf{r}_2(t) = \langle 0, 1 - t, t \rangle.$$

Note that $\mathbf{r}_2(0) = \langle 0, 1, 0 \rangle$, $\mathbf{r}_2(1) = \langle 0, 0, 1 \rangle$, and $\mathbf{r}_2'(t) = \langle 0, -1, 1 \rangle$. Thus,

$$\oint_C yz\,dx + xz\,dy + xy\,dz = \oint_{C_1} yz\,dx + xz\,dy + xy\,dz + \oint_{C_2} yz\,dx + xz\,dy + xy\,dz$$

$$= \oint_0^1 [t(0)(-1) + (1-t)(0)(1) + (1-t)(t)(0)]\,dt$$

$$+ \oint_0^1 [(1-t)(t)(0) + (0)(t)(-1) + (0)(1-t)(1)]\,dt$$

$$= \int_0^1 0\,dt + \int_0^1 0\,dt = 0 + 0 = 0.$$

■

Effects of orientation on line integrals

A single curve C can have infinitely many different parametrizations. What effect does the choice of parametrization have on the value of a line integral over a curve?

Suppose a curve C has a smooth parametrization $\mathbf{r}$ defined over the interval $[a, b]$. If $u(t)$ is a function such that $u(c) = a$, $u(d) = b$, and $u'(t) > 0$, then the new parametrization

$$\mathbf{p}(t) = \mathbf{r}(u(t)), \qquad \text{for } c \leq t \leq d$$

will trace out the curve C in the *same* direction as $\mathbf{r}$. We say $\mathbf{p}$ **preserves** the original orientation of C.

If $u(t)$ is a function such that $u(c) = b$, $u(d) = a$, and $u'(t) < 0$, then the new parametrization

$$\mathbf{q}(t) = \mathbf{r}(u(t)), \qquad \text{for } c \leq t \leq d$$

will trace out the curve in the *opposite* direction of $\mathbf{r}$. We say $\mathbf{q}$ **reverses** the original orientation of C. The effect of such a change in parametrization is different for line integrals of *scalar* and *vector* fields, as documented in the following theorem.

Theorem 17.1

Hypothesis 1: Curve C has smooth parametrization $\mathbf{r}$ defined on $[a, b]$.

Hypothesis 2: f is a differentiable scalar field defined at each point of C, and $\mathbf{F}$ is a differentiable vector field defined at each point of C.

Hypothesis 3: $\mathbf{p}$ is a smooth parametrization of C defined on $[c, d]$.

Conclusion 1: If $\mathbf{p}$ preserves the orientation of C given by $\mathbf{r}$, then

$$\int_a^b f(\mathbf{r}(t))\|\mathbf{r}'(t)\|\, dt \;=\; \int_c^d f(\mathbf{p}(t))\|\mathbf{p}'(t)\|\, dt$$

and

$$\int_a^b \mathbf{F}(\mathbf{r}(t)) \cdot \mathbf{r}'(t)\, dt \;=\; \int_c^d \mathbf{F}(\mathbf{p}(t)) \cdot \mathbf{p}'(t)\, dt.$$

Conclusion 2: If $\mathbf{q}$ reverses the orientation of C given by $\mathbf{r}$, then

$$\int_a^b f(\mathbf{r}(t))\|\mathbf{r}'(t)\|\, dt \;=\; \int_c^d f(\mathbf{q}(t))\|\mathbf{q}'(t)\|\, dt$$

and

$$\int_a^b \mathbf{F}(\mathbf{r}(t)) \cdot \mathbf{r}'(t)\, dt \;=\; -\int_c^d \mathbf{F}(\mathbf{q}(t)) \cdot \mathbf{q}'(t)\, dt.$$

In other words, $\oint_C f\, ds$ does not depend on the the parametrization of C, regardless of the orientation. However, the value of $\oint_C \mathbf{F} \cdot d\mathbf{s}$ changes sign when the orientation of C is reversed.

EXERCISES

Consider the straight line segment C connecting the point P_1 to the point P_2. Find a path $\mathbf{r}$ starting at P_1 and traveling along C to P_2, and use it to find

$$\oint_C f\, ds \quad \text{and} \quad \oint_C \mathbf{F} \cdot d\mathbf{s}$$

in each of exercises 1-6.

	P_1	P_2	$f(x,y,z)$	$\mathbf{F}(x,y,z)$
1.	$(1,1,1)$	$(2,2,2)$	$xy + yz$	$\langle x, y^2, z^2 \rangle$
2.	$(0,1,0)$	$(1,0,0)$	$xz + yz$	$\langle x^3, y, x^2 \rangle$
3.	$(0,-1,1)$	$(0,1,-1)$	$xy + z^2$	$\langle x^2, y^3, z \rangle$
4.	$(-1,-1,-1)$	$(1,1,1)$	$x^2 + z^2$	$\langle xy, y^2, yz \rangle$
5.	$(1,2,3)$	$(3,2,1)$	$y^2 + z^2$	$\langle x^2, xy, xz \rangle$

6. $(1, -1, 0)$ $(0, 0, 1)$ $x^2 + y^2$ $\langle xz, yz, z^2 \rangle$

In each of exercises 7-10, calculate

$$\oint_C \mathbf{F} \cdot \mathbf{T} \, ds,$$

where C is the curve parametrized by the given position function $\mathbf{r}$, and $\mathbf{F}$ is the given vector field.

7. $\mathbf{r} : [1, 2] \longrightarrow \mathbb{R}^3,$ where $\mathbf{r}(t) = \langle \dfrac{t^3}{3}, \dfrac{t^2}{\sqrt{2}}, t \rangle$

   $\mathbf{F}(x, y, z) = (\ln(xyz))\,\mathbf{i}$ $+$ $xyz\,\mathbf{j}$ $+$ $(x^2 + y^2 + z^2)\,\mathbf{k}$

8. $\mathbf{r} : [1, 2] \longrightarrow \mathbb{R}^3,$ where $\mathbf{r}(t) = \langle \dfrac{t^3}{3}, t^2, 4t \rangle$

   $\mathbf{F}(x, y, z) = (y^2 z^2)\,\mathbf{i}$ $+$ $\cos(x^2 + z^2)\,\mathbf{j}$ $+$ $\ln(xyz)\,\mathbf{k}$

9. $\mathbf{r} : [0, \pi] \longrightarrow \mathbb{R}^3,$ where $\mathbf{r}(t) = \langle \cos 2t, \ \sin 2t, \ t \rangle$

   $\mathbf{F}(x, y, z) = x^2 + y^2\,\mathbf{i}$ $+$ $z\,\mathbf{j}$ $+$ $\arctan(y/x)\,\mathbf{k}$

10. $\mathbf{r} : [1, e] \longrightarrow \mathbb{R}^3,$ where $\mathbf{r}(t) = \langle t^2, 2t, \ln t \rangle$

   $\mathbf{F}(x, y, z) = \tan(y^2 + z^2)\,\mathbf{i}$ $+$ $xy^2 z^3\,\mathbf{j}$ $+$ $e^{xyz}\,\mathbf{k}$

For exercises 11-22, consider the scalar field

$$f(x, y) = x^2 y^3,$$

the vector field

$$\mathbf{F}(x, y) = y\mathbf{i} - x\mathbf{j},$$

and the unit quarter circle C parametrized in three different ways:

$$\mathbf{r}(t) = \langle \cos t, \sin t \rangle \quad \text{where } 0 \le t \le \pi/2,$$

$$\mathbf{p}(t) = \langle \cos \pi t, \sin \pi t \rangle \quad \text{where } 0 \le t \le 1/2,$$

$$\mathbf{q}(t) = \langle \sin(t/2), \cos(t/2) \rangle \quad \text{where } 0 \le t \le \pi.$$

11.   Which of the parametrizations have the same orientation? Which have opposite orientation?

12.   If t represents time in seconds, and each of the parametrizations describes the path of a different particle, which particle moves slowest? Which moves fastest? Which particle would collide with the other two, and when?

13.   Find $\oint_C f \, ds$ using the parametrization $\mathbf{r}$.

14. Find $\oint_C f\, ds$ using the parametrization p.

15. Find $\oint_C f\, ds$ using the parametrization q.

16. Does the orientation of the curve affect the value of $\oint_C f\, ds$?

17. Find $\oint_C \mathbf{F} \cdot ds$ using the parametrization r.

18. Find $\oint_C \mathbf{F} \cdot ds$ using the parametrization p.

19. Find $\oint_C \mathbf{F} \cdot ds$ using the parametrization q.

20. Does the orientation of the curve affect the value of $\oint_C \mathbf{F} \cdot ds$?

21. The quarter circle C has endpoints at $(1,0)$ and $(0,1)$. Find $\oint_{C'} f\, ds$, where C' is the straight line segment connecting $(1,0)$ to $(0,1)$. Is the value different than over the quarter circle C?

22. Find $\oint_{C'} \mathbf{F} \cdot ds$, where C' is the straight line segment connecting $(1,0)$ to $(0,1)$. Is the value different than over the quarter circle C?

---

23. Thinking of a scalar field f as measuring the mass density along a wire in the shape of the curve C, explain why it makes sense that $\oint_C f\, ds$ *does not depend* on the orientation of C.

24. Thinking of a vector field F as measuring gravitational force at each point along the curve C, explain why it makes sense that $\oint_C f\, ds$ *does depend* on the orientation of C.

25. Suppose that $\mathbf{F}(\mathbf{r}(t))$ is orthogonal to $\mathbf{r}'(t)$ at every point $\mathbf{r}(t)$ along the curve C. Show that

$$\oint_C \mathbf{F} \cdot ds = 0.$$

26. Suppose that $\mathbf{F}(\mathbf{r}(t))$ points in the same direction as $\mathbf{r}'(t)$ at every point $\mathbf{r}(t)$ along the curve C. Show that

$$\oint_C \mathbf{F} \cdot ds = \oint_C \|\mathbf{F}\|\, ds.$$

17.4 CONSERVATIVE FIELDS AND POTENTIALS

In the calculus of scalar-valued functions of a single variable, we have two *Fundamental Theorems* that state the intimate relationship between derivatives and integrals. Let's recall what those important theorems say.

First Fundamental Theorem of Calculus: If f is continuous on $[a, b]$ then f has an antiderivative given by

$$A(x) = \int_a^x f(t)\,dt, \qquad a \leq x \leq b.$$

In other words, we are guaranteed that $A'(x) = f(x)$.

Second Fundamental Theorem of Calculus: If F is *any* antiderivative of f (in other words, $F'(x) = f(x)$) then

$$\int_a^b f(x)\,dx = F(b) - F(a).$$

This is really quite amazing: the value of the definite integral of f over the entire interval $[a, b]$ only depends on the value of its antiderivative at the two endpoints of the interval!

In this section, we are going to discuss what could be called the fundamental theorem of calculus for *line integrals*. First, we need a little terminology.

Conservative fields–independence of path

Definition 8

> A vector field $\mathbf{F}$ is said to be **conservative** in a region if for any piecewise smooth curve C in that region, the value of line integral
>
> $$\oint_C \mathbf{F} \cdot d\mathbf{s}$$
>
> depends only on the endpoints of the curve C.

If a vector field $\mathbf{F}$ is conservative, then for two different piecewise smooth curves C_1 and C_2 starting at the same point A and ending at the same point B, we have

$$\oint_{C_1} \mathbf{F} \cdot d\mathbf{s} = \oint_{C_2} \mathbf{F} \cdot d\mathbf{s}$$

(see Figure 17.13).

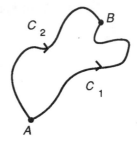

Figure 17.13 $\oint_{C_1} \mathbf{F} \cdot ds = \oint_{C_2} \mathbf{F} \cdot ds$ for a conservative vector field $\mathbf{F}$.

Let's examine this illustration a bit. Suppose we reverse the orientation of the parametrization of C_2 (so that the curve runs from B to A). If we call this curve $-C_2$ for convenience, then for a conservative vector field $\mathbf{F}$, we must have

$$\oint_{C_1} \mathbf{F} \cdot ds = - \oint_{-C_2} \mathbf{F} \cdot ds.$$

The two curves C_1 and $-C_2$ together form a piecewise smooth closed curve C (see Figure 17.14), and we must have

$$\oint_C \mathbf{F} \cdot ds = \oint_{C_1} \mathbf{F} \cdot ds \ + \ \oint_{-C_2} \mathbf{F} \cdot ds = 0.$$

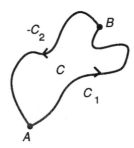

Figure 17.14 C_1 and $-C_2$ form a closed curve.

Conversely, starting with any piecewise smooth closed curve C, we could always break it into two pieces like this. We can conclude the following:

The vector field F is conservative in a region if and only if

$$\oint_C \mathbf{F} \cdot ds = 0$$

for every closed curve C in that region.

Let's put these remarks in a physical context. If $\mathbf{F}$ is a *conservative force field*, then the work performed in moving from point A to a point B depends only on the two points, and *not* on the particular path we travel. Moreover, no matter what distance we travel along the way, if we arrive at our original starting point (a closed curve), then *zero* work is performed.

EXAMPLE 27 Show that the vector field $\mathbf{F}(x, y) = y\mathbf{i} - x\mathbf{j}$ is *not conservative*.

Solution To show that $\mathbf{F}$ is not conservative, we need find only one example of a pair of curves C_1 and C_2 with the same initial point A and terminal point B, but with

$$\oint_{C_1} \mathbf{F} \cdot ds \neq \oint_{C_2} \mathbf{F} \cdot ds.$$

Suppose we connect $A = (1, 0)$ and $B = (0, 1)$ with the two curves illustrated in Figure 17.15.

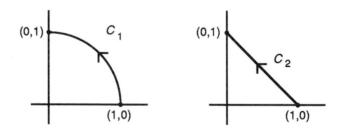

Figure 17.15 Two curves connecting $A = (1, 0)$ to $B = (0, 1)$.

1) The curve C_1 is parametrized by the path

$$\mathbf{r}_1(t) = \langle \cos t, \sin t \rangle \qquad \text{for } 0 \leq t \leq \pi/2.$$

This path traces out the unit quarter circle in the first quadrant.

2) The curve C_2 is parametrized by the path

$$\mathbf{r}_2(t) = \langle 1 - t, t \rangle \qquad \text{for } 0 \leq t \leq 1.$$

This path traces out the straight line segment connecting $(1, 0)$ to $(0, 1)$.

Now, let's calculate the two line integrals:

1) For C_1, we have $\mathbf{r}'_1(t) = \langle -\sin t, \cos t \rangle$, so

$$\oint_{C_1} \mathbf{F} \cdot d\mathbf{s} = \int_0^{\pi/2} \langle \sin t, -\cos t \rangle \cdot \langle -\sin t, \cos t \rangle \, dt$$

$$= \int_0^{\pi/2} (-\sin^2 t - \cos^2 t) \, dt$$

$$= \int_0^{\pi/2} (-1) \, dt = -t \Big]_0^{\pi/2} = -\pi/2.$$

2) For C_2, we have $\mathbf{r}'_2(t) = \langle -1, 1 \rangle$, so

$$\oint_{C_2} \mathbf{F} \cdot d\mathbf{s} = \int_0^1 \langle t, t-1 \rangle \cdot \langle -1, 1 \rangle \, dt$$

$$= \int_0^1 (-t + (t-1)) \, dt$$

$$= \int_0^1 (-1) \, dt = -t \Big]_0^1 = -1.$$

Since $-\pi/2 \neq -1$, we can see that the value of the line integrals differ, although C_1 and C_2 have the same initial and terminal points. We conclude that $\mathbf{F}$ is *not* conservative. ∎

This example illustrates a way one can tell that a vector field $\mathbf{F}$ is not conservative:

 If we can find a single example of a pair of curves C_1 and C_2 having the same initial and terminal endpoints, but

$$\oint_{C_1} \mathbf{F} \cdot d\mathbf{s} \neq \oint_{C_2} \mathbf{F} \cdot d\mathbf{s},$$

then $\mathbf{F}$ is *not* conservative.

However, even if one chooses two such curves and finds that

$$\oint_{C_1} \mathbf{F} \cdot d\mathbf{s} = \oint_{C_2} \mathbf{F} \cdot d\mathbf{s},$$

then we cannot conclude that $\mathbf{F}$ is conservative. The two values may be equal by coincidence, and we have no way of checking all of the infinitely many different pairs of curves connecting the same two points. There are simpler ways of detecting that a field $\mathbf{F}$ is conservative, as we will see.

Potentials

If we start with a differentiable scalar field f, we can obtain its gradient vector field ∇f. Now, let's reverse the situation: suppose we start with a vector field $\mathbf{F}$. Can we find a scalar field having $\mathbf{F}$ as its gradient? If so, we call this scalar field a **potential** for $\mathbf{F}$.

It's traditional to use the Greek letter phi (φ) to denote a potential scalar field, and using this notation we have the following definition.

Definition 9

A continuous vector field $\mathbf{F}$ has a **potential** φ if and only if

$$\nabla \varphi = \mathbf{F}.$$

If $\mathbf{F}$ is a 2-dimensional vector field with

$$\mathbf{F}(x,y) = \langle P(x,y), Q(x,y) \rangle$$

then $\varphi(x,y)$ is a potential for $\mathbf{F}$ if and only if

$$\frac{\partial \varphi}{\partial x} = P(x,y) \qquad \text{and} \qquad \frac{\partial \varphi}{\partial y} = Q(x,y).$$

Similarly, if $\mathbf{F}$ is a 3-dimensional vector field

$$\mathbf{F}(x,y,z) = P(x,y,z)\mathbf{i} + Q(x,y,z)\mathbf{j} + R(x,y,z)\mathbf{k}$$

then $\varphi(x,y,z)$ is a potential for $\mathbf{F}$ if and only if

$$\frac{\partial \varphi}{\partial x} = P(x,y,z), \qquad \frac{\partial \varphi}{\partial y} = Q(x,y,z), \qquad \frac{\partial \varphi}{\partial z} = R(x,y,z).$$

The potential φ is a kind of "antiderivative" for the vector field $\mathbf{F}$. Just as a continuous function of one variable has infinitely many antiderivatives (remember that an arbitrary constant can be added), if a vector field $\mathbf{F}$ has one potential φ, then it has infinitely many (again by simply adding on an arbitrary constant).

EXAMPLE 28   Determine a potential, if one exists, for the vector field

$$\mathbf{F}(x, y, z) = \langle 3x^2 y, x^3 + y^3, 2z \rangle.$$

Solution   We need a function $\varphi(x, y, z)$ such that

$$\nabla \varphi = \langle 3x^2 y, x^3 + y^3, 2z \rangle,$$

so we must have

$$\frac{\partial \varphi}{\partial x} = 3x^2 y, \qquad \frac{\partial \varphi}{\partial y} = x^3 + y^4, \qquad \frac{\partial \varphi}{\partial z} = 2z.$$

The first requirement $\dfrac{\partial \varphi}{\partial x} = 3x^2 y$ forces the function φ to be of the form

$$\varphi(x, y, z) = x^3 y + (\text{ expression in terms of } y \text{ and } z).$$

The second requirement $\dfrac{\partial \varphi}{\partial y} = x^3 + y^3$ forces the function φ to be of the form

$$\varphi(x, y, z) = x^3 y + \frac{y^4}{4} + (\text{ expression in terms of } x \text{ and } z).$$

The third requirement $\dfrac{\partial \varphi}{\partial z} = 2z$ forces $\varphi(x, y, z)$ to be of the form

$$\varphi(x, y, z) = z^2 + (\text{ expression in terms of } x \text{ and } y).$$

We note that any scalar field of the form

$$\varphi(x, y, z) = x^3 y + \frac{y^4}{4} + z^2 + C \qquad (C \text{ a constant})$$

satisfies all three requirements, and we can check that

$$\frac{\partial \varphi}{\partial x} = 3x^2 y, \qquad \frac{\partial \varphi}{\partial y} = x^3 + y^3, \qquad \frac{\partial \varphi}{\partial z} = 2z.$$

■

Potentials and line integrals of conservative vector fields enjoy a relationship very similar to that between antiderivatives and definite integrals of continuous single variable functions. If $\mathbf{F}$ has potential function in a region, then $\mathbf{F}$ will be conservative. Here's why:

Suppose $\mathbf{F} = \nabla \varphi$ and C is a smooth curve connecting points A and B. If $\mathbf{r}$ is a parametrization of the curve with $\mathbf{r}(a) = A$ and $\mathbf{r}(b) = B$.

$$\oint_C \mathbf{F} \cdot d\mathbf{s} = \int_a^b \mathbf{F}(\mathbf{r}(t)) \cdot \mathbf{r}'(t) \, dt$$

$$= \int_a^b \nabla \varphi(\mathbf{r}(t)) \cdot \mathbf{r}'(t) \, dt.$$

Examine the integrand closely. This is simply the derivative of

$$\varphi(\mathbf{r}(t))$$

since

$$\frac{d}{dt}\Big[\varphi(\mathbf{r}(t))\Big] = \nabla\varphi(\mathbf{r}(t)) \cdot \mathbf{r}'(t).$$

Since $\varphi(\mathbf{r}(t))$ is an antiderivative of $\nabla\varphi(\mathbf{r}(t)) \cdot \mathbf{r}'(t)$, we have

$$\int_a^b \nabla\varphi(\mathbf{r}(t)) \cdot \mathbf{r}'(t) = \varphi(\mathbf{r}(b)) - \varphi(\mathbf{r}(a)) = \varphi(B) - \varphi(A).$$

So the integral's value only depends on the two endpoints $\mathbf{r}(a) = A$ and $\mathbf{r}(b) = B$ and not on any of the other points $\mathbf{r}(t)$ $(a < t < b)$.

EXAMPLE 29   Find $\oint_C \mathbf{F} \cdot \mathbf{T} \, ds$ where

$$\mathbf{F}(x, y, z) = \langle 3x^2 y, x^3 + y^3, 2z \rangle,$$

and C is parametrized by

$$\mathbf{r}(t) = \langle \cos t, \sin t, \tan t \rangle \qquad \text{for } 0 \le t \le \pi/4.$$

Solution   In the last example, we saw that $\mathbf{F}$ has a potential function

$$\varphi(x, y, z) = x^3 y + \frac{y^4}{4} + z^2.$$

The initial point and terminal points of the curve C are:

$$\mathbf{r}(0) = \langle 1, 0, 0 \rangle \qquad \text{and} \qquad \mathbf{r}(\pi/4) = \langle \sqrt{2}/2, \sqrt{2}/2, 1 \rangle.$$

Therefore,

$$\oint_C \mathbf{F} \cdot \mathbf{T} \, ds = \varphi(\sqrt{2}/2, \sqrt{2}/2, 1) - \varphi(1, 0, 0)$$

$$= (\frac{1}{4} + \frac{1}{4} + 1) - (0 + 0 + 0) = \frac{3}{2}.$$

Knowing that $\mathbf{F}$ had a potential φ greatly simplified our task.   ∎

The fundamental theorem for line integrals

In fact, the *only* conservative vector fields are those that have potentials:

Theorem 17.2

> Vector field $\mathbf{F}$ is conservative if and only if $\mathbf{F} = \nabla\varphi$ for a scalar field φ.

Paraphrased, the theorem states that the line integral of $\mathbf{F}$ will depend only on the endpoints of the path if and only if $\mathbf{F}$ is the *gradient* of a scalar field φ.

When can we tell that a vector field is conservative? The fundamental theorem for line integrals tells us that the following statements are all equivalent:

$$\oint_C \mathbf{F} \cdot ds \text{ depends only on the endpoints of } C.$$

$$\oint_C \mathbf{F} \cdot ds = 0 \text{ for every closed curve } C.$$

$$\mathbf{F} = \nabla\varphi \text{ for some scalar potential } \varphi.$$

To determine that $\mathbf{F}$ is conservative, we need to demonstrate a potential φ. Here are some clues that can be used to tell whether or not to bother looking.

How to tell whether a 2-dimensional vector field is conservative:

If $\mathbf{F}$ is a C^1-differentiable vector field

$$\mathbf{F}(x, y) = \langle P(x, y), Q(x, y) \rangle$$

then P and Q have continuous partial derivatives. Now, if $\mathbf{F}$ has a potential φ, then

$$\frac{d\varphi}{dx} = P(x, y) \quad \frac{d\varphi}{dy} = Q(x, y)$$

and φ is C^2-differentiable. Let's look at the mixed second-order partials of φ:

$$\frac{\partial^2\varphi}{\partial y\,\partial x} = \frac{\partial P}{\partial y} \quad \frac{\partial^2\varphi}{\partial x\,\partial y} = \frac{\partial Q}{\partial x}.$$

These must match. In other words, for $\mathbf{F}$ to have a potential we must have

$$\frac{\partial P}{\partial y} = \frac{\partial Q}{\partial x}$$

or equivalently

$$\frac{\partial Q}{\partial x} - \frac{\partial P}{\partial y} = 0.$$

This won't *guarantee* that $\mathbf{F}$ has a potential but if $\dfrac{\partial Q}{\partial x} - \dfrac{\partial P}{\partial y} \neq 0$, then $\mathbf{F}$ is definitely *not* conservative.

How to tell whether a 3-dimensional vector field is conservative:

We can use the same type of reasoning for a 3-dimensional vector field $\mathbf{F}$. If $\mathbf{F}(x, y, z) = P(x, y, z)\mathbf{i} + Q(x, y, z)\mathbf{j} + R(x, y, z)\mathbf{k}$ is C^1-differentiable, then we'll need

$$\frac{\partial P}{\partial y} = \frac{\partial Q}{\partial x} \quad \frac{\partial P}{\partial z} = \frac{\partial R}{\partial x} \quad \frac{\partial Q}{\partial z} = \frac{\partial R}{\partial y}$$

in order for all the second-order partials of φ to match. But this also means that

$$\mathbf{curl}\ \mathbf{F} = \left(\frac{\partial R}{\partial y} - \frac{\partial Q}{\partial z}\right)\mathbf{i} + \left(\frac{\partial P}{\partial z} - \frac{\partial R}{\partial x}\right)\mathbf{j} + \left(\frac{\partial Q}{\partial x} - \frac{\partial P}{\partial y}\right)\mathbf{k} = \mathbf{0}.$$

Again, this doesn't guarantee that $\mathbf{F}$ is conservative, but if $\mathbf{curl}\ \mathbf{F} \neq 0$ then $\mathbf{F}$ is definitely *not* conservative.

EXAMPLE 30   Show that the vector field $\mathbf{F}(x, y) = y\,\mathbf{i} - x\,\mathbf{j}$ is *not conservative*.

Solution   We have $P(x, y) = y$ and $Q(x, y) = -x$. We simply note that

$$\frac{\partial Q}{\partial x} = -1 \neq 1 = \frac{\partial P}{\partial y},$$

so $\mathbf{F}$ cannot have a potential function, and therefore is not conservative. We noted this in an earlier example by directly showing that $\mathbf{F}$ was not path-independent for the purposes of line integrals.   ■

EXAMPLE 31   We note that for $\mathbf{F}(x, y, z) = \langle 3x^2y, x^3 + y^3, 2z\rangle$, we have

$$\mathbf{curl}\ \mathbf{F} = \nabla \times \mathbf{F}$$

$$= \begin{vmatrix} \mathbf{i} & \mathbf{j} & \mathbf{k} \\ \partial/\partial x & \partial/\partial y & \partial/\partial z \\ 3x^2y & x^3 + y^3 & 2z \end{vmatrix}$$

$$= (0 - 0)\mathbf{i} - (0 - 0)\mathbf{j} + (3x^2 - 3x^2)\mathbf{k}$$

$$= \langle 0, 0, 0\rangle.$$

This does not by itself prove that $\mathbf{F}$ is conservative. However, recall that earlier, we successfully found a potential function for $\mathbf{F}$:

$$\varphi(x, y, z) = x^3y + \frac{y^4}{4} + z^2.$$

■

Relationship between curl and line integrals

A closed oriented curve lying in a plane determines a particular normal direction by use of a "right hand rule." If you place your right hand so that your fingers point in the direction of the curve, with your palm facing the interior of the plane region enclosed by the curve, then your thumb will extend in the normal direction determined by the curve.

Figure 17.16 illustrates this right hand rule for three curves in planes parallel to the coordinates planes. For each of the curves C_1, C_2, C_3, we have indicated the unit normal vector n pointing in the direction determined by the right hand rule.

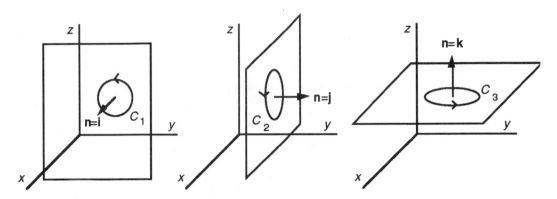

Figure 17.16 Right hand rule for closed plane curves.

Given a vector field

$$\mathbf{F}(x, y, z) = P(x, y, z)\mathbf{i} + Q(x, y, z)\mathbf{j} + R(x, y, z)\mathbf{k},$$

we can express curl F in terms of line integrals along closed curves. Recall that if F represents a velocity vector field, we can think of

$$\oint_C \mathbf{F} \cdot \mathbf{T} \, ds$$

as giving us the circulation along the curve C. Now, if C is a very tiny closed plane curve around a point (x_0, y_0, z_0), and we measure the area ΔS enclosed by the curve, then the component of curl $\mathbf{F}(x_0, y_0, z_0)$ in the direction determined by the curve is approximately the circulation divided by the area enclosed. In other words, if n is a unit normal vector determined by the curve C, then

$$\text{curl } \mathbf{F}(x_0, y_0, z_0) \cdot \mathbf{n} \Delta S \approx \oint_C \mathbf{F} \cdot \mathbf{T} \, ds.$$

The approximation becomes better and better as $\Delta S \to 0$. In fact,

$$\text{curl } \mathbf{F}(x_0, y_0, z_0) \cdot \mathbf{n} = \lim_{\Delta S \to 0} \frac{1}{\Delta S} \oint_C \mathbf{F} \cdot \mathbf{T} \, ds.$$

EXAMPLE 32   Suppose (x_0, y_0, z_0) is a point enclosed by the curves C_1, C_2, C_3 in the planes $x = x_0$, $y = y_0$, and $z = z_0$, respectively. If the curves are oriented as shown in Figure 17.16, the unit normal vectors are **i, j,** and **k**. The discussion above implies that

$$\frac{1}{\Delta S} \oint_{C_1} \mathbf{F} \cdot \mathbf{T} \, ds \approx \frac{\partial R}{\partial y} - \frac{\partial Q}{\partial z},$$

$$\frac{1}{\Delta S} \oint_{C_2} \mathbf{F} \cdot \mathbf{T} \, ds \approx \frac{\partial P}{\partial z} - \frac{\partial R}{\partial x},$$

$$\frac{1}{\Delta S} \oint_{C_3} \mathbf{F} \cdot \mathbf{T} \, ds \approx \frac{\partial Q}{\partial x} - \frac{\partial P}{\partial y}.$$

The components of **curl F** are sometimes defined as the limits of these approximations. ∎

EXERCISES

Consider the vector field

$$\mathbf{F}(x, y, z) = ye^{xy}\mathbf{i} + \left(xe^{xy} + \frac{1}{y}\right)\mathbf{j}.$$

Which of the statements in exercises 1-8 are true, and which are false? Explain your answer in each case.

1.   **F** is a conservative vector field.

2.   div **F** = 0.

3.   curl **F** = 0.

4. $\oint_C ye^{xy} \, dx + \left(xe^{xy} + \frac{1}{y}\right) dy = 0$ for any simple, closed, smooth curve C.

5.   **F** = grad f for some function f.

6.   **F** is irrotational at every point.

7.   div(**curl F**) = 0.

8.   curl (div **F**) = 0.

Suppose $\mathbf{F}(x, y, z) = \langle yz, xz, xy \rangle$. *Which of the statements in exercises 9-13 are true and which are false? Explain your reasons.*

9. $\mathbf{F} = \nabla \varphi$ for some scalar field φ.

10. $\nabla \times \mathbf{F} = 0$.

11.   **F** is conservative.

12.   **F** is irrotational at every point (x, y, z) .   13. $\oint_C \mathbf{F} \cdot \mathbf{T} \, ds = 0$ for every simple, smooth, closed curve C.

Determine whether or not the following vector fields in exercises 14-17 are conservative (explain your reasoning). If so, find a potential function for the vector field.

14. $\mathbf{F}(x, y) = (e^x \sin y + 8x)\mathbf{i} + (e^x \cos y + 1)\mathbf{j}.$

15. $\mathbf{G}(x, y) = (\ln y \cos x)\mathbf{i} + (y \sin x)\mathbf{j}.$

16. $\mathbf{H}(x, y) = \langle e^x - \cos y, e^y - x \sin y \rangle.$

17. $\mathbf{L}(x, y, z) = \frac{y^4}{4x}\mathbf{i} + y^3 \ln x \mathbf{j} + \cos z \mathbf{k}.$

18. Find the work done in moving a particle directly from $(-2, 0)$ to $(2, 0)$ along the x-axis by the force $\mathbf{F}(x, y) = \langle x^2, -y \rangle.$

19. Find the work done in moving a particle along the semicircle

$$x^2 + y^2 = 4, \qquad y \geq 0,$$

from $(-2, 0)$ to $(2, 0)$ by the same force $\mathbf{F}(x, y) = \langle x^2, -y \rangle.$

20. Consider

$$\mathbf{F}(x, y) = P(x, y)\mathbf{i} + Q(x, y)\mathbf{j} = \frac{-y}{x^2 + y^2}\mathbf{i} + \frac{x}{x^2 + y^2}\mathbf{j}.$$

Show that

$$\frac{\partial Q}{\partial x} = \frac{\partial P}{\partial y},$$

but that the integral

$$\oint_C P(x, y)\, dx + Q(x, y)\, dy \neq 0$$

over the closed curve C, where C is the unit circle parametrized by $\mathbf{r}(t) = \langle \cos t, \sin t \rangle$ for $0 \leq t \leq 2\pi$. What has gone wrong?

21. Show that the gravitational force field described in Example 1 has a potential

$$\varphi(x, y, z) = \frac{GMm}{\sqrt{x^2 + y^2 + z^2}}.$$

This is called the *Newtonian potential*.

17.5 PARAMETRIZED SURFACES

A position function $\mathbf{r}(t) = \langle x(t), y(t), z(t) \rangle$ parametrizes a *curve* in space ($\mathbb{R}^3$). We can think of a particle constrained to move along this curve as having "one degree of freedom" represented by the single parameter t.

In contrast, we think of a particle constrained to move on a *surface* as having *two* degrees of freedom of movement. For example, when we move on the surface of the earth, both our *latitude* and *longitude* can change. Accordingly, a parametrization of a surface requires two parameters, say u and v.

Definition 10

A surface S in space $\mathbb{R}^3$ is the image of a position function

$$\mathbf{r} : D \longrightarrow \mathbb{R}^3$$

$$\mathbf{r}(u, v) = \langle x(u, v), y(u, v), z(u, v) \rangle$$

where D is some domain in the uv-plane.

The parameters u and v provide a sort of "coordinate system" for the surface. You have seen maps of regions of the earth where the latitude and longitude lines appear to be horizontal and vertical, respectively. This flat map corresponds to the uv-plane where u is latitude and v is longitude.

The position function $\mathbf{r}$ in this case associates or *maps* each ordered pair (u, v) to a corresponding point on the earth's surface. This correspondence allows us to use the coordinates (u, v) to chart our position and navigate about the surface.

To calculate the effects of a scalar or vector field on a surface, we'll need the navigational system provided by a parametrization. In this section we'll illustrate some techniques for parametrizing surfaces.

Parametrizing a function graph

To parametrize a surface S in space we need to describe two things explicitly:

1) the coordinate functions x, y, z of the position function $\mathbf{r}$

$$\mathbf{r}(u, v) = \langle x(u, v), y(u, v), z(u, v) \rangle$$

2) the domain D of points (u, v) corresponding to points $\mathbf{r}(u, v)$ on the surface S.

If our surface S is that of a function graph

$$z = f(x, y)$$

over some domain in the xy-plane, then it is an easy matter to parametrize the surface in terms of u and v. Simply let

$$x = u$$

$$y = v$$

$$z = f(u, v)$$

and let D be the domain of the function f in the uv-plane.

EXAMPLE 33 Parametrize the upper hemisphere of the unit sphere: $\{(x, y, z) : x^2+y^2+z^2 = 1, \quad z \geq 0\}$.

Solution The surface can be described as the graph of the function $z = \sqrt{1 - x^2 - y^2}$ over the unit disk (the unit circle and its interior) in the xy-plane.

Hence, our parametrization is

$$\mathbf{r}(u, v) = \langle u, v, \sqrt{1 - u^2 - v^2} \rangle$$

for $D = \{(u, v) : u^2 + v^2 \leq 1\}$. Figure 17.17 illustrates the parametrization $\mathbf{r}$ and its domain D. ■

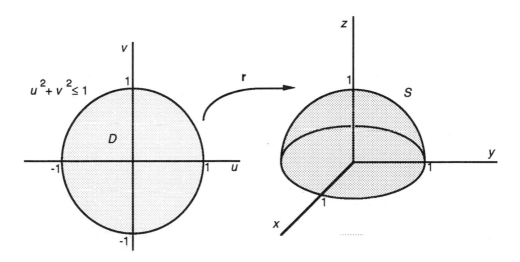

Figure 17.17 $\mathbf{r}(u, v) = \langle u, v, \sqrt{1 - u^2 - v^2} \rangle$ for D in the uv-plane.

Any surface having an equation where one of the coordinates x, y, z is expressed in terms of the other two can be parametrized in a similar way.

Using cylindrical or spherical coordinates to parametrize a surface

Cylindrical or spherical coordinates can also prove useful in parametrizing a surface, if the equation of the surface allows us to express each coordinate x, y and z in terms of two cylindrical or two spherical parameters. We'll illustrate by parametrizing the surface of the previous example in two more ways.

EXAMPLE 34 Parametrize the upper hemisphere of the unit sphere using cylindrical coordinates.

Solution Since $x = r\cos\theta$ and $y = r\sin\theta$ and $z = \sqrt{1 - x^2 - y^2} = \sqrt{1 - r^2}$, we have all three coordinates in terms of the parameters r and θ:

$$\mathbf{p}(r, \theta) = \langle r\cos\theta, r\sin\theta, \sqrt{1 - r^2}\rangle$$

where $0 \le r \le 1$ and $0 \le \theta \le 2\pi$. The domain of our parametrization is actually a rectangle in the $r\theta$-plane (see Figure 17.18). ∎

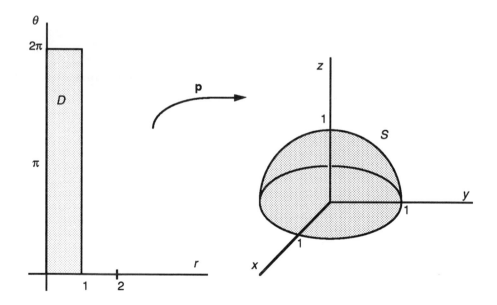

Figure 17.18 $\mathbf{p}(r, \theta) = \langle r\cos\theta, r\sin\theta, \sqrt{1 - r^2}\rangle$ for D in the $r\theta$-plane.

EXAMPLE 35 Parametrize the upper hemisphere of the unit sphere using spherical coordinates .

Solution Since $\rho = 1$ on the unit sphere, we have

$$x = \rho \sin \varphi \cos \theta = \sin \varphi \cos \theta$$

$$y = \rho \sin \varphi \sin \theta = \sin \varphi \sin \theta$$

$$z = \rho \cos \varphi = \cos \varphi.$$

Hence all three coordinates are in terms of θ and φ

$$\mathbf{q}(\theta, \varphi) = \langle \sin \varphi \cos \theta, \sin \rho \sin \theta, \cos \varphi \rangle$$

where $0 \le \theta \le 2\pi$ and $0 \le \varphi \le \frac{\pi}{2}$. The domain of this parametrization is a rectangle in the $\theta\phi$-plane (see Figure 17.19). ■

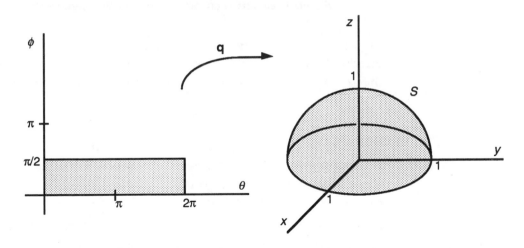

Figure 17.19 $\mathbf{q}(\theta, \varphi) = \langle \sin \varphi \cos \theta, \sin \rho \sin \theta, \cos \varphi \rangle$ for D in the $\theta\phi$-plane.

Graphing parametrized surfaces

A parametrized surface can be plotted using a technique similar to the slicing method discussed in Chapter 14. Suppose S is parametrized by

$$\mathbf{r}(u, v) = \langle x(u, v), y(u, v), z(u, v) \rangle$$

over some domain D in the uv-plane. If we set u or v equal to some constant value c, we obtain a parametrized *curve* in space.

For example, if we set $u = 3$, then

$$\mathbf{r}(3, v) = \langle x(3, v), y(3, v), z(3, v) \rangle$$

will trace out a curve in space for values of the single remaining parameter v. Similarly, if we set $v = -2$, then

$$\mathbf{r}(u, -2) = \langle x(u, -2), y(u, -2), z(u, -2) \rangle$$

traces out another curve in space for values of the single parameter u.

These curves are simply the images of vertical ($u = c$) and horizontal ($v = c$) grid lines in the uv-plane (see Figure 17.20).

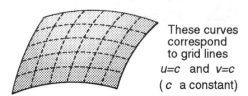

These curves
correspond
to grid lines
$u=c$ and $v=c$
(c a constant)

Figure 17.20 Image curves of grid lines on a parametrized surface S.

If a machine plots several image curves (or their piecewise linear approximations) of uv-grid lines, the result resembles a wireframe model of the surface. Indeed, for a function graph $z = f(x, y)$ parametrized by

$$\mathbf{r}(u, v) = \langle u, v, f(u, v) \rangle,$$

the image curves of the grid lines of the domain D correspond exactly to x-slices and y-slices of the function graph. For other parametrizations, these image curves of the grid lines of the domain D need not correspond to cross-sections of the surface S.

EXAMPLE 36   Describe the image curves of horizontal and vertical grid lines of the domain D in each of the parametrizations of the upper unit hemisphere of the three previous examples.

Solution   For the parametrization

$$\mathbf{r}(u, v) = \langle u, v, \sqrt{1 - u^2 - v^2} \rangle,$$

the image curve of $u = c$ for $-1 \le c \le 1$ is a vertical cross-section of the hemisphere cut parallel to the yz-plane. The image curve of $v = c$ for $-1 \le c \le 1$ is a vertical cross-section of the hemisphere cut parallel to the xz-plane. Each of these curves is in the shape of the upper half of an ellipse.

For the parametrization in cylindrical coordinates

$$\mathbf{p}(r, \theta) = \langle r \cos \theta, r \sin \theta, \sqrt{1 - r^2} \rangle,$$

the image curve of $r = c$ for $0 \leq r \leq 1$ is a circular horizontal cross-section of the hemisphere cut parallel to the xy-plane. The image curve of $\theta = c$ for $0 \leq \theta \leq 2\pi$ is a unit quarter circle extending from the base of the hemisphere (the unit circle in the xy-plane) up to the "North Pole" $(0, 0, 1)$.

For the parametrization in spherical coordinates

$$\mathbf{q}(\theta, \varphi) = \langle \sin \varphi \cos \theta, \sin \rho \sin \theta, \cos \varphi \rangle,$$

the image curve of $\theta = c$ is the same as for cylindrical coordinates. The image curve of $\varphi = c$ for $0 \leq \varphi \leq \pi/2$ is a circular horizontal cross-section of the hemisphere cut parallel to the xy-plane. The values of θ and ϕ are the equivalent of longitude and latitude lines. ∎

Smooth surfaces, normal vectors, and tangent planes

If S is a surface parametrized by a C^1-differentiable position function

$$\mathbf{r}(u, v) = \langle x(u, v), y(u, v), z(u, v) \rangle,$$

then two vectors can be computed at each point $\mathbf{r}(u, v)$:

$$\mathbf{T}_u = \frac{\partial \mathbf{r}}{\partial u} = \left\langle \frac{\partial x}{\partial u}, \frac{\partial y}{\partial u}, \frac{\partial z}{\partial u} \right\rangle$$

and

$$\mathbf{T}_v = \frac{\partial \mathbf{r}}{\partial v} = \left\langle \frac{\partial x}{\partial v}, \frac{\partial y}{\partial v}, \frac{\partial z}{\partial v} \right\rangle.$$

If these vectors are nonzero, then they are tangent to the surface S at the point $\mathbf{r}(u, v)$ (see Figure 17.21).

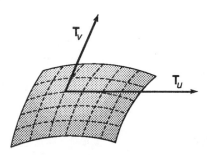

Figure 17.21 $\mathbf{T}_u = \partial \mathbf{r}/\partial u$ and $\mathbf{T}_v = \partial \mathbf{r}/\partial v$ are tangent vectors to S.

The vector

$$\mathbf{T}_u \times \mathbf{T}_v$$

is called the **fundamental vector product** of the parametrization r. Every point at which $T_u \times T_v \neq 0$ is called a **smooth point**. A point at which r is not C^1-differentiable or where $T_u \times T_v = 0$ is called a **singular point**.

At a smooth point, the vector $T_u \times T_v$ is normal to the surface S (see Figure 17.22).

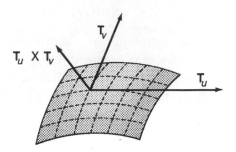

Figure 17.22 $T_u = \partial r / \partial u$ and $T_v = \partial r / \partial v$ are tangent vectors to S.

We can use the fundamental vector product to find the equation of the tangent plane to the surface at a smooth point. Recall, if (x_0, y_0, z_0) is a point in a plane and $\langle a, b, c \rangle$ is a vector normal to the plane, then for any point (x, y, z) in the plane we have $\langle a, b, c \rangle \cdot \langle x - x_0, y - y_0, z - z_0 \rangle = 0$, so

$$a(x - x_0) + b(y - y_0) + c(z - z_0) = 0.$$

EXAMPLE 37 Find T_u, T_v, the fundamental vector product $T_u \times T_v$, and the equation of the tangent plane to the surface parametrized by

$$r(u, v) = \langle uv, u - v, \frac{u}{v} \rangle$$

at the point $r(2, -1)$.

Solution First, we calculate $T_u = \dfrac{\partial r}{\partial u} = \langle v, 1, \frac{1}{v} \rangle$ and $T_v = \dfrac{\partial r}{\partial v} = \langle u, -1, \frac{-u}{v^2} \rangle$.

The fundamental vector product is

$$T_u \times T_v = \begin{vmatrix} i & j & k \\ v & 1 & 1/v \\ u & -1 & -u/v^2 \end{vmatrix}$$

$$= \langle \frac{-u}{v^2} + \frac{1}{v}, \frac{u}{v} + \frac{u}{v}, -v - u \rangle$$

$$= \langle \frac{v - u}{v^2}, \frac{2u}{v}, -v - u \rangle.$$

At $(u, v) = (2, -1)$, we have

$$T_u = \langle -1, 1, -1 \rangle \qquad T_v == \langle 2, -1, -2 \rangle \qquad T_u \times T_v = \langle -3, 4, -1 \rangle.$$

The equation of the tangent plane to the surface at $\mathbf{r}(2, -1) = (-2, 3, -2)$ is

$$-3(x + 2) + 4(y - 3) - (z + 2) = 0,$$

or, after expanding and collecting constant terms,

$$-3x + 4y - z - 20 = 0.$$

■

If the surface S is the graph of a function $z = f(x, y)$ then we can parametrize the surface as $\mathbf{r}(u, v) = \langle u, v, f(u, v) \rangle$. If f is C^1-differentiable, we have

$$\mathbf{T}_u = \frac{\partial \mathbf{r}}{\partial u} = \langle 1, 0, \frac{\partial f}{\partial u} \rangle$$

$$\mathbf{T}_v = \frac{\partial \mathbf{r}}{\partial v} = \langle 0, 1, \frac{\partial f}{\partial v} \rangle$$

$$\mathbf{T}_u \times \mathbf{T}_v = \frac{\partial \mathbf{r}}{\partial u} \times \frac{\partial \mathbf{r}}{\partial v} = \left\langle -\frac{\partial f}{\partial u}, -\frac{\partial f}{\partial v}, 1 \right\rangle.$$

EXAMPLE 38 Find a *unit* normal vector and the equation of the tangent plane to the graph of $z = x^2 y^2$ at the point $(-1, 2, 4)$.

Solution We can parametrize the surface with

$$\mathbf{r}(u, v) = \langle u, v, u^2 v^2 \rangle,$$

so

$$\mathbf{T}_u \times \mathbf{T}_v = \langle -2uv^2, -2u^2v, 1 \rangle.$$

At the point $(-1, 2, 4)$ we have $u = -1$ and $v = 2$, so

$$\mathbf{T}_u \times \mathbf{T}_v \Big|_{(u,v)=(-1,2)} = \langle 8, -4, 1 \rangle,$$

and a unit normal vector is given by

$$\frac{\langle 8, -4, 1 \rangle}{\sqrt{8^2 + (-4)^2 + 1^2}} = \langle \frac{8}{9}, -\frac{4}{9}, \frac{1}{9} \rangle.$$

An equation of the tangent plane is

$$8(x + 1) - 4(y - z) + (z - 4) = 0.$$

■

EXAMPLE 39   A surface S is parametrized by

$$\mathbf{r}(u,v) = (u^2 - v^2, u + v, u^2 + 4v)$$

over $D = \{(u,v) : \ 0 \le u \le 1, \ 0 \le v \le 1\}$. Find $\mathbf{T}_u$, $\mathbf{T}_v$, the fundamental vector product $\mathbf{T}_u \times \mathbf{T}_v$, and the equation of the tangent plane to the surface at the point $\mathbf{r}(0, 1/2) = (-1/4, 1/2, 2)$.

Solution   We have

$$\mathbf{T}_u = \langle 2u, 1, 2u \rangle, \qquad \mathbf{T}_v = \langle -2v, 1, 4 \rangle,$$

so that

$$\mathbf{T}_u \Big|_{(0,1/2)} = \langle 0, 1, 0 \rangle, \qquad \mathbf{T}_v \Big|_{(0,1/2)} = \langle -1, 1, 4 \rangle.$$

The fundamental vector product at this point is

$$(\mathbf{T}_u \times \mathbf{T}_v) \Big|_{(0,1/2)} = \begin{vmatrix} \mathbf{i} & \mathbf{j} & \mathbf{k} \\ 0 & 1 & 0 \\ -1 & 1 & 4 \end{vmatrix} = 4\mathbf{i} - 0\mathbf{j} + 1\mathbf{k} = \langle 4, 0, 1 \rangle.$$

The equation of tangent plane is

$$4(x + 1/4) + 0(y - 1/2) + 1(z - 2) = 0,$$

or, after collecting terms:

$$4x + z - 1 = 0.$$

■

EXAMPLE 40   Find the equation of the tangent plane to $z = 3x^2 + 8xy$ at the point $(1, 0, 0)$.

Solution   First, we need to parametrize the surface. If we let $x = u$, $y = v$, then $z = 3u^2 + 8uv$, so

$$\mathbf{r}(u,v) = \langle u, v, 3u^2 + 8uv \rangle.$$

We can compute

$$\mathbf{T}_u = \langle 1, 0, 6u + 8v \rangle, \qquad \mathbf{T}_v = \langle 0, 1, 8u \rangle.$$

At $(u,v) = (0,1)$, we have $\mathbf{r}(1,0) = \langle 1, 0, 3 \rangle$, and

$$\mathbf{T}_u = \langle 1, 0, 6 \rangle, \qquad \mathbf{T}_v = \langle 0, 1, 8 \rangle.$$

The fundamental vector product is

$$\mathbf{T}_u \times \mathbf{T}_v = \begin{vmatrix} \mathbf{i} & \mathbf{j} & \mathbf{k} \\ 1 & 0 & 6 \\ 0 & 1 & 8 \end{vmatrix} = -6\mathbf{i} - 8\mathbf{j} + 1\mathbf{k}.$$

The equation of the tangent plane is $-6x - 8y + z + 3 = 0$. ∎

EXAMPLE 41 Find the equation of the tangent plane to $x^3 + 3xy + z^2 = 2$ at the point $(1, 1/3, 0)$.

Solution First, we need to parametrize the surface. Let $x = u$ and $z = v$. Then

$$3xy = 2 - z^2 - x^3$$

and

$$y = \frac{2}{3x} - \frac{z^2}{3x} - \frac{x^2}{3} = \frac{2}{3u} - \frac{v^2}{3u} - \frac{u^2}{3}.$$

Now we can write the parametrization as

$$\mathbf{r}(u, v) = \langle u, \frac{2}{3u} - \frac{v^2}{3u} - \frac{u^2}{3}, v \rangle.$$

We can compute

$$\mathbf{T}_u = \langle 1, -\frac{2}{3u^2} + \frac{v^2}{3u^2} - \frac{2u}{3}, 0 \rangle$$

$$\mathbf{T}_v = \langle 0, \frac{2v}{3u}, 1 \rangle.$$

At $(u, v) = (1, 0)$, we have

$$\mathbf{r}(1, 0) = \langle 1, 1/3, 0 \rangle, \quad \mathbf{T}_u = \langle 1, -4/3, 0 \rangle, \quad \mathbf{T}_v = \langle 0, 0, 1 \rangle.$$

The fundamental vector product is

$$\mathbf{T}_u \times \mathbf{T}_v = \begin{vmatrix} \mathbf{i} & \mathbf{j} & \mathbf{k} \\ 1 & -4/3 & 0 \\ 0 & 0 & 1 \end{vmatrix}$$

$$= -4/3\mathbf{i} - 1\mathbf{j} + 0\mathbf{k} = \langle -4/3, -1, 0 \rangle.$$

An equation for the tangent plane at $(1, 1/3, 0)$ is

$$-4x/3 - y + 5/3 = 0 \qquad \text{or} \; -4x - 3y + 5 = 0.$$

∎

EXERCISES

Parametrize the surfaces described in exercises 1-6 as the image of

$$\mathbf{r} : D \longrightarrow \mathbb{R}^3$$

where D is a subset of $\mathbb{R}^2$ (the uv-plane). Sketch the region D.

1. $z = x^3 + y^2$; $x^2 + y^2 \leq 1$.

2. $x = z^2 y^2$; $0 \leq z \leq 1$; $0 \leq y \leq z$.

3. $y = \sqrt{x + z}$; $0 \leq x \leq z$; $0 \leq z \leq 1$.

4. $x^3 + y^3 + z = 0$; $x^2 + y^2 \leq 4$.

5. $x + 2y^2 - z^2 = 3$; $y \geq 0$; $z \geq 0$; $y^2 + z^2 \leq 1$.

6. $x^2 + y + z^3 = 10$; $x \geq 0$; $z \leq 0$; $x^2 + z^2 \leq 4$.

Suppose a surface S is parametrized by $\mathbf{r}$ over the domain D as indicated in each of exercises 7-13. Find the following:

$$\mathbf{T}_u, \qquad \mathbf{T}_v, \qquad \mathbf{T}_u \times \mathbf{T}_v,$$

and the equation of the tangent plane to the surface at the indicated point.

7. $\mathbf{r}(u, v) = \langle v \cos u, v \sin u, v \rangle$; $\mathbf{r}(\pi, -\pi)$.

8. $\mathbf{r}(u, v) = \langle u \cos v, u \sin v, u \rangle$; $\mathbf{r}(0, \pi/2)$.

9. $\mathbf{r}(u, v) = \langle \cos u \sin u, v \rangle$; $\mathbf{r}(0, 1)$.

10. $\mathbf{r}(u, v) = \langle \cos v \sin v, u \rangle$; $\mathbf{r}(1, \pi/4)$.

11. $\mathbf{r}(u, v) = \langle u + v, u - v, v - u \rangle$; $\mathbf{r}(-1, 2)$.

12. $\mathbf{r}(u, v) = \langle u - v, v - u, u + v \rangle$; $\mathbf{r}(1, -1)$.

13. $\mathbf{r}(u, v) = \langle u^2, v^2, u^2 + v^2 \rangle$; $\mathbf{r}(1, 1)$.

---

14. Find a unit normal vector to the surface in exercise 1 at the origin $(0, 0, 0)$.

15. Find a unit normal vector to the surface in exercise 2 at the point $(1/16, 1/2, 1/2)$.

16. Find a unit normal vector to the surface in exercise 3 at the point $(1/8, 1/2, 1/8)$.

17. Parametrize the surface in exercise 1 using cylindrical coordinates.

18. Parametrize the surface in exercise 4 using cylindrical coordinates.

19. Find the equation of the tangent plane to the surface in exercise 5 at the point $(5/2, 1/2, 0)$.

20. Find the equation of the tangent plane to the surface in exercise 6 at the point $(-1, 8, 1)$.

17.6 SURFACE INTEGRALS

A *surface integral* is to a parametrized surface S what a *line integral* is to a parametrized curve C. In this section we'll define surface integrals, explain some of their physical interpretations, and illustrate how they are computed.

Finding the surface area of a parametrized surface

In the last chapter, we noted how the double integral

$$\iint_D 1 \, dA$$

simply measures the area of the region D.

Suppose

$$\mathbf{r} : D \longrightarrow \mathbb{R}^3,$$

where

$$\mathbf{r}(u, v) = \langle x(u, v), y(u, v), z(u, v) \rangle$$

is a parametrization of a surface S. How might we go about measuring the *surface area* of S?

If we consider a fine rectangular grid partition of the domain D in the uv-plane, we could "follow" its image to the surface S.

If the vertical grid lines of D are spaced Δu apart and the horizontal lines are spaced Δv apart, then each tiny rectangle of the original grid has area

$$\Delta A = \Delta u \Delta v.$$

Now, we want to examine the image of one of these grid rectangles under the parametrization $\mathbf{r}$ (see Figure 17.23).

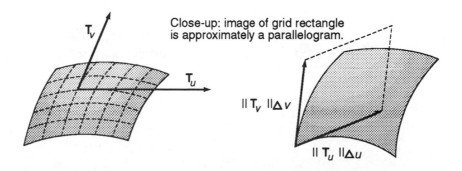

Figure 17.23 Approximating surface area using the image grid.

Provided the parametrization **r** is smooth and the partition grid of D is fine enough, the images of the sides of the rectangle will be approximately straight line segments (but not necessarily at right angles to each other). This suggests that we could approximate the image of the rectangle R with a *parallelogram*. Now, the grid lines correspond to holding either u constant (the vertical lines) or v constant (the horizontal lines). Hence the sides of the parallelogram are approximated by

$$\|\frac{\partial \mathbf{r}}{\partial u}\Delta u\| = \|\mathbf{T}_u\|\Delta u \quad \text{and} \quad \|\frac{\partial \mathbf{r}}{\partial v}\Delta v\| = \|\mathbf{T}_v\|\Delta v.$$

The area of this tiny parallelogram is

$$\Delta S \approx \|\mathbf{T}_u\|\, \|\mathbf{T}_v\|\sin\theta\Delta u\Delta v,$$

where θ is the angle between $\mathbf{T}_u$ and $\mathbf{T}_v$.

But the length of the fundamental vector product is

$$\|\mathbf{T}_u \times \mathbf{T}_v\| = \|\mathbf{T}_u\|\, \|\mathbf{T}_v\|\sin\theta,$$

so we can write

$$\Delta S \approx \|\mathbf{T}_u \times \mathbf{T}_v\|\Delta u\Delta v.$$

If we add up these parallelogram surface area approximations we obtain an approximation of the entire surface area of S. The finer the partition, the better the approximation, and we are led to defining the surface area of S as follows:

Definition 11

> If S is a surface with smooth parametrization $\mathbf{r} : D \longrightarrow \mathbb{R}^3$, then the **surface area** of S is
>
> $$\iint\limits_{S} dS = \iint\limits_{D} \|\mathbf{T}_u \times \mathbf{T}_v\|\, du\, dv.$$

The double integral symbol $\iint$ reminds us that we are integrating over a surface requiring two parameters. The notation dS indicates a *surface area element* (just as ds indicates an *arc length element*).

EXAMPLE 42 Suppose the surface S is parametrized by

$$\mathbf{r}(u, v) = \langle u\cos v, u\sin v, u^2 \rangle; \qquad 0 \leq u \leq 1, \quad 0 \leq v \leq 2\pi.$$

Find the surface area of S.

Solution First, we need to find the length of the fundamental vector product:

$$\mathbf{T}_u = \langle \cos v, \sin v, 2u \rangle \qquad \mathbf{T}_v = \langle -u\sin v, u\cos v, 0 \rangle$$

and

$$\mathbf{T}_u \times \mathbf{T}_v = \begin{vmatrix} \mathbf{i} & \mathbf{j} & \mathbf{k} \\ \cos v & \sin v & 2u \\ -u\sin v & u\cos v & 0 \end{vmatrix}$$

$$= (-2u^2\cos v)\mathbf{i} - (2u^2\sin v)\mathbf{j} + (u\cos^2 v + u\sin^2 v)\mathbf{k}$$

$$= \langle -2u^2\cos v, -2u^2\sin v, u \rangle.$$

$$\|\mathbf{T}_u \times \mathbf{T}_v\| = \sqrt{4u^4\cos^2 v + 4u^4\sin^2 v + u^2}$$

$$= \sqrt{4u^4 + u^2} = u\sqrt{4u^2 + 1} \qquad (\text{since } u \geq 0).$$

The surface area of S is

$$\iint_D \|\mathbf{T}_u \times \mathbf{T}_v\| \, du\, dv = \int_0^{2\pi} \int_0^1 u\sqrt{4u^2 + 1}\, du\, dv.$$

If we substitute $t = 4u^2 + 1$, so that $dt = 8u\, du$, we have

$$\int u\sqrt{4u^2 + 1}\, du = \int \frac{1}{8}\sqrt{t}\, dt = \frac{1}{12} t^{3/2} = \frac{1}{12}(4u^2 + 1)^{3/2}.$$

Therefore,

$$\int_0^{2\pi} \int_0^1 u\sqrt{4u^2 + 1}\, du\, dv = \int_0^{2\pi} \frac{1}{12}(4u^2 + 1)^{3/2} \Big]_0^1 \, dv$$

$$= \frac{5^{3/2}}{12} \int_0^{2\pi} dv = \frac{5\sqrt{5}\pi}{6}.$$

■

EXAMPLE 43 Verify that the surface area of a sphere of radius R is $4\pi R^2$ by means of a surface integral.

Solution We'll use spherical coordinates to parametrize the surface of the sphere:

$$\mathbf{r} : D \longrightarrow \mathbb{R}^3$$

where $D = \{(\theta, \varphi): \ 0 \le \theta \le 2\pi; \ -\frac{\pi}{2} \le \varphi \le \frac{\pi}{2}\}$, and

$$\mathbf{r}(\theta, \varphi) = \langle R\cos\theta\sin\varphi, \ R\sin\theta\sin\varphi, \ R\cos\varphi \rangle.$$

We compute

$$\mathbf{T}_\theta = \langle -R\sin\theta\sin\varphi, \ R\cos\theta\sin\varphi, \ 0 \rangle$$

and

$$\mathbf{T}_\varphi = \langle R\cos\theta\cos\varphi, \ R\sin\theta\cos\varphi, \ -R\sin\varphi \rangle.$$

Hence,

$$\mathbf{T}_\theta \times \mathbf{T}_\varphi = \begin{vmatrix} \mathbf{i} & \mathbf{j} & \mathbf{k} \\ -R\sin\theta\sin\varphi & R\sin\theta\cos\varphi & 0 \\ R\cos\theta\cos\varphi & R\sin\theta\cos\varphi & -R\sin\varphi \end{vmatrix}$$

$$= (-R^2\cos\theta\sin^2\varphi)\mathbf{i} - (R^2\sin\theta\sin^2\varphi)\mathbf{j}$$

$$\quad + R^2(-\sin^2\theta\sin\varphi\cos\varphi - \cos^2\theta\sin\varphi\cos\varphi)\mathbf{k}$$

$$= R^2\langle -\cos\theta\sin^2\varphi, \ -\sin\theta\sin^2\varphi, \ -\sin\varphi\cos\varphi \rangle.$$

Therefore,

$$\|\mathbf{T}_\theta \times \mathbf{T}_\varphi\| = R^2\sqrt{\cos^2\theta\sin^4\varphi + \sin^2\theta\sin^4\varphi + \sin^2\varphi\cos^2\varphi}$$

$$= R^2\sqrt{\sin^4\varphi + \sin^2\varphi\cos^2\varphi}$$

$$= R^2\sqrt{\sin^2\varphi(\sin^2\varphi + \cos^2\varphi)}$$

$$= R^2\sqrt{\sin^2\varphi} = R^2|\sin\varphi|.$$

(Notice the absolute value signs.) So, the surface area of the sphere is

$$\iint\limits_D \|\mathbf{T}_\theta \times \mathbf{T}_\varphi\| \, d\theta \, d\varphi = \int_0^{2\pi} \int_{-\frac{\pi}{2}}^{\frac{\pi}{2}} R^2|\sin\varphi| \, d\varphi \, d\theta.$$

Since R^2 is a constant, we can factor it outside the integral signs. To handle the absolute value $|\sin\varphi|$, we split the inner integral over two intervals:

$$R^2\left(\int_0^{2\pi}\int_{-\frac{\pi}{2}}^0 (-\sin\varphi)\,d\varphi\,d\theta + \int_0^{2\pi}\int_0^{\frac{\pi}{2}}(\sin\varphi)\,d\varphi\,d\theta\right)$$

$$= R^2\left(\int_0^{2\pi}\cos\varphi\bigg]_{-\frac{\pi}{2}}^0 d\theta + \int_0^{2\pi}(-\cos\varphi)\bigg]_0^{\frac{\pi}{2}}d\theta\right)$$

$$= R^2\left(\int_0^{2\pi}1\,d\theta + \int_0^{2\pi}1\,d\theta\right)$$

$$= R^2(2\pi + 2\pi) = 4\pi R^2.$$

■

Surface integral of a scalar field

The surface integral of a scalar field $f(x,y,z)$ over a surface S is written

$$\iint_S f\,dS = \iint_D f(\mathbf{r}(u,v))\|\mathbf{T}_u \times \mathbf{T}_v\|\,du\,dv.$$

For a physical interpretation of the surface integral $\displaystyle\iint_S f\,dS$, consider f as measuring the mass per unit of area of a thin sheet of material. If this density function is integrated over the surface, we obtain

$$\text{the total mass of the surface } = \iint_S f\,dS,$$

while

$$\text{the total area of the surface } = \iint_S 1\,dS.$$

EXAMPLE 44 A sheet of material S corresponds to the part of the plane $x = z$ inside the cylinder $x^2 + y^2 = 1$ (units in centimeters). The mass per square centimeter of the material is given by the density function $f(x,y,z) = x^2$ grams per square centimeter. Find the total mass of the sheet.

Solution First, we'll need to parametrize the surface S. If we let $x = u$ and $y = v$, then $z = x = u$ and we can parametrize S as

$$\mathbf{r} : D \longrightarrow \mathbb{R}^3$$

$$\mathbf{r}(u,v) = (u,v,u)$$

$$D = \{(u,v): \quad u^2 + v^2 \le 1\}.$$

Now we compute

$$\mathbf{T}_u = \langle 1, 0, 1 \rangle \qquad \mathbf{T}_v = \langle 0, 1, 0 \rangle$$

and

$$\mathbf{T}_u \times \mathbf{T}_v = \langle -1, 0, 1 \rangle \qquad \| \mathbf{T}_u \times \mathbf{T}_v \| = \sqrt{2}.$$

The mass of the sheet is given by

$$\iint\limits_S f\, dS = \sqrt{2} \iint\limits_D u^2\, du\, dv.$$

Now, D is a unit disk in the uv-plane, so we'll make a change to polar coordinates by

$$u = r \cos \theta, \qquad v = r \sin \theta, \qquad dA = r\, dr\, d\theta.$$

We have

$$\sqrt{2} \iint\limits_D u^2\, du\, dv = \sqrt{2} \int_0^{2\pi} \int_0^1 r^2 \cos^2 \theta\, dr\, d\theta$$

$$= \sqrt{2} \int_0^{2\pi} \frac{r^4}{4} \cos^2 \theta \Big]_0^1 d\theta$$

$$= \frac{\sqrt{2}}{4} \int_0^{2\pi} \cos^2 \theta\, d\theta$$

$$= \frac{\sqrt{2}}{4} \int_0^{2\pi} \frac{1 + \cos 2\theta}{2}\, d\theta$$

$$= \frac{\sqrt{2}}{8} \theta + \frac{\sin 2\theta}{16} \Big]_0^{2\pi}$$

$$= \left(\frac{\sqrt{2}\pi}{4} + 0 \right)(-(0+0)) = \frac{\sqrt{2}\pi}{4} \approx 1.11.$$

Hence, the sheet has a mass of approximately 1.11 grams. ∎

EXAMPLE 45   Find $\displaystyle\iint\limits_S x\, dS$ where S is the part of the plane $x + y + z = 1$ in the first octant $(x \geq 0,\ y \geq 0,\ z \geq 0)$.

Solution   The surface S is a triangle with vertices $(1, 0, 0)$, $(0, 0, 1)$ and $(0, 1, 0)$. If we solve for z, then

$$z = 1 - x - y$$

and the parametrization we use is:

$$\mathbf{r} : D \longrightarrow \mathbb{R}^3, \qquad \mathbf{r}(u, v) = \langle u, v, 1 - u - v \rangle,$$

where D is shown in Figure 17.24.

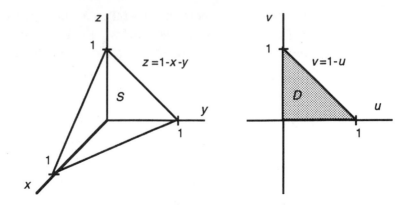

Figure 17.24 Surface S and its domain of parametrization D.

We compute

$$\mathbf{T}_u = \langle 1, 0, -1 \rangle \qquad \mathbf{T}_v = \langle 0, 1, -1 \rangle$$

$$\mathbf{T}_u \times \mathbf{T}_v = \langle 1, 1, 1 \rangle \qquad \|\mathbf{T}_u \times \mathbf{T}_v\| = \sqrt{3}.$$

We have

$$\iint_S x \, dS = \iint_D u \|\mathbf{T}_u \times \mathbf{T}_v\| \, du \, dv$$

$$= \int_0^1 \int_0^{1-u} u\sqrt{3} \, dv \, du$$

$$= \int_0^1 uv\sqrt{3} \Big]_0^{1-u} du$$

$$= \sqrt{3} \int_0^1 (u - u^2) \, du$$

$$= \sqrt{3} \Big(\frac{u^2}{2} - \frac{u^3}{3} \Big) \Big]_0^1$$

$$= \sqrt{3} \Big(\frac{1}{2} - \frac{1}{3} \Big) = \frac{\sqrt{3}}{6}.$$

■

Oriented surfaces

The fundamental vector product $\mathbf{T}_u \times \mathbf{T}_v$ provides a normal vector to a surface S parametrized by $\mathbf{r}(u, v)$. If we divide the fundamental vector product by its length, we obtain what is called the **principal unit normal vector**

$$\mathbf{n}(u, v) = \frac{\mathbf{T}_u \times \mathbf{T}_v}{\|\mathbf{T}_u \times \mathbf{T}_v\|}.$$

The direction of $\mathbf{n}(u, v)$ establishes an *orientation* to the surface, much like the direction of a tangent vector establishes an orientation to a curve.

We generally think of a surface as having two sides, like a sheet of paper. Choosing one of the two sides distinguishes a particular orientation to the surface, like designating one side of a piece of paper as the "front" and the other as the "back." The surface shown in Figure 17.25 has an orientation with the principal unit normal pointing outward at each point. A different parametrization of the same surface might *reverse* the orientation, with the principal unit normal pointing inward.

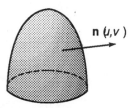

Figure 17.25 The principal unit normal $\mathbf{n}(u, v)$ orients a surface.

Remarkably, there exist surfaces that are not orientable. The Möbius band is perhaps the most famous example of such a surface—it has only one side! You can make your own Möbius band by taking a rectangular strip of paper, twisting it once, and then gluing the opposite ends together (see Figure 17.26).

Figure 17.26 The Möbius band is a non-orientable surface.

 Try this: make a Möbius band and draw a continuous line lengthwise along its center. What happens? Now try cutting the Möbius band "in half" along this line.

Surface integral of a vector field

If we think of a vector field $\mathbf{F}$ as representing the velocity field of a fluid and a surface S as a porous membrane permitting the flow of fluid through it, then one might ask about the total volume flow rate through the surface.

If $\mathbf{r}$ parametrizes the surface, then at any smooth point $\mathbf{r}(u, v)$ on the surface we can calculate the velocity vector of the fluid flow

$$\mathbf{F}(\mathbf{r}(u, v))$$

as well as a normal vector to the surface by means of the fundamental vector product

$$\mathbf{T}_u \times \mathbf{T}_v.$$

We form the **principal unit normal vector** to the surface at this point as

$$\mathbf{n}(u, v) = \frac{\mathbf{T}_u \times \mathbf{T}_v}{\|\mathbf{T}_u \times \mathbf{T}_v\|}.$$

Hence, the component of fluid flow perpendicular to the surface at $\mathbf{r}(u, v)$ is

$$\mathbf{F} \cdot \mathbf{n} = \mathbf{F}(\mathbf{r}(u, v)) \cdot \mathbf{n}(u, v).$$

This dot product is a scalar field that we can integrate over the surface to obtain the total volume rate of fluid flow through the surface in the direction of the principal unit normal $\mathbf{n}$.

Definition 12

If $\mathbf{F}$ is a C^1-differentiable vector field and S is a surface with a smooth parametrization $\mathbf{r} : D \longrightarrow \mathbb{R}^3$, then the surface integral of the normal component of $\mathbf{F}$ over S with respect to this parametrization is denoted

$$\iint_S \mathbf{F} \cdot \mathbf{n}\, dS \quad \text{or} \quad \iint_S \mathbf{F} \cdot d\mathbf{S},$$

where either notation indicates the double integral to be computed:

$$\iint_D \mathbf{F}(\mathbf{r}(u, v)) \cdot \mathbf{n}(u, v)\|\mathbf{T}_u \times \mathbf{T}_v\|\, du\, dv = \iint_D \mathbf{F}(\mathbf{r}(u, v)) \cdot (\mathbf{T}_u \times \mathbf{T}_v)\, du\, dv.$$

This surface integral value is called the **flux** of $\mathbf{F}$ over the surface S.

The last equation in the definition follows from noting that

$$\mathbf{n}(u, v)\|\mathbf{T}_u \times \mathbf{T}_v\| = \frac{\mathbf{T}_u \times \mathbf{T}_v}{\|\mathbf{T}_u \times \mathbf{T}_v\|}\|\mathbf{T}_u \times \mathbf{T}_v\| = \mathbf{T}_u \times \mathbf{T}_v.$$

The value of the flux of a vector field depends on the orientation of the surface, indicated by the direction of the principal unit normal vector. If q is a smooth parametrization of S reversing the orientation, then

$$\iint_S \mathbf{F} \cdot \mathbf{n} \, dS$$

changes sign (since the principal unit normal vector now points in the opposite direction).

If you parametrize a surface S for the purpose of calculating the flux of a vector field, but find that your parametrization has the opposite of the desired orientation, then simply remember to reverse the sign at the end of the calculation.

EXAMPLE 46   Calculate the *outward* flux of $\mathbf{F}(x, y, z) = \langle x, y, z \rangle$ over the upper unit hemisphere

$$S = \{(x, y, z) : \ x^2 + y^2 + z^2 = 1; \ z \geq 0\}.$$

Solution   We can express S as the graph of $z = \sqrt{1 - x^2 - y^2}$ (positive square root since this is the upper hemisphere), so one parametrization of the surface, using cylindrical coordinates, is

$$\mathbf{r} : D \longrightarrow \mathbb{R}^3$$

$$\mathbf{r}(r, \theta) = \langle r \cos \theta, \ r \sin \theta, \ \sqrt{1 - r^2} \rangle$$

$$D = \{(r, \theta) : \ 0 \leq r \leq 1; \ 0 \leq \theta \leq 2\pi\}.$$

We compute

$$\mathbf{T}_r = \langle \cos \theta, \ \sin \theta, \ \frac{-r}{\sqrt{1 - r^2}} \rangle \qquad \mathbf{T}_\theta = \langle -r \sin \theta, \ r \cos \theta, \ 0 \rangle$$

and

$$\mathbf{T}_r \times \mathbf{T}_\theta = \langle \frac{r^2 \cos \theta}{\sqrt{1 - r^2}}, \frac{r^2 \sin \theta}{\sqrt{1 - r^2}}, r \rangle.$$

Does this normal vector have the desired direction? Since $0 \leq r \leq 1$, we can see that the z-component of $\mathbf{T}_r \times \mathbf{T}_\theta$ is nonnegative. Thus, the normal vector must point up and out of the hemisphere, as desired. Now, we can compute

$$\iint_S \mathbf{F} \cdot d\mathbf{S} = \iint_D \mathbf{F}(\mathbf{r}(r, \theta)) \cdot (\mathbf{T}_r \times \mathbf{T}_\theta) \, dr \, d\theta$$

$$= \int_0^{2\pi} \int_0^1 \langle r\cos\theta,\ r\sin\theta,\ \sqrt{1-r^2}\rangle \cdot \langle \frac{r^2\cos\theta}{\sqrt{1-r^2}}, \frac{r^2\sin\theta}{\sqrt{1-r^2}}, r\rangle \, dr\, d\theta$$

$$= \int_0^{2\pi} \int_0^1 \left(\frac{r^3\cos^2\theta}{\sqrt{1-r^2}} + \frac{r^3\sin^3\theta}{\sqrt{1-r^2}} + r\sqrt{1-r^2} \right) dr\, d\theta$$

$$= \int_0^{2\pi} \int_0^1 \left(\frac{r^3(\cos^2\theta + \sin^2\theta) + r(1-r^2)}{\sqrt{1-r^2}} \right) dr\, d\theta$$

$$= \int_0^{2\pi} \int_0^1 \frac{r}{\sqrt{1-r^2}} \, dr\, d\theta$$

$$= \int_0^{2\pi} (-\sqrt{1-r^2})\Big]_0^1 \, d\theta = \int_0^{2\pi} [(0) - (-1)] \, d\theta$$

$$= \int_0^{2\pi} 1 d\theta = 2\pi.$$

■

EXAMPLE 47 A fluid is flowing at a constant rate straight upwards, so that its velocity vector field is

$$\mathbf{F}(x,y,z) = \mathbf{k}.$$

Calculate the outward flux of $\mathbf{F}$ over the cone $z = \sqrt{x^2 + y^2}$, for $x^2 + y^2 \le 1$.

Solution We can parametrize the cone with

$$\mathbf{r} : D \longrightarrow \mathbb{R}^3,$$

where $D = \{(u,v)|u^2 + v^2 \le 1\}$, and

$$\mathbf{r}(u,v) = \langle u, v, \sqrt{u^2 + v^2}\rangle.$$

We have

$$\mathbf{T}_u = \langle 1, 0, \frac{u}{\sqrt{u^2+v^2}}\rangle, \qquad \mathbf{T}_v = \langle 0, 1, \frac{v}{\sqrt{u^2+v^2}}\rangle,$$

so

$$\mathbf{T}_u \times \mathbf{T}_v = \begin{vmatrix} \mathbf{i} & \mathbf{j} & \mathbf{k} \\ 1 & 0 & \frac{u}{\sqrt{u^2+v^2}} \\ 0 & 1 & \frac{v}{\sqrt{u^2+v^2}} \end{vmatrix} = \frac{-u}{\sqrt{u^2 + v^2}}\mathbf{i} + \frac{-v}{\sqrt{u^2 + v^2}}\mathbf{j} + 1\mathbf{k}.$$

Since the z-component is positive, this normal vector points up and into the cone, rather than outward. Hence, the outward flux is

$$-\iint_S \mathbf{F} \cdot d\mathbf{S} = -\iint_D \mathbf{k} \cdot (\mathbf{T}_u \times \mathbf{T}_v)\, du\, dv$$

$$= -\iint_D \langle 0, 0, 1 \rangle \cdot \langle \frac{-u}{\sqrt{u^2 + v^2}}, \frac{-v}{\sqrt{u^2 + v^2}}, 1 \rangle\, du\, dv$$

$$= -\iint_D du\, dv = -\pi.$$

(Since D is a unit circle, it has area π.)   ∎

EXERCISES

Suppose a surface S is parametrized by $\mathbf{r}$ over the domain $\mathbf{D}$ as indicated in each of exercises 1-6. Find $\|\mathbf{T}_u \times \mathbf{T}_v\|$, the surface area of S, and the value of $\iint_S f\, dS$ where $f(x, y, z)$ is the scalar field indicated.

1. $\mathbf{r}(u, v) = \langle v \cos u, v \sin u, v \rangle$ $D = [0, \frac{\pi}{2}] \times [0, 1]$.

   $f(x, y, z) = z$.

2. $\mathbf{r}(u, v) = \langle u \cos v, u \sin v, u \rangle$; $D = [0, 1] \times [0, \frac{\pi}{2}]$.

   $f(x, y, z) = z^2$

3. $\mathbf{r}(u, v) = \langle \cos u, \sin u, v \rangle$; $D = [0, \pi] \times [0, 1]$.

   $f(x, y, z) = x + y + z$.

4. $\mathbf{r}(u, v) = \langle \cos v, \sin v, u \rangle$; $D = [0, 1] \times [0, \pi]$.

   $f(x, y, z) = x^2 + y^2 + z^2$.

5. $\mathbf{r}(u, v) = \langle u + v, u - v, v - u \rangle$; $D = [1, 2] \times [0, 1]$.

   $f(x, y, z) = x^2 + y^2$.

6. $\mathbf{r}(u, v) = \langle u - v, v - u, u + v \rangle$; $D = [0, 1] \times [1, 2]$.

   $f(x, y, z) = y^2 + z^2$.

---

7.   Imagine a unit circle in the xz-plane whose center is at a distance $R > 1$ from the z-axis. If this circle is rotated about the z-axis, it generates a surface S in the shape of a torus or "doughnut." This surface has a parametrization given by

$$\mathbf{r} : D \longrightarrow \mathbb{R}^3,$$

where $D = [0, 2\pi] \times [0, 2\pi]$ and

$$\mathbf{r}(\theta, \varphi) = \langle (R + \cos \varphi) \cos \theta, (R + \cos \varphi) \sin \theta, \sin \varphi \rangle.$$

Show that the surface area of the torus is $4\pi^2 R$.

---

8. Consider the left hemisphere of the unit sphere; in other words,

$$x^2 + y^2 + z^2 = 1, \qquad y \le 0.$$

Parametrize this hemisphere in terms of suitable variables u and v. Be sure to state the domain D of the parametrization and sketch it in the uv-plane.

9. Use your parametrization to verify that the surface area of the hemisphere is 2π.

---

10. Explain, without performing the calculation, why

$$\iint\limits_S (x + y + z)\, dS = 0,$$

if $S = \{(x, y, z) :\ x^2 + y^2 + z^2 = 1\}$ is the unit sphere.

11. A fluid is flowing at a constant rate given by the constant velocity vector field

$$\mathbf{F}(x, y, z) = -\langle \frac{\sqrt{2}}{2}, 0, \frac{\sqrt{2}}{2} \rangle.$$

Find the flux of $\mathbf{F}$ over the cone $z = \sqrt{x^2 + y^2}$, for $x^2 + y^2 \le 1$.

17.7 THE FUNDAMENTAL THEOREMS OF VECTOR CALCULUS

In this final section, we discuss some of the fundamental theorems of vector calculus. For each of these theorems, you should draw an analogy with the Second Fundamental Theorem of Calculus for a continuous single variable function f:

$$\text{If } F'(x) = f(x), \text{ then } \int_a^b f(x)dx = F(b) - F(a).$$

This theorem tells us that the integral of f over the entire interval $[a, b]$ depends only on the values of a related function (the antiderivative F) on the *boundary* of the interval, namely, the two endpoints a and b.

Earlier in this chapter, we saw that a similar statement holds true for line integrals of *conservative* vector fields $\mathbf{F}$:

$$\text{If } \nabla\varphi(x) = \mathbf{F}(x), \text{ then } \int_C \mathbf{F} \cdot \mathbf{T}\, ds = \varphi(B) - \varphi(A),$$

where A and B are the two endpoints of the curve C. Recall, we called φ a *potential* for **F**.

In this section we'll see three of the fundamental theorems of vector calculus applied to vector fields in $\mathbb{R}^2$ and $\mathbb{R}^3$. Each theorem carries a similar message, relating the integral of a "derivative" of a function over a region to the values of the original function on the *boundary* of the region.

Terminology

To set the stage for the theorems, we'll need to discuss some terminology regarding curves and surfaces.

A parametrized surface S is said to be **simple** if its parametrization **r** is *one-to-one*. Like a simple curve, a simple surface never "crosses" itself in the sense that no point on the surface can be the image of two different points in the domain D. In terms of its parametrization **r**, if $(u_1, v_1) \neq (u_2, v_2)$ are two points in the domain D, then we must have

$$\mathbf{r}(u_1, v_1) \neq \mathbf{r}(u_2, v_2).$$

A **closed** surface is a surface that has no boundary. For example, a sphere is a closed surface. Note that for an *oriented* closed surface, it makes sense to refer to the principal unit normal vector as being either *inward* or *outward*, depending on the specific parametrization.

Suppose a simple surface S has a parametrization $\mathbf{r}(u, v)$ with domain D that is enclosed by a simple closed curve C^* in the uv-plane (as shown in Figure 17.27).

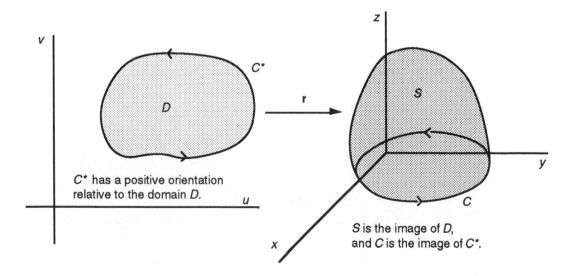

Figure 17.27 Domain D enclosed by a simple closed curve C^*.

If the closed curve C^* is oriented counterclockwise (so D is always to the left of the curve as we travel around it), then we say C^* is **positively oriented** with respect to D. The image of the curve C^* forms the boundary C of the surface S. If C is given the orientation "inherited" from C^*, then we say C is **positively oriented** with respect to the orientation of S. Figure 17.28 illustrates a surface S with its principal unit normal vector **n** and a positively oriented boundary C.

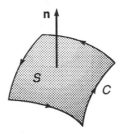

Figure 17.28 Surface S with positively oriented boundary C.

Stokes' Theorem

Stokes' Theorem relates a surface integral involving the *curl* of a vector field over a surface S with the line integral of its tangential component around the boundary of S. It is named in honor of George G. Stokes (1819-1903), an Irish mathematician.

In terms of the notation for surface and line integrals, the statement of Stokes' Theorem can be stated precisely as follows:

Theorem 17.3

Stokes' Theorem.
Hypothesis 1: **F** is a C^1-differentiable vector field in some region of $\mathbb{R}^3$ containing a simple smooth surface S.
Hypothesis 2: C is the piecewise smooth boundary of S with a positive orientation.
Conclusion: We have

$$\iint\limits_{S} (\nabla \times \mathbf{F}) \cdot \mathbf{n} \, dS = \oint_{C} \mathbf{F} \cdot \mathbf{T} \, ds.$$

Reasoning   Let's try to understand why Stokes' Theorem makes sense. First, let's consider the line integral

$$\oint \mathbf{F} \cdot \mathbf{T} \, ds$$

around the positively oriented boundary C of S as shown in Figure 17.28.

Now, suppose we subdivide S into 4 subregions as shown in Figure 17.29, with boundaries C_1, C_2, C_3, and C_4, each positively oriented.

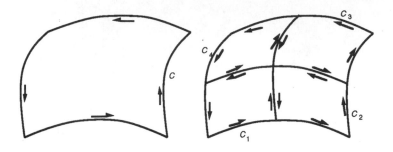

Figure 17.29 Subdividing S into 4 subregions.

If we traverse each of the four resulting curves C_1, C_2, C_3, and C_4, note that we end up traversing each of the interior curve segments twice, once in each direction. On the other hand, those segments along the original boundary only get traversed once, and always in the original direction of the boundary C. This means that if we sum up the line integrals of $\mathbf{F} \cdot \mathbf{T}$ along the four curves, the parts along the interior curve segments will cancel each other out, and we'll find

$$\oint_C \mathbf{F} \cdot \mathbf{T} \, ds = \oint_{C_1} \mathbf{F} \cdot \mathbf{T} \, ds + \oint_{C_2} \mathbf{F} \cdot \mathbf{T} \, ds + \oint_{C_3} \mathbf{F} \cdot \mathbf{T} \, ds + \oint_{C_4} \mathbf{F} \cdot \mathbf{T} \, ds.$$

The same is true if we subdivide S into many small subregions as shown in Figure 17.30.

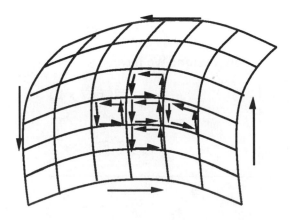

Figure 17.30 All the interior curve segments are traversed in both directions.

All of the interior curve segments are still traversed twice, once in each direction, but those along the original boundary are traversed only once. If we sum up all the line integrals over all these tiny closed curves C_i the result is equal to our original line integral around the surface boundary C:

$$\oint_C \mathbf{F} \cdot \mathbf{T}\,ds = \sum_{i=1}^{N} \oint_{C_i} \mathbf{F} \cdot \mathbf{T}\,ds.$$

We noted earlier (in Section 17.4) the relationship between the curl of a vector field at a point and the line integral about a tiny closed planar curve C surrounding that point. Precisely, we have

$$\text{curl } \mathbf{F} \cdot \mathbf{n} = \lim_{\Delta S \to 0} \frac{1}{\Delta S} \oint_C \mathbf{F} \cdot \mathbf{T}\,ds$$

where C is a curve lying in a plane surrounding the point, ΔS is the area enclosed by curve C, and n is the unit normal to the plane determined by C.

Examine Figure 17.30 again. As we subdivide the surface S finer and finer, the individual closed curves become approximately planar.

If ΔS is the approximate surface area enclosed by each C_i, and we choose a point on the surface within each grid curve C_i, then each of the line integrals $\oint_{C_i} \mathbf{F} \cdot \mathbf{T}\,ds$ is approximated by

$$\oint_{C_i} \mathbf{F} \cdot \mathbf{T}\,ds \approx \text{curl } \mathbf{F} \cdot \mathbf{n}\Delta S.$$

Summing these, we obtain

$$\oint_C \mathbf{F} \cdot \mathbf{T}\,ds \approx \sum_{i=1}^{N} \text{curl } \mathbf{F} \cdot \mathbf{n}\,\Delta S$$

The hypotheses of the theorem guarantee that the approximation becomes better and better as $N \to \infty$. Hence

$$\oint_C \mathbf{F} \cdot \mathbf{T}\,ds = \lim_{N \to \infty} \sum_{i=1}^{N} \text{curl } \mathbf{F} \cdot \mathbf{n}\Delta S = \iint_S (\nabla \times \mathbf{F}) \cdot \mathbf{n}\,dS,$$

and this is exactly the conclusion of Stokes' Theorem. $\square$

Let's illustrate Stoke's Theorem with an example.

EXAMPLE 48   Suppose **F** is the vector field defined by

$$\mathbf{F}(x, y, z) = (x^2 + y - 4)\mathbf{i} + 3xy\mathbf{j} + (2xz + z^2)\mathbf{k},$$

and S is the hemisphere of radius 4 defined as the set of points

$$S = \{(x, y, z) : x^2 + y^2 + z^2 = 16, z \geq 0\}.$$

Verify that Stokes' Theorem holds:

$$\iint_S \nabla \times \mathbf{F} \cdot \mathbf{n}\, dS = \int_C \mathbf{F} \cdot \mathbf{T}\, ds,$$

where C is the positively oriented boundary of S.

Solution   To verify Stokes' Theorem, we essentially need to calculate both sides of the equation

$$\iint_S (\nabla \times \mathbf{F}) \cdot \mathbf{n}\, dS = \oint_C \mathbf{F} \cdot \mathbf{T}\, ds,$$

and check that they are equal.

First, we'll calculate the left-hand side of the equation. We have

$$\nabla \times \mathbf{F}(x, y, z) = \begin{vmatrix} \mathbf{i} & \mathbf{j} & \mathbf{k} \\ \dfrac{\partial}{\partial x} & \dfrac{\partial}{\partial y} & \dfrac{\partial}{\partial z} \\ x^2 + y - 4 & 3xy & 2xz + z^2 \end{vmatrix}$$

$$= 0\mathbf{i} - (2z - 0)\mathbf{j} + (3y - 1)\mathbf{k}$$

$$= \langle 0, -2z, 3y - 1 \rangle.$$

The hemisphere is the graph of $z = \sqrt{16 - x^2 - y^2}$ (positive square root since $z \geq 0$) for $x^2 + y^2 \leq 16$, so we can use the parametrization

$$\mathbf{r} : D \longrightarrow \mathbb{R}^3$$

$$\mathbf{r}(u, v) = \langle u,\ v,\ \sqrt{16 - u^2 - v^2} \rangle,$$

where $D = \{(u, v) :\ u^2 + v^2 \leq 16\}$. Figure 17.31 illustrates the surface S, with its boundary C, and the domain D of the parametrization.

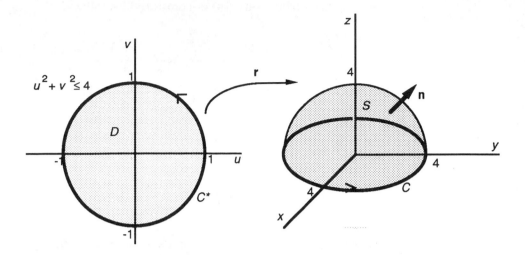

Figure 17.31 Parametrizing a sphere of radius 4.

We compute

$$\mathbf{T}_u = \langle 1, 0, \frac{-u}{\sqrt{16 - u^2 - v^2}} \rangle, \qquad \mathbf{T}_v = \langle 0, 1, \frac{-v}{\sqrt{16 - u^2 - v^2}} \rangle,$$

$$\mathbf{T}_u \times \mathbf{T}_v = \begin{vmatrix} \mathbf{i} & \mathbf{j} & \mathbf{k} \\ 1 & 0 & \frac{-u}{\sqrt{16-u^2-v^2}} \\ 0 & 1 & \frac{-v}{\sqrt{16-u^2-v^2}} \end{vmatrix}$$

$$= \left\langle \frac{u}{\sqrt{16 - u^2 - v^2}}, \frac{v}{\sqrt{16 - u^2 - v^2}}, 1 \right\rangle.$$

This means we have

$$\nabla \times \mathbf{F}(\mathbf{r}(u, v)) \cdot (\mathbf{T}_u \times \mathbf{T}_v)$$

$$= \langle 0, -2\sqrt{16 - u^2 - v^2}, 3v - 1 \rangle \cdot \left\langle \frac{u}{\sqrt{16 - u^2 - v^2}}, \frac{v}{\sqrt{16 - u^2 - v^2}}, 1 \right\rangle$$

$$= 0 - 2v + (3v - 1) = v - 1.$$

Hence, the surface integral is computed as follows:

$$\iint\limits_S (\nabla \times \mathbf{F}) \cdot \mathbf{n} \, dS = \int_{-4}^4 \int_{-\sqrt{16-v^2}}^{\sqrt{16-v^2}} (\nabla \times \mathbf{F})(\mathbf{r}(u,v)) \cdot (\mathbf{T}_u \times \mathbf{T}_v) \, du \, dv$$

$$= \int_{-4}^4 \int_{-\sqrt{16-v^2}}^{\sqrt{16-v^2}} (v-1) \, du \, dv$$

$$= \int_{-4}^4 (uv - u) \Big]_{-\sqrt{16-v^2}}^{\sqrt{16-v^2}} dv$$

$$= \int_{-4}^4 \left(2v\sqrt{16-v^2} - 2\sqrt{16-v^2}\right) dv$$

$$= \frac{-(16-v^2)^{\frac{3}{2}}}{3} \Big]_{-4}^4 - 2 \int_{-4}^4 \sqrt{16-v^2}) \, dv$$

$$= 0 - (\text{area of circle of radius } 4)$$

$$= -16\pi.$$

Now, let's calculate the right-hand side of the equation. We can parametrize C with

$$\mathbf{r} : [0, 2\pi] \longrightarrow \mathbb{R}^3 \qquad \mathbf{r}(t) = \langle 4\cos t, 4\sin t, 0 \rangle.$$

The velocity vector is

$$\mathbf{r}'(t) = \langle -4\sin t, 4\cos t, 0 \rangle,$$

so

$$\oint_C \mathbf{F} \cdot \mathbf{T} \, ds = \int_0^{2\pi} \mathbf{F}(\mathbf{r}(t)) \cdot \mathbf{r}'(t) \, dt$$

$$= \int_0^{2\pi} \langle 16\cos^2 t + 4\sin t - 4, \; 48\cos t \sin t, \; 0 \rangle \cdot \langle -4\sin t, \; 4\cos t, \; 0 \rangle \, dt$$

$$= \int_0^{2\pi} (-64\cos^2 t \sin t - 16\sin^2 t - 16\sin t + 192\cos^2 t \sin t) \, dt$$

$$= 128 \int_0^{2\pi} \cos^2 t \sin t \, dt - 16 \int_0^{2\pi} \sin^2 t \, dt + 16 \int_0^{2\pi} \sin t \, dt$$

$$= 128 \left(\frac{-\cos^3 t}{3}\right) \Big]_0^{2\pi} - 16\left(\frac{t}{2} - \frac{\sin 2t}{4}\right)\Big]_0^{2\pi} - 16\cos t \Big]_0^{2\pi}$$

$$= 0 - 16(\pi) - 0 = -16\pi.$$

This matches exactly our original computation.   ∎

Stokes' Theorem tells us that the curl of a C^1-differentiable vector field has the property that for all surfaces "capping" the same closed boundary curve C, the surface integral has the same value

$$\oint_C \mathbf{F} \cdot \mathbf{T} \, ds.$$

(See Figure 17.32.)

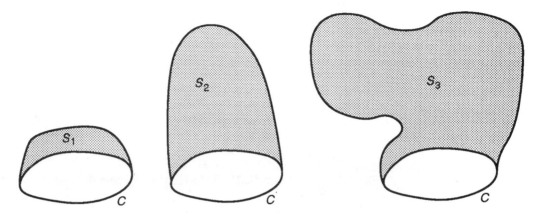

Figure 17.32 Stoke s' Theorem: $\iint_S \nabla \mathbf{F} \times \mathbf{n} \, dS$ is the same for all three surfaces capping C.

This is entirely consistent with our previous observation regarding conservative vector fields. Recall, if $\mathbf{F}$ is a conservative vector field, then

$$\oint_C \mathbf{F} \cdot \mathbf{T} \, ds = 0$$

for every closed curve C. If $\mathbf{F}$ is C^1-differentiable, then this also means **curl** $\mathbf{F} = 0$. But then

$$\iint_S \mathbf{curl} \, \mathbf{F} \cdot \mathbf{n} \, dS = \iint_S \mathbf{0} \cdot \mathbf{n} \, dS = 0$$

for any surface S capping the closed curve C.

EXAMPLE 49   Suppose

$$\mathbf{F}(x, y, z) = \langle -2xz, x, y^2 \rangle.$$

Find the value of the surface integral

$$\iint_S \nabla \times \mathbf{F} \cdot \mathbf{n} \, dS,$$

for each of the surfaces shown in Figure 17.33, where the principal unit normal in each case points outward.

The boundary of each surface below is the circle of radius 2 in the *xy*-plane, centered at the origin.

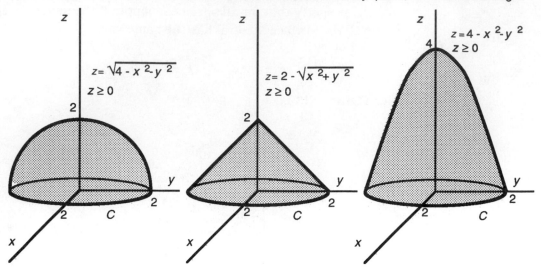

Figure 17.33 Three surfaces capping the same curve.

Solution   Each of the surfaces shown has the same boundary, namely the circle of radius 2 in the xy-plane. This circle is parametrized by

$$\mathbf{r}(t) = \langle 2\cos t, \ 2\sin t, \ 0 \rangle \qquad \text{for } 0 \le t \le 2\pi.$$

The velocity vector is

$$\mathbf{r}'(t) = \langle -2\sin t, \ 2\cos t, \ 0 \rangle.$$

Hence, if S is any surface capping the positively oriented boundary C, we must have

$$\iint\limits_{S} \nabla \times \mathbf{F} \cdot \mathbf{n}\, dS = \oint_{C} \mathbf{F} \cdot \mathbf{T}\, ds$$

$$= \int_{0}^{2\pi} \langle 0, \ 2\cos t, \ 4\sin^2 t \rangle \cdot \langle -2\sin t, 2\cos t, 0 \rangle\, dt$$

$$= \int_{0}^{2\pi} 4\cos^2 t\, dt$$

$$= \int_{0}^{2\pi} 2 + 2\cos 2t\, dt$$

$$= 2t + \sin 2t \, \Big]_{0}^{2\pi}$$

$$= 4\pi.$$

This is the value of the surface integral for each of the surfaces shown. ■

Green's Theorem

It's also possible to apply Stokes' Theorem to 2-dimensional vector fields. This special case of Stokes' Theorem is usually called **Green's Theorem**.

If $\mathbf{F}(x,y) = \langle P(x,y),\ Q(x,y) \rangle$ is a 2-dimensional vector field, then we can consider it as a 3-dimensional vector field

$$\mathbf{F}(x,y,z) = \langle P(x,y),\ Q(x,y),\ 0 \rangle$$

just by restricting our attention to the xy-plane. In this case,

$$\text{curl }\mathbf{F} = \left\langle 0,\ 0,\ \frac{\partial Q}{\partial x} - \frac{\partial P}{\partial y} \right\rangle.$$

A smooth simple curve that lies entirely in the xy-plane encloses a plane region D, that can be parametrized simply by

$$\mathbf{r}(x,y) = \langle x,\ y,\ 0 \rangle.$$

(See Figure 17.34.)

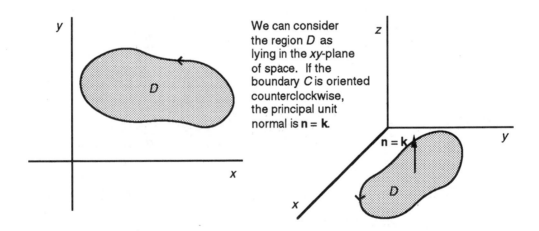

We can consider the region D as lying in the xy-plane of space. If the boundary C is oriented counterclockwise, the principal unit normal is $\mathbf{n} = \mathbf{k}$.

Figure 17.34 Considering a plane region in space.

If we traverse the boundary C of D in the counterclockwise direction, then the principal unit normal vector to D will just be $\mathbf{k} = \langle 0,\ 0,\ 1 \rangle$ at every point. Now,

$$\text{curl }\mathbf{F} \cdot \mathbf{k} = \langle 0,\ 0,\ \frac{\partial Q}{\partial x} - \frac{\partial P}{\partial y} \rangle \cdot \langle 0,\ 0,\ 1 \rangle = \frac{\partial Q}{\partial x} - \frac{\partial P}{\partial y},$$

and the surface integral in this case is simply the double integral

$$\iint\limits_{D} \frac{\partial Q}{\partial x} - \frac{\partial P}{\partial y}\ dx\,dy.$$

Hence, Stokes' Theorem tells us that

$$\iint_D \frac{\partial Q}{\partial x} - \frac{\partial P}{\partial y}\, dx\, dy = \oint_C P\, dx + Q\, dy.$$

This particular application of Stokes' Theorem is known as **Green's Theorem** (as are several other theorems in vector calculus), stated more precisely below:

Theorem 17.4

Green's Theorem:
Hypothesis 1: C is a simple piecewise smooth curve in the xy-plane oriented counterclockwise and D is the region enclosed by C.
Hypothesis 2: $\mathbf{F}(x, y) = \langle P(x, y), Q(x, y) \rangle$ is a C^1-differentiable vector field defined on D.
Conclusion:

$$\iint_D \left(\frac{\partial Q}{\partial x} - \frac{\partial P}{\partial y} \right) dx\, dy = \oint_C P\, dx + Q\, dy.$$

$\square$

EXAMPLE 50   Verify Green's Theorem for $\mathbf{F}(x, y) = \langle xy, \ x + y \rangle$, where C is the boundary of the unit square $[0, 1] \times [0, 1]$ (oriented counterclockwise).

Solution   To calculate $\oint_C P\, dx + Q\, dy$ will require splitting the curve C into 4 smooth parts as shown in Figure 17.35.

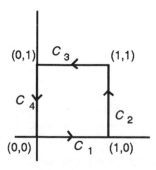

Figure 17.35 The boundary of the square can be split into four smooth curves.

We have the following parametrizations and line integral computations:

$$C_1 : \qquad \mathbf{r}_1(t) = \langle t, 0 \rangle \qquad \text{for } 0 \le t \le 1, \qquad \text{so } \mathbf{r}_1'(t) = \langle 1, 0 \rangle,$$

and

$$\oint_{C_1} P\,dx + Q\,dy = \int_0^1 \langle 0, 0 + t \rangle \cdot \langle 1, 0 \rangle \, dt = \int_0^1 0 \, dt = 0.$$

$$C_2 : \qquad \mathbf{r}_2(t) = \langle 1, t \rangle \qquad \text{for } 0 \le t \le 1, \qquad \text{so } \mathbf{r}_2'(t) = \langle 0, 1 \rangle,$$

and

$$\oint_{C_2} P\,dx + Q\,dy = \int_0^1 \langle t, 1 + t \rangle \cdot \langle 0, 1 \rangle \, dt = \int_0^1 (1 + t) \, dt = t + \frac{t^2}{2} \Big]_0^1 = \frac{3}{2}.$$

$$C_3 : \qquad \mathbf{r}_3(t) = \langle 1 - t, 1 \rangle \qquad \text{for } 0 \le t \le 1, \qquad \text{so } \mathbf{r}_3'(t) = \langle -1, 0 \rangle,$$

and

$$\oint_{C_3} P\,dx + Q\,dy = \int_0^1 \langle (1-t)1, 1 - t + 1 \rangle \langle -1.0 \rangle \, dt = \int_0^1 (t - 1) \, dt = \frac{t^2}{2} - t \Big]_0^1 = -\frac{1}{2}.$$

$$C_4 : \qquad \mathbf{r}_4(t) = \langle 0, 1 - t \rangle \qquad \text{for } 0 \le t \le 1, \qquad \text{so } \mathbf{r}_1'(t) = \langle 0, -1 \rangle,$$

and

$$\oint_{C_4} P\,dx + Q\,dy = \int_0^1 \langle 0(1-t),\ 0 + (1-t) \rangle \cdot \langle 0, -1 \rangle \, dt = \int_0^1 (t - 1) \, dt$$

$$= \frac{t^2}{2} - t \Big]_0^1 = -\frac{1}{2}.$$

Summing up the four pieces, we have

$$\oint_C P\,dx + Q\,dy = 0 + \frac{3}{2} - \frac{1}{2} - \frac{1}{2} = \frac{1}{2}.$$

Now, by Green's Theorem we should achieve the same result by calculating

$$\iint_D \left(\frac{\partial Q}{\partial x} - \frac{\partial P}{\partial y} \right) dx\,dy,$$

where D is the unit square $[0, 1] \times [0, 1]$. We have

$$\frac{\partial Q}{\partial x} = \frac{\partial}{\partial x}(x + y) = 1 \qquad \frac{\partial P}{\partial y} = \frac{\partial}{\partial y}(xy) = x.$$

Hence,

$$\iint\limits_{D} \left(\frac{\partial Q}{\partial x} - \frac{\partial P}{\partial y} \right) dx\,dy = \int_0^1 \int_0^1 (1 - x)\,dx\,dy$$

$$= \int_0^1 x - \frac{x^2}{2} \bigg]_0^1 dy$$

$$= \int_0^1 \frac{1}{2}\,dy = \frac{y}{2} \bigg]_0^1 = \frac{1}{2}.$$

We see that the results agree.   ∎

Calculating area using Green's Theorem

Green's Theorem provides an alternative for calculating the area of a region D enclosed by a closed piecewise curve C (oriented counterclockwise).

If $P(x,y)$ and $Q(x,y)$ are any two C^1-differentiable functions such that $\frac{\partial Q}{\partial x} - \frac{\partial P}{\partial y} = 1$, then the area of D is

$$\iint\limits_{D} 1\,dx\,dy = \iint\limits_{D} \left(\frac{\partial Q}{\partial x} - \frac{\partial P}{\partial y} \right) dx\,dy = \oint_C P\,dx + Q\,dy.$$

Thus, we can calculate the area of a region by means of a line integral around its boundary, provided we make suitable choices for $P(x,y)$ and $Q(x,y)$. Here are three possiblities:

If D is a region enclosed by a closed piecewise simple curve C (oriented counterclockwise) then each of the following line integrals has a value equal to the area of D.

$$\textbf{area of } D = \oint_C x\,dy = -\oint_C y\,dx = \frac{1}{2} \oint_C x\,dy - y\,dx.$$

EXAMPLE 51   Find the area of the ellipse whose boundary has equation

$$\frac{x^2}{a^2} + \frac{y^2}{b^2} = 1$$

using a line integral.

Solution   The boundary curve C can be parametrized with counterclockwise orientation with

$$\mathbf{r}(t) = \langle a \cos t, \ b \sin t \rangle \qquad 0 \le t \le 2\pi.$$

We have $\mathbf{r}'(t) = \langle -a \sin t, \ b \cos t \rangle$. Using a line integral to measure the area we have

$$
\begin{aligned}
\text{area of } D &= \frac{1}{2} \oint_C x \, dy - y \, dx \\
&= \frac{1}{2} \int_0^{2\pi} \langle -b \sin t, \ a \cos t \rangle \cdot \langle -a \sin t, \ b \cos t \rangle \\
&= \frac{1}{2} \int_0^{2\pi} (ab \sin^2 t + ab \cos^2 t) \, dt \\
&= \frac{ab}{2} \int_0^{2\pi} 1 \, dt \\
&= \frac{ab}{2}(2\pi) = \pi ab.
\end{aligned}
$$

$\blacksquare$

The divergence theorem—Gauss' Theorem

Stokes' Theorem relates the integral of the normal component of curl $\mathbf{F}$ over a surface to the integral of the tangential component of $\mathbf{F}$ over the closed boundary curve of the surface. **Gauss' Theorem**, also known as the **divergence theorem**, is analogous to Stokes' Theorem—it relates the integral of div $\mathbf{F}$ over a 3-dimensional region R to the outward flux of $\mathbf{F}$ over the closed boundary surface S of the region. Here is the precise statement of the theorem:

Theorem 17.5

The Divergence Theorem (Gauss' Theorem).
Hypothesis 1: R is a region of $\mathbb{R}^3$ bounded by a closed simple surface S, oriented so the principal unit normal **n** points outward.
Hypothesis 2: **F** is a C^1-differentiable vector field defined on R.
Conclusion:

$$\iint\limits_S \mathbf{F} \cdot \mathbf{n}\, dS = \iiint\limits_R \nabla \cdot \mathbf{F}\, dV.$$

In other words, the integral of the divergence of **F** over the region R has the same value as the total outward flux of **F** over its boundary surface S.

Reasoning   The reasoning behind the divergence theorem is quite similar to that behind Stokes' Theorem. Imagine that we subdivide the region R into several subregions and consider the surface integral of the outward normal component of **F** over the boundary surface of each subregion (see Figure 17.36).

Close-up of two interior subregions.

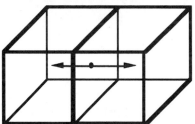

Within the region, the outward normals of the faces of the adjacent interior subregions point in exactly the opposite directions. The only normals that do not cancel out are those along the exterior surface of the region as shown.

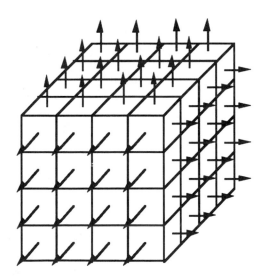

Figure 17.36 Subdividing a region R.

Two adjacent regions within the region will share a face, but their outward normals will point in opposite directions. Therefore, when we take the sum of all these surface integrals, the only parts that do not cancel belong to the faces that make up the original boundary of the region R.

Now, as we partition our region R into finer and finer subregions the surface integral of the outward flux of F over each tiny closed surface S_i has a value

$$\iint_{S_i} \mathbf{F} \cdot \mathbf{n}\, dS \approx \text{div } \mathbf{F} \Delta V,$$

where div F is evaluated at a point within the subregion and ΔV is the volume of the subregion. This approximation becomes better and better as the partition is made finer and finer, and we obtain

$$\iint_S \mathbf{F} \cdot \mathbf{n}\, dS = \lim_{N \to \infty} \sum_{i=1}^{N} \iint_{S_i} \mathbf{F} \cdot \mathbf{n}\, dS = \lim_{N \to \infty} \sum_{i=1}^{N} \nabla \cdot \mathbf{F} \Delta V = \iiint_R \nabla \cdot \mathbf{F}\, dV.$$

This is simply the statement of the divergence theorem. $\qquad\square$

EXAMPLE 52 Verify the divergence theorem for

$$\mathbf{F}(x, y, z) = x\mathbf{i} + y\mathbf{j} + z\mathbf{k}$$

for the cube

$$R = [-1, 1] \times [-1, 1] \times [-1, 1].$$

Solution First, we'll calculate the volume integral. The divergence of **F** is

$$\nabla \cdot \mathbf{F} = 1 + 1 + 1 = 3,$$

and

$$\int_{-1}^{1} \int_{-1}^{1} \int_{-1}^{1} 3\, dx\, dy\, dz = 3(\text{volume of the cube}) = 3 \cdot 8 = 24.$$

Now, we'll calculate the surface integral. Figure 17.37 illustrates the region R, and the principal unit normals to each face of the cube.

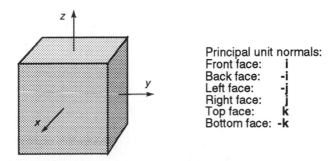

Figure 17.37 The region R and its principal unit normal vectors.

Let's calculate $\mathbf{F} \cdot \mathbf{n}$ for each face of the cube:

front face: $\langle x, y, z \rangle \cdot \langle 1, 0, 0 \rangle = x = 1$

(since $x = 1$ over the entire front face of the cube)

back face: $\langle x, y, z \rangle \cdot \langle -1, 0, 0 \rangle = -x = 1$

(since $x = -1$ over the entire back face of the cube)

left face: $\langle x, y, z \rangle \cdot \langle 0, -1, 0 \rangle = -y = 1$

right face: $\langle x, y, z \rangle \cdot \langle 0, 1, 0 \rangle = y = 1$

top face: $\langle x, y, z \rangle \cdot \langle 0, 0, 1 \rangle = z = 1$

bottom face: $\langle x, y, z \rangle \cdot \langle 0, 0, -1 \rangle = -z = 1.$

The front face has a surface area of 4, hence

$$\iint\limits_{front} \mathbf{F} \cdot \mathbf{n} \, dS = \iint\limits_{front} 1 \, dS = 4.$$

Similarly, the surface integral over each of the other faces has the same value 4. Thus,

$$\iint\limits_{S} \mathbf{F} \cdot \mathbf{n} \, dS = 6 \cdot 4 = 24.$$

We can see that the surface integral of $\mathbf{F} \cdot \mathbf{n}$ over the surface S of the cube has the same value as the volume integral of $\nabla \cdot \mathbf{F}$ over the region R. ■

EXERCISES

For exercises 1-5, consider

$$\mathbf{F}(x, y, z) = \langle y, -x, z - x^2 - y^2 \rangle \text{ and}$$
$$\mathbf{G}(x, y, z) = \langle -2y, 2x, -2 \rangle.$$

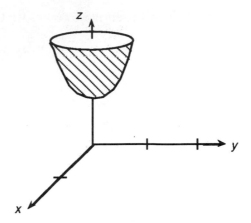

The surface S pictured above has parametrization $\mathbf{r} : D \to \mathbb{R}^3$, *where*

$$\mathbf{r}(u, v) = (u, v, u^2 + v^2 + 1) \quad \text{and} \quad D = \{(u, v) : u^2 + v^2 \le 1\}.$$

(Note: D is the unit disk in the uv plane.) C is the boundary of S, and has positively oriented parametrization

$$\mathbf{g} : [0, 2\pi] \to \mathbb{R}^3 \quad \text{and} \quad \mathbf{g}(t) = \langle \cos t, \sin t, 2 \rangle.$$

1. Express the surface area of S in terms of a double integral.

2. Find $\displaystyle\iint_S \mathbf{G} \cdot \mathbf{n} \, dS$.

3. Find $\displaystyle\oint_C \mathbf{F} \cdot \mathbf{T} \, ds$.

4. Find $\nabla \times \mathbf{F}$.

5. Does Stokes' Theorem hold for $\mathbf{F}$, S, and C? Explain why or why not.

Keeping in mind the usefulness of Green's Theorem, evaluate the integrals in exercises 6-10.

6. Use Green's Theorem to evaluate

$$\oint_C (4y + \sqrt{\sin^3 x}) \, dx + (6x - \sec^3 y) \, dy$$

where C is a circle centered at the origin of radius 4.

7. Find

$$\oint_C e^x \, dx + xy \, dy$$

where C is the path that starts at $(0, 0)$, follows the x-axis to $(2, 0)$ and then returns to $(0, 0)$ along the parabola $y = 2x - x^2$.

8. Use Green's theorem to evaluate the line integral

$$\oint_C (2^x - e^y)\, dx + (4x^2 - e^y)\, dy$$

where C is the path consisting of the straight line segment from $(0,0)$ to $(1,0)$, the straight line segment from $(1,0)$ to $(1,1)$, the straight line segment from $(1,1)$ to $(0,1)$, and the straight line segment from $(0,1)$ to $(0,0)$.

9. Use Green's theorem to transform the integral

$$\int_0^1 \int_0^x y\, dy\, dx$$

into a line integral, then evaluate the line integral.

10. If C is the boundary of $[0,2] \times [1,2]$, then find $\oint_C y^2\, dx - x^2\, dy$.

---

11. Find

$$\iint_S \mathbf{F} \cdot \mathbf{n}\, dS,$$

where S is the surface bounding the tetrahedron formed by the coordinate planes and the plane $x + y + z = 1$, $\mathbf{n}$ is the outward unit normal to S, and $\mathbf{F} = (2x + e^y)\mathbf{i} + (\ln x - 2y)\mathbf{j} + (\cos(xy) + z)\mathbf{k}$. You may use the fact that the volume of the tetrahedron is $\frac{1}{6}$.

12. Referring to Figure 17.33, verify the result of Example 49 by calculating

$$\iint_S \nabla \times \mathbf{F} \cdot \mathbf{n}\, dS$$

for each surface, where $\mathbf{F}(x, y, z) = \langle -2xz,\ x,\ y^2 \rangle$.

13. Referring to Example 51, verify that each of the line integrals

$$\oint_C x\, dy \quad \text{and} \quad -\oint_C y\, dx$$

also yield the area πab enclosed by the ellipse C having equation

$$\frac{x^2}{a^2} + \frac{y^2}{b^2} = 1.$$

14. Verify the divergence theorem for the *solid* hemisphere $\{(x, y, z) : x^2 + y^2 + z^2 \le 1, z \ge 0\}$, where $\mathbf{F}(x, y, z) = \langle x,\ y,\ z \rangle$. (Note that the outward principal unit normal is $\mathbf{n} = \langle x,\ y,\ z \rangle$ along the spherical portion of the surface, and $\mathbf{n} = -\mathbf{k}$ along the flat bottom of the surface.)

15. Use Gauss' Theorem to calculate the outward flux of

$$\mathbf{F}(x, y, z) = \langle 2x,\ 3y\, 4z \rangle$$

over the closed cylindrical surface

$$S = \{(x, y, z): \ x^2 + y^2 = 1, \ -1 \le z \le 2\}.$$

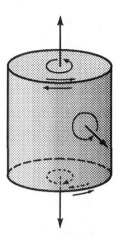

16. Justify the following statement: The outward flux of an incompressible C^1-differentiable vector field $\mathbf{F}$ is 0 over any simple, closed surface.

ANSWERS TO SELECTED EXERCISES

Chapter 12

12.1

1.

vector	initial point	terminal point
n	$(-1,1)$	$(-1,-1)$
p	$(-1,1)$	$(-1,4)$
q	$(-1,1)$	$(2,1)$
u	$(2,4)$	$(2,1)$
v	$(2,1)$	$(5,1)$
w	$(-1,1)$	$(5,-1)$

3. $\|\mathbf{m}\| = \sqrt{13}$, $\|\mathbf{w}\| = 2\sqrt{10}$ 5. p, q, and u; p, v, and u   7. 6   9. $(9,-6)$ 11. $\overrightarrow{PQ} = \langle -3,-5,-1 \rangle$.   13. $\overrightarrow{PQ} = \langle -2.5,3.4,4.1 \rangle$.   15. $\overrightarrow{PQ} = \langle 3,5,1 \rangle$.   17.   11) $T = (5.5,1.6,-5.1)$ 12) $T = (-7.5,-1.4,-4.1)$ 13) $T = (5,-6.8,-8.2)$ 14) $T = (2.5,-3.4,-4.1)$ 15) $T = (-.5,-8.4,-3.1)$ 19. $\langle -2\sqrt{3},2 \rangle$ 21. $\langle 0,0 \rangle$ 23. **m:** $\sqrt{13}$ units at $\approx 236.3°$, **n:** 2 units due South $(180°)$, **p:** 3 units due North $(0°)$, **q:** 3 units due East $(90°)$, **u:** 3 units due South $(180°)$, **v:** 3 units due East $(90°)$ **w:** $2\sqrt{10}$ units at $\approx 108.4°$

12.2

1. $\mathbf{p}+\mathbf{q}+\mathbf{u}=\mathbf{q}$ 3. $\mathbf{q}-\mathbf{v}=0$ 5. $\mathbf{w}-\mathbf{n}=2\mathbf{q}$ 7. listing clockwise: $\mathbf{w}-\mathbf{v}, -\mathbf{v}, -\mathbf{w}, \mathbf{v}-\mathbf{w}$ 9. $2\mathbf{v}=\langle -8,-6 \rangle$ 11. $\mathbf{v}+\mathbf{w}=\langle -1.28,.14 \rangle$ 13. $\mathbf{w}-\mathbf{v}=\langle 6.72,6.14 \rangle$ 15. $\frac{\mathbf{w}}{5}-4\mathbf{v}=\langle 16.544,12.628 \rangle$ 17. $2\mathbf{v}=4\mathbf{i}-6\mathbf{j}+2\mathbf{k}$ 19. $\mathbf{v}+\mathbf{w}=2.5\mathbf{i}-2\mathbf{j}-6\mathbf{k}$ 21. $\mathbf{w}-\mathbf{v}=-1.5\mathbf{i}+4\mathbf{j}-8\mathbf{k}$ 23. $\frac{\mathbf{w}}{5}-4\mathbf{v}=-7.9\mathbf{i}+12.2\mathbf{j}-5.4\mathbf{k}$

12.3

1. $\mathbf{p} \cdot \mathbf{n} = -6$ 3. $\mathbf{u} \cdot \mathbf{p} = -9$ 5. $(\mathbf{m} \cdot \mathbf{u})\mathbf{u} = \langle 0, -2 \rangle$ where $\mathbf{u}$ is the unit vector in the direction $\mathbf{u}$.   7. $\mathbf{v} \cdot \mathbf{w} = \frac{1}{2}$ 9. $\mathbf{v} \cdot \mathbf{w} = -20.3$ 11. $\|\mathbf{v}\| = 5$ 13. $\|\mathbf{v} + \mathbf{w}\| \approx 1.29$ 15. $\|\mathbf{v} - \mathbf{w}\| \approx 9.10$ 17. $\langle -\frac{4}{5}, -\frac{3}{5} \rangle$ 19. $\theta \approx 167.8°$ 21. $(\mathbf{v} \cdot \mathbf{u})\mathbf{u} \approx \langle -3.20, -3.69 \rangle$ where $\mathbf{u}$ is the unit vector in the direction $\mathbf{w}$.   23. $\mathbf{v} \cdot \mathbf{w} = -9$ 25. $\|\mathbf{v}\| \approx 3.74$ 27. $\|\mathbf{v} + \mathbf{w}\| \approx 6.80$ 29. $\|\mathbf{v} - \mathbf{w}\| \approx 9.07$ 31. $\approx \langle .53, -.80, .27 \rangle$ 33. $\theta \approx 109.8°$

35. $(\mathbf{v} \cdot \mathbf{u})\mathbf{u} \approx \langle -8.96, -.18, 1.25 \rangle$

37. no; for example, let $\mathbf{u} = \langle 3, 2 \rangle$ and $\mathbf{v} = \langle -2, 3 \rangle$.

39. Since $\mathbf{v} - \mathrm{comp}_\mathbf{u}\mathbf{v} = \mathbf{v} - (\mathbf{v} \cdot \mathbf{u})\mathbf{u}$, we have

$$\mathbf{u} \cdot (\mathbf{v} - \mathrm{comp}_\mathbf{u}\mathbf{v}) = \mathbf{u} \cdot \mathbf{v} - (\mathbf{v} \cdot \mathbf{u})(1) = 0 \Rightarrow \mathbf{u} \perp (\mathbf{v} - \mathrm{comp}_\mathbf{u}\mathbf{v})$$

41. Assuming θ is expressed such that $0 \leq \theta < 2\pi$, we have

$$(r, \theta) = (\|\mathbf{v}\|, \arccos(\mathbf{v} \cdot \mathbf{i}/\|\mathbf{v}\|)) \qquad \text{if } 0 < \theta \leq \pi,$$

$$(r, \theta) = (\|\mathbf{v}\|, 2\pi - \arccos(\mathbf{v} \cdot \mathbf{i}/\|\mathbf{v}\|)) \qquad \text{if } \pi < \theta < 2\pi.$$

43. For $\mathbf{v}$: $\cos^2 \alpha + \cos^2 \beta = \frac{25}{25} = 1$;   for $\mathbf{w}$, $\cos^2 \alpha + \cos^2 \beta = 1$.

45. For $\mathbf{v}$, $\alpha \approx 57.7°$, $\beta \approx 143.3°$, $\gamma \approx 74.5°$; for $\mathbf{w}$, $\alpha \approx 86°$, $\beta \approx 81.9°$, $\gamma \approx 170.9°$

47. Let $\mathbf{u} = \langle u_1, u_2, u_3 \rangle$ be any vector in $\mathbb{R}^3$, and let α, β, and γ be the direction angles.

$$\cos^2 \alpha + \cos^2 \beta + \cos^2 \gamma = \frac{u_1^2}{\|\mathbf{u}\|^2} + \frac{u_2^2}{\|\mathbf{u}\|^2} + \frac{u_3^2}{\|\mathbf{u}\|^2} = \frac{u_1^2 + u_2^2 + u_3^2}{\|\mathbf{u}\|^2} = \frac{\|\mathbf{u}\|^2}{\|\mathbf{u}\|^2} = 1.$$

12.4

1. $U = \begin{bmatrix} 5 & -3 \\ 4 & 4 \\ 1 & -2 \end{bmatrix}$ 3. $X = \begin{bmatrix} 4 & 2 \\ 9 & 2 \\ -5 & -11 \end{bmatrix}$ 5. Associativity for addition

7. $W = \begin{bmatrix} -10 & 6 \\ -8 & -8 \\ -2 & 4 \end{bmatrix}$ 9. scalar distributivity   11. $F = \begin{bmatrix} 10 & 5 \\ 23 & 1 \end{bmatrix}$

13. $I_3 C = C$, $C I_3 = C$ 15. $\det(E) = 0$ 17. $\det(I_3) = 1$ 19. $\det(O_n) = 0$ for any n 21. $\det(P) = -11$ 23. $\det(P) = 33$ 25. $\det(P) = -11a$ 27. $\det(P) = 0$ 29. $\det(P) = 0$

12.5

1. $\mathbf{b} \times \mathbf{a} = 5\mathbf{i} = 5\mathbf{k}$ 3. $(\mathbf{b} - \mathbf{a}) \times (\mathbf{c} - \mathbf{a}) = -10\mathbf{i} - 5\mathbf{k}$ 5. $\langle x(t), y(t), z(t) \rangle = \langle 3t, t, -t \rangle$ 7. $\langle x(t), y(t), z(t) \rangle = \langle 2 + t, -1 - 3t, 1 - 2t \rangle$ (other answers are possible)   9. Area ≈ 8.66.   11. The volume of the parallelipiped determined by the three vectors is not zero. Therefore, all three vectors do not lie in the same plane.   13. $8x + 31y - 9z - 69 = 0$ 15. $\approx 70.9°$ 17. False   19. False   21. False   23. True   25. False

Chapter 13

13.1

Here is the image curve for exercises 1-4:

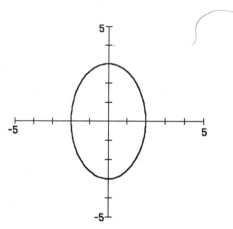

Here is the image curve for exercises 5-8:

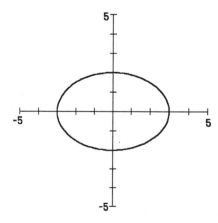

9. The orientation of the curves in problems 1,4,5, and 8 is counterclockwise. The orientation of the curves in problems 2,3,6, and 7 is clockwise. The conic section traced out by the position functions in problems 1-8 is an ellipse.

Here is the image curve for exercises 11 and 12:

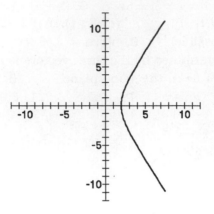

Here is the image curve for exercises 13 and 14:

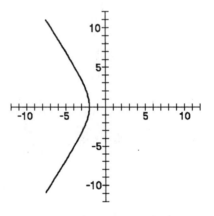

Here is the image curve for exercises 15 and 16:

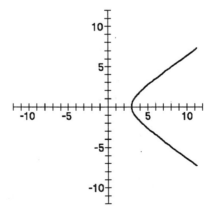

Here is the image curve for exercises 17 and 18:

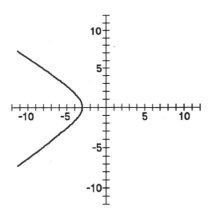

19. The orientation of the curves in problems 11,14,15, and 18 is clockwise. The orientation of the curves in problems 12,13,16, and 17 is counterclockwise. The conic section traced out by each position function is an hyperbola.

21.

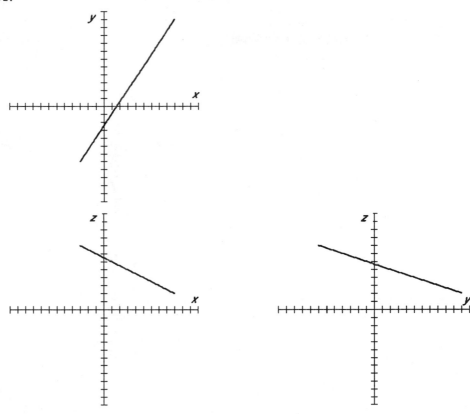

23.

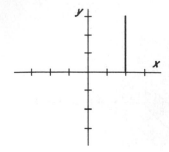

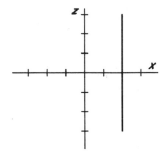

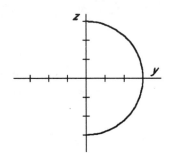

25.

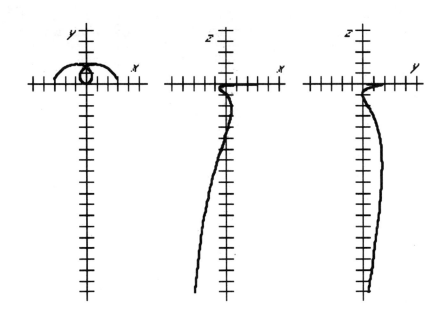

27.

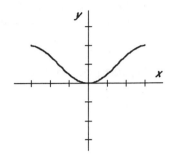

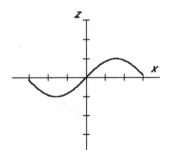

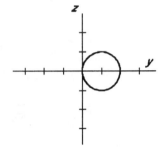

29.

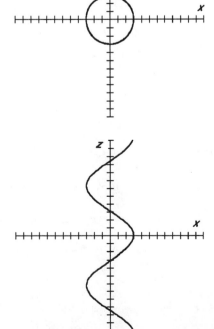

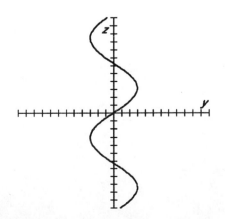

13.2

1. $\mathbf{v}(t) = \langle 6t, \frac{1}{2\sqrt{t}} \rangle$, $\mathbf{v}(2) = \langle 12, \frac{1}{2\sqrt{2}} \rangle$ 3. $\mathbf{T}(t) = \frac{\langle 12t^{\frac{3}{2}}, 1 \rangle}{\sqrt{144t^3 + 1}}$,

$\mathbf{T}(2) \approx \langle .9996, .0295 \rangle$ 5. $\langle x(t), y(t) \rangle = \langle 12 + 12t, \sqrt{2} + \frac{1}{2\sqrt{2}}t \rangle$

7. $\mathbf{a}(t) = \langle -9\sin(3t), -9\cos(3t), \frac{3}{2}t^{-\frac{1}{2}} \rangle$, $\mathbf{a}(1) \approx \langle -1.27, 8.91, 1.5 \rangle$

9. speed ≈ 4.24 11. $\mathbf{v}(t) = \langle t, 2t^{\frac{1}{2}}, 2 \rangle$, $\mathbf{v}(1) = \langle 1, 2, 2 \rangle$

13. $\mathbf{T}(t) = \frac{\langle t, 2t^{\frac{1}{2}}, 2 \rangle}{|t+2|}$, $\mathbf{T}(1) = \langle \frac{1}{3}, \frac{2}{3}, \frac{2}{3} \rangle$

15. $\langle x(t), y(t), z(t) \rangle = \langle \frac{1}{2} + t, \frac{4}{3} + 2t, 2 + 2t \rangle$

17. $\mathbf{a}(t) = \langle -\pi^2 \sin(\pi t), -\frac{1}{t^2}, e^t \rangle$, $\mathbf{a}(1) = \langle 0, -1, e \rangle$

19. speed $= \sqrt{\pi^2 + 1 + e^2} \approx 4.27$

21. $t = e$ 23. $t = 1, t = 2$ 25. $\langle x(t), y(t), z(t) \rangle = \langle -\pi t, t, e + et \rangle$

13.3

1. $\mathbf{r}'(t) = \langle 6t, \frac{1}{2\sqrt{t}} \rangle$, $\mathbf{r}'(\frac{1}{2}) = \langle 3, \frac{1}{\sqrt{2}} \rangle$ 3. $\theta \approx 20°$ 5. $\mathbf{T}(t) = \frac{\langle 12t^{\frac{3}{2}}, 1 \rangle}{\sqrt{144t^3 + 1}}$,

$\mathbf{T}(\frac{1}{2}) = \frac{\langle 3\sqrt{2}, 1 \rangle}{\sqrt{19}} \approx \langle .97, .23 \rangle$ 7. $\|\mathbf{r}'(t)\| = \sqrt{\frac{144t^3 + 1}{4t}}$

9. $\frac{d}{dt}(f(t)\mathbf{r}(t)) = \langle 6t - 12t^3, (t^{-1/2} - 5t^{3/2})/2 \rangle$, computed either directly or by use of the scalar product rule.

11. $\mathbf{v}_1(t) = \langle 3\cos(3t), -3\sin(3t), 3t^{\frac{1}{2}} \rangle$, $\mathbf{v}_2(t) = \langle t, 2t^{\frac{1}{2}}, 2 \rangle$

13. $\mathbf{T}_1(t) = \frac{\langle 3\cos(3t), -3\sin(3t), 3t^{\frac{1}{2}} \rangle}{3\sqrt{1+t}}$, $\mathbf{T}_2(t) = \frac{\langle t, 2t^{\frac{1}{2}}, 2 \rangle}{t+2}$

15. $\mathbf{a}(t) = \langle 6\cos(3t^2) - 36t^2 \sin(3t^2), -6\sin(3t^2) - 36t^2\cos(3t^2), 12t \rangle$

17. $\frac{d}{dt}(\mathbf{r}_1 \cdot \mathbf{r}_2) = \sin(3t)(t - 4t^{\frac{3}{2}}) + \cos(3t)(\frac{3}{2}t^2 + 2t^{\frac{1}{2}}) + 10t^{\frac{3}{2}}$, computed either directly or by applying the dot product rule.

19. $\|\mathbf{r}_1\| = \sqrt{1 + 4t^3}$, $\frac{d}{dt}(\|\mathbf{r}_1\|) = \frac{6t^2}{\sqrt{1 + 4t^3}}$,

$\|\mathbf{r}_2\| = \frac{t\sqrt{9t^2 + 64t + 144}}{6}$, $\frac{d}{dt}(\|\mathbf{r}_2\|) = \frac{3t^2 + 16t + 24}{\sqrt{9t^2 + 64t + 144}}$

21. Computed either directly or by the cross product rule, we have

$$\frac{d}{dt}(\mathbf{r}_1 \times \mathbf{r}_2) = (y_1' z_2 + y_1 z_2' - y_2' z_1 - y_2 z_1')\mathbf{i}$$
$$+ (-x_1' z_2 - x_1 z_2' + x_2' z_1 + x_2 z_1')\mathbf{j}$$
$$+ (x_1' y_2 + x_1 y_2' - x_2' y_1 - x_2 y_1')\mathbf{k}.$$

23. Let $\mathbf{r}(t) = \langle \sin(t), \cos(t) \rangle$.   **25.** Let $\mathbf{r}(t) = \langle x(t), y(t), z(t) \rangle$, then

$$\frac{d\mathbf{T}}{dt} = \left(\frac{y'(y'x'' - x'y'') + z'(z'x'' - x'z'')}{((x')^2 + (y')^2 + (z')^2)^{3/2}} \right)\mathbf{i}$$

$$+ \left(\frac{x'(x'y'' - y'x'') + z'(z'y'' - y'z'')}{((x')^2 + (y')^2 + (z')^2)^{3/2}} \right)\mathbf{j}$$

$$+ \left(\frac{x'(x'z'' - z'x'') + y'(y'z'' - z'y'')}{((x')^2 + (y')^2 + (z')^2)^{3/2}} \right)\mathbf{k},$$

and $\dfrac{d\mathbf{T}}{dt} \times \mathbf{T} == \dfrac{\mathbf{r}''(t)}{\|\mathbf{r}'(t)\|} \times \mathbf{T}(t)$.

13.4

1. $\mathbf{a}(t) = \langle 6t, \frac{1}{2\sqrt{t}} \rangle$ **3.** $\mathbf{r}(4) - \mathbf{r}(1) = \langle 64, \frac{16}{3} \rangle - \langle 1, \frac{2}{3} \rangle = \langle 63, \frac{14}{3} \rangle$. Therefore, the net distance is $= \|\langle 63, \frac{14}{3}, \rangle\| \approx 63.17$.   **5.** $\mathbf{r}(t) = \langle t^3 - 3, \frac{2}{3}t^{\frac{3}{2}} + \frac{7}{3} \rangle$, $\mathbf{r}(4) = \langle 61, \frac{23}{3} \rangle$ **7.** $s(t) = \int_1^t (9u^4 + u)^{\frac{1}{2}} du$ **9.** $\int \mathbf{r}(t)dt = \langle \frac{-\cos(3t)}{3}, \frac{\sin(3t)}{3}, \frac{4}{5}t^{\frac{5}{2}} \rangle + \mathbf{C}$ **11.** $\mathbf{r}(2) - \mathbf{r}(0) = \langle \sin(6), \cos(6), 2^{\frac{5}{2}} \rangle - \langle 0, 1, 0 \rangle = \langle \sin(6), \cos(6) - 1, 2^{\frac{5}{2}} \rangle$. Therefore, the net distance is $\|\langle \sin(6), \cos(6) - 1, 2^{\frac{5}{2}} \rangle\| \approx 5.66$ **13.** $s(t) = 2[(1 + t)^{\frac{3}{2}} - 1]$ **15.** $\mathbf{v}(t) = \langle \frac{t^3}{3} + 1, \frac{t^4}{4} + 2, t^2 + 3 \rangle$, terminal velocity $= \langle \frac{11}{3}, 6, 7 \rangle$ **17.** $\mathbf{r}(2) - \mathbf{r}(0) = \langle \frac{7}{3}, \frac{28}{5}, \frac{20}{3} \rangle - \langle -1, 0, -2 \rangle = \langle \frac{10}{3}, \frac{28}{5}, \frac{26}{3} \rangle$. Therefore, the net distance is $\|\langle \frac{10}{3}, \frac{28}{5}, \frac{26}{3} \rangle\| \approx 10.84$ **19.** total distance $= 10\pi$; reparametrization by arc length: $s : [0, 10\pi] \to \mathbb{R}^3$, $s \mapsto \langle 3\sin(\frac{s}{5}), 3\cos(\frac{s}{5}), \frac{4s}{5} \rangle$

13.5

1. $\mathbf{T}(t) = \frac{\langle 12t^{\frac{3}{2}}, 1 \rangle}{\sqrt{144t^3 + 1}}$, $\mathbf{T}(2) = \frac{\langle 24\sqrt{2}, 1 \rangle}{\sqrt{1153}} \approx \langle .99957, .02945 \rangle$

3. $\kappa(2) \approx .0018$, $\rho(2) \approx 544$

5. $\mathbf{N}(t) = \frac{\langle -t^{\frac{1}{2}}(6(1+t)\sin(3t) + \cos(3t)), -t^{\frac{1}{2}}(6(1+t)\cos(3t) - \sin(3t)), 1 \rangle}{\sqrt{36t(1+t)^2 + t + 1}}$

$\mathbf{N}(1) = \langle \frac{-12\sin(3) - \cos(3)}{\sqrt{146}}, \frac{-12\cos(3) + \sin(3)}{\sqrt{146}}, \frac{1}{\sqrt{146}} \rangle \approx \langle -.058, .995, .083 \rangle$

7. approximate equation: $.71x - .017y + .70z - 1.52 = 0$

9. $\mathbf{N}(t) = \frac{\langle 2\sqrt{t}, -t+2, -2\sqrt{t} \rangle}{t+2}$, $\mathbf{N}(1) = \langle \frac{2}{3}, \frac{1}{3}, \frac{-2}{3} \rangle$ **11.** $\frac{2}{3}x - \frac{2}{3}y + \frac{1}{3}z - \frac{1}{9} = 0$

13. $\mathbf{N}(t) = \frac{\langle 2t, -2t, -2t^2 + 1 \rangle}{2t^2 + 1}$, $\mathbf{N}(1) = \langle \frac{2}{3}, -\frac{2}{3}, -\frac{1}{3} \rangle$.

15. $-\frac{1}{3}x - \frac{2}{3}y + \frac{2}{3}z - 1 = 0$.   **17.** Let $\mathbf{B} \cdot \mathbf{B} = 1$. Then, differentiate with respect to s to obtain

$$\frac{d\mathbf{B}}{ds} \cdot \mathbf{B} + \mathbf{B} \cdot \frac{d\mathbf{B}}{ds} = 0 \implies 2(\mathbf{B} \cdot \frac{d\mathbf{B}}{ds}) = 0 \implies \mathbf{B} \cdot \frac{d\mathbf{B}}{ds} = 0.$$

This implies $\dfrac{d\mathbf{B}}{ds}$ is orthogonal to $\mathbf{B}$.

19. $\dfrac{d\mathbf{B}}{dt} = \dfrac{d\mathbf{B}}{ds}\left(\dfrac{ds}{dt}\right) \implies \dfrac{d\mathbf{B}}{ds} = \dfrac{d\mathbf{B}/dt}{ds/dt}$ where $\dfrac{ds}{dt} = \|\mathbf{v}\|$.

4) $\frac{d\mathbf{B}}{ds} \approx .479\mathbf{N}$;   8) $\frac{d\mathbf{B}}{ds} = -\frac{1}{9}\mathbf{N}$;   12) $\frac{d\mathbf{B}}{ds} = \frac{2}{3}\mathbf{N}$.

Centers and radii of circles are approximate in exercises 21-30.

21. $\kappa \approx .024852$, $r \approx 40.238$, $(x + 37.544)^2 + (y - 14.480)^2 = (40.238)^2$

23. $\kappa = 12$, $r = \frac{1}{12}$, $(x + 2)^2 + (y - 15.917)^2 = (.083333)^2$.

25. $\kappa \approx .15086$, $r \approx 6.6287$, $(x + 5.5)^2 + (y - 2.3)^2 = (6.6287)^2$.

27. $\kappa \approx .43196$, $r \approx 2.3150$, $(x + .46833)^2 + (y - 2.5)^2 = (2.3150)^2$.

29. $\kappa \approx 0$.

31. $\mathbf{v}(t) = \langle x'(t), y'(t) \rangle$, and

$$\mathbf{T} = \frac{\langle x'(t), y'(t) \rangle}{\sqrt{(x'(t))^2 + (y'(t))^2}} = \frac{x'(t)}{\sqrt{(x'(t))^2 + (y'(t))^2}}\mathbf{i} + \frac{y'(t)}{\sqrt{(x'(t))^2 + (y'(t))^2}}\mathbf{j}.$$

Therefore,

$$\frac{d\mathbf{T}}{dt} = \frac{y'(y'x'' - x'y'')}{((x')^2 + (y')^2)^{\frac{3}{2}}}\ \mathbf{i} + \frac{x'(x'y'' - y'x'')}{((x')^2 + (y')^2)^{\frac{3}{2}}}\ \mathbf{j},$$

and we have

$$\left\|\frac{d\mathbf{T}}{dt}\right\| = \frac{|x'y'' - y'x''|}{((x')^2 + (y')^2)^{\frac{3}{2}}}.$$

33. $v(t)\mathbf{T}(t) = v(t)\dfrac{\mathbf{v}(t)}{v(t)} = \mathbf{v}(t)$

35. $\kappa(t)v(t)\mathbf{N}(t) = \left(\dfrac{\|d\mathbf{T}/dt\|}{\|\mathbf{v}(t)\|}\right)\|\mathbf{v}(t)\|\left(\dfrac{d\mathbf{T}/dt}{\|d\mathbf{T}/dt\|}\right) = \dfrac{d\mathbf{T}}{dt}$.

37.

$$\mathbf{a}(t) \times \mathbf{v}(t) = \left(\frac{dv}{dt}\mathbf{T}(t) + \kappa(t)v^2(t)\mathbf{N}(t)\right) \times (v(t)\mathbf{T}(t))$$

$$= \frac{dv}{dt}\mathbf{T}(t) \times v(t)\mathbf{T}(t) + \kappa(t)v^2(t)\mathbf{N}(t) \times v(t)\mathbf{T}(t)$$

$$= v(t)\frac{dv}{dt}\mathbf{T}(t) \times \mathbf{T}(t) + \kappa(t)v^3(t)\mathbf{N}(t) \times \mathbf{T}(t)$$

$$= 0 + \kappa(t)v^3(t)\mathbf{N}(t) \times \mathbf{T}(t)$$

$$= \kappa(t)v^3(t)\mathbf{N}(t) \times \mathbf{T}(t).$$

39. $\dfrac{\mathbf{a}(t) \times \mathbf{v}(t)}{v^3(t)} = \dfrac{\kappa(t)v^3(t)\mathbf{N}(t) \times \mathbf{T}(t)}{v^3(t)} = \kappa(t)\mathbf{N}(t) \times \mathbf{T}(t) = \kappa(t)\mathbf{B}(t)$.

Chapter 14

14.1

1. A vector parallel to the line of intersection is $\langle 1, 0, 1 \rangle$. The line of intersection can be described by $\langle x(t), y(t), z(t) \rangle = \langle 0, 0, 0 \rangle + t\langle 1, 0, 1 \rangle$.   3. Domain $= \{(x, y) : y \neq 0\}$.   5. Hyperbolic cylinder running parallel to the y-axis. The cross-section in the xz plane is shown.   7. Circular cylinder with the y-axis running down its center. The cross section in the xz plane is shown.   9. Square cylinder with the z-axis running down its center. The cross-section in the xy plane is shown.   11. $(x + 2)^2 + (y - 4)^2 + (z + 7)^2 = 49$.   13. when $a = b = c$.   15. It is not the graph of a function.   17. It is not the graph of a function.   19. It is the graph of a function.   21. $\dfrac{y^2}{a^2} + \dfrac{z^2}{b^2} = 1$, $\dfrac{y^2}{a^2} + \dfrac{z^2}{b^2} = 1$ 23. $\dfrac{x^2}{c^2} = \dfrac{y^2}{a^2} + \dfrac{z^2}{b^2} - 1$, $\dfrac{y^2}{c^2} = \dfrac{x^2}{a^2} + \dfrac{z^2}{b^2} - 1$ 25. $cx = \dfrac{y^2}{a^2} + \dfrac{z^2}{b^2}$, $cy = \dfrac{x^2}{a^2} + \dfrac{z^2}{b^2}$

14.2

1. minima at $\approx (1, .6)$, $(1.35, 0)$, $(0, 1.65)$; maxima at $(0, 0)$, $(3, 3)$.

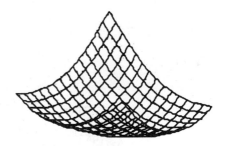

3. minima at $(0, x)$ for $0 \le x = le3$; maxima at $(3, 3)$.

5. minimum at $(0, 3)$; maximum at $(3, 0)$.

7. saddle

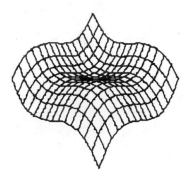

9. saddle

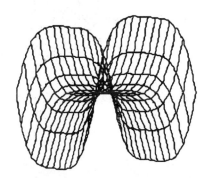

11. saddle

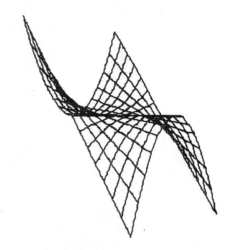

13. trough

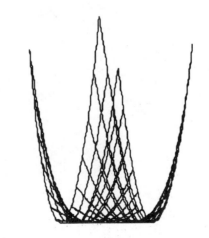

15. uphill   17. uphill   19. downhill   21. downhill   23. N, S

29.

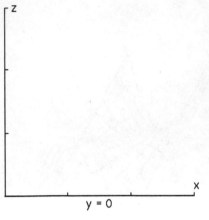

y = 0

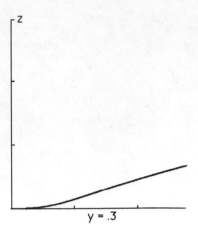

y = .3

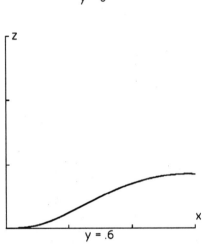

y = .6

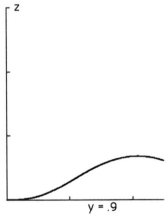

y = .9

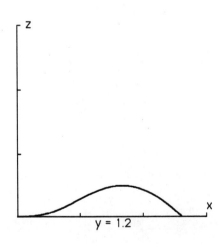

y = 1.2

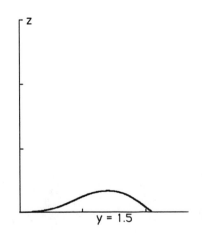

y = 1.5

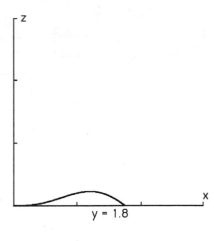

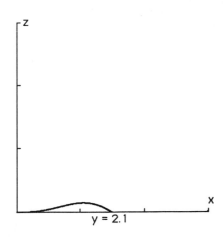

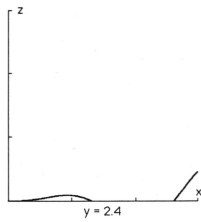

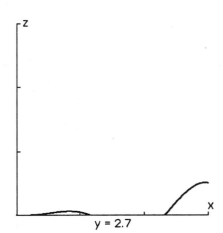

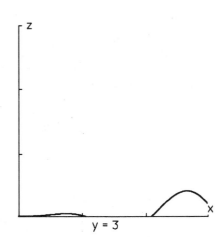

14.3

1. $m_1 = 2$ 3. $z = 2x + \frac{1}{2}y - \frac{3}{2}$ 5. $z = 2(x+3) + \frac{1}{2}(y-2) - \frac{13}{2}$ 7. $0, 1, 0, z = (y-4) + 4$ 9. slopes undefined, plane contains entire z-axis, plane is not the graph of a function   11. Find the discriminant: $A = 3$, $B = 0$, $C = -1$, $D = 3(-1) - 0^2 = -3$. since $D < 0$, a saddle point occurs at $(0,0)$.   13. Find the discriminant: $A = 1$, $B = -2$, $C = 4$, $D = (1)(4) - (-2)^2 = 0$. No strict maximum, minimum, nor saddle point occurs at $(0,0)$. Graph is a trough whose bottom edge lies along the line $x - 2y = 0$. 15. Find the discriminant: $A = 0$, $B = 0$, $C = 0$, $D = 0$. No strict maximum, minimum, no saddle point occurs at $(0,0)$. Graph is the xy plane.   17. $z = (x+1)^2 - (y-3)^2 - 3$.   19. $z = (x+1)^2 + 2(y-3)^2 + 2(x+1) - 12(y-3) - 12$. 21. Find the discriminant: $A = -1$, $B = 4$, $C = -4$, $D = (-1)(-4) - 4^2 = -12$. since $D < 0$, a saddle point occurs at $(-1, 3, 17)$.   Answers may vary for exercises 24-30.   25. $z = -2x^2 - 3(y-1)^2 - 5$.   27. $z = (2x - y + 3)^2 + 5$. 29. $2y + 3z = 6$.

14.4

1. $(\sqrt{5}, \text{arccot}(\frac{1}{2}), -2) \approx (2.24, 1.11, -2)$;

   $(3, \text{arccot}(\frac{1}{2}), \text{arccot}(\frac{-2}{\sqrt{5}})) \approx (3, 1.11, 2.30)$

3. $(-\frac{5\sqrt{3}}{4}, \frac{15}{4}, \frac{-5}{2}) \approx (-2.17, 3.75, -2.5)$; $(5\frac{\sqrt{3}}{2}, \frac{2\pi}{3}, -\frac{5}{2}) \approx (4.33, 2.09, -2.5)$

5. $(-\frac{7}{\sqrt{2}}, \frac{7}{\sqrt{2}}, 6) \approx (-4.95, 4.95, 6)$; $(\sqrt{85}, \frac{3\pi}{4}, \text{arccot}(\frac{6}{7})) \approx (9.22, 2.36, .86)$

7. $(0, 0, -5); (5, 0, \pi)$ 9. $(0, 4, 0); (4, \frac{\pi}{2}, 0)$

11. $(\sqrt{3}, -1, -1) \approx (1.73, -1, -1)$; $(\sqrt{5}, 330°, \text{arccot}(-\frac{1}{2})) \approx (2.24, 330°, 117°)$

13. points on the positive x-axis and points on the z-axis

15. $(0, 0, 0)$ 17. Points on the z-axis have more than one representation in the cylindrical and spherical coordinate systems. (Since $r = 0$, θ can be any angle.)

19. $z = r^2 \cos 2\theta$; $\rho \cos \phi = \rho^2 \sin^2 \phi \cos 2\theta$ or $1 = \rho \tan \phi \sin \phi \cos 2\theta$

21. $z = 5$; $\rho \cos \phi = 5$

23. $r^2 = 9$ or $r = 3$; $\rho^2 \sin^2 \phi = 9$ or $\rho \sin \phi = 3$

25. $z = r^2 \sin^2 \theta$; $\rho \cos \phi = \rho^2 \sin^2 \phi \sin^2 \theta$ or $1 = \rho \tan \phi \sin \phi \sin^2 \theta$

27. $z = \cot \theta$; $\rho \cos \phi = \cot \theta$

29. $z^2 = r^2 + 1$; $\rho^2 \cos^2 \phi = \rho^2 \sin^2 \phi + 1$ or $\rho^2 \cos 2\phi = 1$

Chapter 15

15.1

1. $\frac{\partial f}{\partial x} = \frac{-2y}{(x-y)^2}$; $\quad \frac{\partial f}{\partial x}\Big|_{(1,2)} = -4$; $\quad \frac{\partial f}{\partial y}\frac{2x}{(x-y)^2}$; $\quad \frac{\partial f}{\partial y}\Big|_{(1,2)} = 2$.

3. $\frac{\partial f}{\partial x} = e^{x^2+y^2}(2x)$; $\quad \frac{\partial f}{\partial x}\Big|_{(1,-2)} = 2e^5$; $\quad \frac{\partial f}{\partial y} = e^{x^2+y^2}(2y)$; $\quad \frac{\partial f}{\partial y}\Big|_{(1,-2)} = -4e^5$.

5. $\frac{\partial f}{\partial x} = 2xy\cos(x^2y)$; $\quad \frac{\partial f}{\partial x}\Big|_{(2,0)} = 0$; $\quad \frac{\partial f}{\partial y} = x^2\cos(x^2y)$; $\quad \frac{\partial f}{\partial y}\Big|_{(2,0)} = 4$.

7. $\frac{\partial f}{\partial x} = \left(\frac{-y}{x^2}\right)\sec^2\left(\frac{y}{x}\right)$; $\quad \frac{\partial f}{\partial x}\Big|_{(2,\frac{\pi}{2})} = \frac{-\pi}{4}$; $\quad \frac{\partial f}{\partial y} = \left(\frac{1}{x}\right)\sec^2\left(\frac{y}{x}\right) + \csc^2(y)$; $\quad \frac{\partial f}{\partial y}\Big|_{(2,\frac{\pi}{2})} = 2$.

9. $\frac{\partial f}{\partial x} = \frac{x}{\sqrt{x^2+y^2}}$; $\quad \frac{\partial f}{\partial x}\Big|_{(-3,4)} = -.\overline{6}$; $\quad \frac{\partial f}{\partial y} == \frac{y}{\sqrt{x^2+y^2}}$; $\quad \frac{\partial f}{\partial x}\Big|_{(-3,4)} = .8$.

11. $f_x = \frac{-z}{(x+y)^2}$; $\quad f_x(1,2,3) = -.\overline{3}$; $\quad f_y = \frac{-z}{(x+y)^2}$; $\quad f_y(1,2,3) = -.\overline{3}$;

$f_z = \frac{1}{x+y}$; $f_z(1,2,3) = .\overline{3}$.

13. $f_x = 5$; $\quad f_x(-3,2,4) = 5$; $\quad f_y = -4$; $\quad f_y(-3,2,4) = -4$;

$f_z = 3$; $\quad f_z(-3,2,4) = 3$. $\qquad$ 15. $f_x = \arcsin z$; $\quad f_x(.5,.5,.5) = \arcsin(.5) = \frac{\pi}{6}$;

$f_y = \arctan z$; $\quad f_y(.5,.5,.5) = \arctan(.5) \approx . - .46$; $\quad f_z = \frac{x}{\sqrt{1-z^2}} - \frac{y}{1+z^2}$;

$f_z(.5,.5,.5) = \frac{.5}{\sqrt{1-.5^2}} - \frac{.5}{1+.5^2} \approx .18$

17. (uphill). $\qquad$ 19. (downhill). $\qquad$ 21. (downhill). $\qquad$ 23. (uphill).

25. $\frac{\partial P}{\partial y}\Big|_{(0,0)} \approx 2.47$. $\qquad$ 27. $\frac{\partial P}{\partial y}\Big|_{(.5,-.12)} \approx 3.72$. $\qquad$ 29. $\frac{\partial P}{\partial y}\Big|_{(0,0)} \approx 2.51$.

31. $\frac{\partial P}{\partial y}\Big|_{(-.35,-.42)} \approx 2.76$. $\qquad$ 33. $\frac{\partial T}{\partial y}\Big|_{(0,-3,0)} = 18$.

35. The temperature will decrease at points where the z coordinate is positive regardless of whether we move straight up or straight down.

15.2

1. $Df = \begin{bmatrix} \frac{-2y}{(x-y)^2} & \frac{2x}{(x-y)^2} \end{bmatrix}$; $\quad f(x,y) \approx -3 - 4(x-1) + 2(y-2)$.

3. $Df = \begin{bmatrix} 2xe^{x^2+y^2} & 2ye^{x^2+y^2} \end{bmatrix}$; $\quad f(x,y) \approx e^5 + 2e^5(x-1) - 4e^5(y+2)$.

5. $Df = \begin{bmatrix} 2xy\cos(x^2y) & x^2\cos(x^2y) \end{bmatrix}$; $\quad f(x,y) \approx 4y$.

7. $Df = \begin{bmatrix} \frac{-y}{x^2}\sec^2\left(\frac{y}{x}\right) & \left(\frac{1}{x}\right)\sec^2\left(\frac{y}{x}\right) + \csc^2(y) \end{bmatrix}$; $\quad f(x,y) \approx 1 - \frac{\pi}{4}(x-2) + 2(y-\frac{\pi}{2})$.

9. $Df = \begin{bmatrix} \frac{x}{\sqrt{x^2+y^2}} & \frac{y}{\sqrt{x^2+y^2}} \end{bmatrix}$; $\quad f(x,y) \approx 5 - .6(x+3) + .8(y-4)$.

11. $Df = \left[\begin{array}{ccc} \dfrac{-z}{(x+y)^2} & \dfrac{-z}{x+y)^2} & \dfrac{1}{x+y} \end{array}\right]; \quad f(x,y,z) \approx 1 - .\overline{3}(x-1) - .\overline{3}(y-2) +$
$.\overline{3}(z-3).$

13. $Df = \begin{bmatrix} 5 & -4 & 3 \end{bmatrix}; \quad f(x,y,z) \approx -11 + 5(x+3) - 4(y-2) + 3(z-4).$

15. $Df = \left[\begin{array}{ccc} \arcsin z & \arctan z & \dfrac{x}{\sqrt{1-z^2}} - \dfrac{y}{|z|\sqrt{z^2-1}} \end{array}\right];$
$Df(.5,.5,.5) = \begin{bmatrix} .52 & -,46 & .18 \end{bmatrix};$

$$f(x,y,z) \approx f(.5,.5,.5) + \begin{bmatrix} .52 & -.46 & .18 \end{bmatrix} \begin{bmatrix} x - .5 \\ y - .5 \\ z - .5 \end{bmatrix}$$

or $f(x,y,z) \approx .03 + .52(x - .5) - .46(y - .5) + .18(z - .5).$

17. $DH(2,3) = \begin{bmatrix} -112 & 12 \end{bmatrix}; \quad H(2.1, 3.1) - (-47) \approx -10.$

19. $DH(-1,2) = \begin{bmatrix} .28 & 16 \end{bmatrix}; \quad H(-1.3, 1.8) - 1 \approx -11.6.$

21. $DH(3,-2) = \begin{bmatrix} .\overline{3} & -.75 \end{bmatrix}; \quad H(3.2, -2.1) - (-1.5) \approx .14.$

23. $DH(-2,-1) = \begin{bmatrix} -1 & 2 \end{bmatrix}; \quad H(-2.2, -.9) - 2 \approx .4.$

25. $DT = \left[\begin{array}{ccc} \dfrac{-600x}{1 + x^2 + y^2 + z^2)^2} & \dfrac{-600y}{1 + x^2 + y^2 + z^2)^2} & \dfrac{-600z}{1 + x^2 + y^2 + z^2)^2} \end{array}\right];$
$T(1.1, 1.9, 3.2) - 20 \approx -1.\overline{3}.$

15.3

1. $D_{\mathbf{u}}f(1,2) = 2.$ **3.** $D_{\mathbf{u}}f(2,0) = 4.$ **5.** $D_{\mathbf{u}}f(0,0) = -.8.$

7. $\mathbf{u} \approx \langle -0.58 - .58, .58 \rangle.$ **9.** $\mathbf{u} \approx \langle .71, -.57, .42 \rangle.$

11. $\nabla f = \langle \arcsin z, -\arctan z, \dfrac{x}{\sqrt{1-z^2}} - \dfrac{y}{1+z^2} \rangle;$

$\nabla f(.5,.5,.5) = \langle .52, -.46, .18 \rangle;$

$\mathbf{u} = \dfrac{\langle .52, -.46, .18 \rangle}{\sqrt{(.52)^2 + (-.46)^2 + (.18)^2}} \approx \langle .73, -.64, .25 \rangle.$

13. $D_{\mathbf{u}}T(-1,-3,-2) \approx -9.1.$

15. $\mathbf{u} \approx \langle .97, -.16, -.16 \rangle.$ **17.** $D_{\mathbf{r}}T\Big|_{t=4} \approx -3.53.$

19. $D_{\mathbf{i}}f = \left\langle \dfrac{2x}{1+(x^2+y^2+z^2)^2}, \dfrac{2y}{1+(x^2+y^2+z^2)^2}, \dfrac{2z}{1+(x^2+y^2+z^2)^2} \right\rangle \cdot \langle 1,0,0 \rangle$

$\qquad = \dfrac{2x}{1 + (x^2 + y^2 + z^2)^2} = f_x;$

$D_{\mathbf{j}}f = \left\langle \dfrac{2x}{1+(x^2+y^2+z^2)^2}, \dfrac{2y}{1+(x^2+y^2+z^2)^2}, \dfrac{2z}{1+(x^2+y^2+z^2)^2} \right\rangle \cdot \langle 0,1,0 \rangle$

$\qquad = \dfrac{2y}{1 + (x^2 + y^2 + z^2)^2} = f_y;$

$D_{\mathbf{k}}f = \left\langle \dfrac{2x}{1+(x^2+y^2+z^2)^2}, \dfrac{2y}{1+(x^2+y^2+z^2)^2}, \dfrac{2z}{1+(x^2+y^2+z^2)^2} \right\rangle \cdot \langle 0,0,1 \rangle$

$\qquad = \dfrac{2z}{1 + (x^2 + y^2 + z^2)^2} = f_z.$

21. $f(x,y,z) \approx \frac{\pi}{4} + x - 1$.   23. $D_{\mathbf{u}}H(-1,2) \approx 8.49$; $\mathbf{u} \approx \langle .87, -.5 \rangle$.   25. $D_{\mathbf{u}}H(-2,-1) \approx -31.11$; $\mathbf{u} \approx \langle .50, -.87 \rangle$.   27. $D_{\mathbf{u}}H(-1,2) \approx -.18$; $\mathbf{u} \approx \langle -.89, -.45 \rangle$.   29. $D_{\mathbf{u}}H(-2,-1) \approx -.71$; $\mathbf{u} \approx \langle -.45, .89 \rangle$.

Section 15.4

1. $f_x = 2x$; $f_x(0,0) = 0$; $f_y = 2y$; $f_y(0,0) = 0$; $f_{xx} = 2$; $f_{xx}(0,0) = 2$; $f_{xy} = 0$; $f_{xy}(0,0) = 0$; $f_{yy} = 2$; $f_{yy}(0,0) = 2$; $f(x,y) \approx \frac{1}{2}[2x^2 + 2y^2]$ or $f(x,y) \approx x^2 + y^2$; $A = 1, B = 0, C = 1$; $D = (1)(1) = 1$; $D > 0, A > 0 \Rightarrow$ minimum at $(0,0) \Rightarrow$ bowl up.

3. $f_x = 3x^2$; $f_x(0,0) = 0$; $f_y = -2y$; $f_y(0,0) = 0$; $f_{xx} = 6x$; $f_{xx}(0,0) = 0$; $f_{xy} = 0$; $f_{xy}(0,0) = 0$; $f_{yy} = -2$; $f_{yy}(0,0) = -2$; $f(x,y) \approx \frac{1}{2}(-2)y^2 = -y^2$; $A = 0, B = 0, C = -1$; $D = 0$ (trough).

5. $f_x = y$; $f_x(0,0) = 0$; $f_y = x$; $f_y(0,0) = 0$; $f_{xx} = 0$; $f_{xx}(0,0) = 0$; $f_{xy} = 1$; $f_{xy}(0,0) = 1$; $f_{yy} = 0$; $f_{yy}(0,0) = 0$; $f(x,y) \approx \frac{1}{2}(2)xy = xy$; $A = C = 0, B = \frac{1}{2}, D = \frac{-1}{4}$; $D < 0 \Rightarrow$ a saddle point at $(0,0)$.

7. $f_x = y^3$; $f_x(0,0) = 0$; $f_y = 3x^2$; $f_y(0,0) = 0$; $f_{xx} = 0$; $f_{xx}(0,0) = 0$; $f_{xy} = 3y^2$; $f_{xy}(0,0) = 0$; $f_{yy} = 6xy$; $f_{yy}(0,0) = 0$; $f(x,y) \approx 0$; $D = 0 \Rightarrow$

9. The slice of $z = f(x,y)$ with $y = a$ is increasing and concave up. The slice of $z = f(x,y)$ with $x = b$ is decreasing and concave up. The y-slopes are decreasing as we move from slice to slice in the direction of increasing x.

11. $f_x(x,y) = \dfrac{-z}{(x+y)^2}$; $f_y(x,y) = \dfrac{-z}{(x+y)^2}$;

$f_z(x,y) = \frac{1}{x+y}$; $f_{xy}(x,y) = \dfrac{2z}{(x+y)^3} = f_{yx}(x,y)$;

$f_{xz}(x,y) = \frac{-1}{(x+y)^2} = f_{zx}(x,y)$; $f_{yz}(x,y) = \frac{-1}{(x+y)^2} = f_{zy}(x,y)$;

$f_{xx}(x,y) = \dfrac{2z}{(x+y)^3} = f_{yy}(x,y)$; $f_{zz}(x,y) = 0$.

13. $f_x(x,y) = 5$; $f_y(x,y) = -4$; $f_z = 3$;   all second order partial derivatives are 0.

15. $f_x(x,y) = \arcsin z$; $f_y(x,y) = \arctan z$; $f_z(x,y) = \dfrac{x}{\sqrt{1-z^2}} - \dfrac{y}{1+z^2}$;

$f_{xy}(x,y) = 0 = f_{yx}(x,y)$; $f_{xz}(x,y) = \dfrac{1}{\sqrt{1-z^2}} = f_{zx}(x,y)$;

$f_{yz}(x,y) = \dfrac{-1}{1+z^2} = f_{zy}(x,y)$; $f_{xx}(x,y) = f_{yy}(x,y) = 0$;

$f_{zz}(x,y) = \dfrac{xz}{(1-z^2)^{\frac{3}{2}}} + \dfrac{-2z}{1+z^2}$.

17. $f_{xx}(x,y) = 0$; $f_{xy}(x,y) = 1 - \dfrac{1}{y^2}$; $f_{yy}(x,y) = \dfrac{2x}{y^3}$;

$f(x,y) \approx -3.\overline{3} + 3.\overline{3}(x+1) - .\overline{8}(y-3) + .\overline{8}(x+1)(y-3) - .\overline{037}(y-3)^2$.

19. $f_{xx}(x,y) = 2y\cos(x^2 y) - 4x^2 y^2 \sin(x^2 y)$;

$f_{xy}(x,y) = 2x\cos(x^2 y - 2x^3 y\sin(x^2 y)$; $f_{yy}(x,y) = -x^4 \sin(x^2 y)$;

$$f(x, y) \approx 4y + 4(x - 2)y.$$

15.5

1. a saddle point at $(-.5, -.5)$.

3.

point	A	B	C	Δ	classification
$(0,0)$	0	3	0	-9	saddle point
$(1,1)$	-6	3	-6	27	relative maximum

5. a saddle point at $(0, 0)$.

7.

point	A	B	C	Δ	classification
$(0, 1)$	-16	0	6	-96	saddle point
$(0, -1)$	-16	0	-6	96	relative maximum
$(2, 1)$	32	0	6	192	relative minimum
$(2, -1)$	32	0	-6	-192	saddle point
$(-2, 1)$	32	0	6	192	relative minimum
$(-2, -1)$	32	0	-6	-192	saddle point

9. A relative minimum occurs at $(0, 0)$.   11. $(0, 0)$ is a saddle point.

13. $(\frac{\pi}{2} + n\pi, 0)$ (n any integer) is a saddle point.   15. A local minimum occurs at $(0, 0)$.

17. Absolute minimum value is -2 and absolute maximum value is 1.

19. Absolute minimum value is 0 and absolute maximum value is 2.

21. Absolute minimum value is 0 and absolute maximum value is 4.

23. Absolute minimum value is -1 and absolute maximum value is 2.

25. Absolute minimum value is 0 and absolute maximum value is 1.

26. $(\frac{9}{7}, \frac{6}{7}, \frac{3}{7})$ is the point closest to the origin.

28. $(2, 2, 2)$, $(2, -2, -2)$, $(-2, -2, 2)$, $(-2, 2, -2)$ are the points closest to the origin.

30. $(2, 2, 2)$ and $(-2, -2, -2)$ are the points closest to the origin.

32. Box is a cube with each side of length $\sqrt[3]{6000} \approx 18.17$ cm;   Surface area $= 2(\sqrt[3]{6000})^2 + 2(\sqrt[3]{6000})^2 + 2(\sqrt[3]{6000})^2 = 6(6000)^{\frac{2}{3}} \approx 1981.16$ cm^2.

34. Minimize $C = 2xy + 2xz + 2yz$ subject to $xyz = 6000$;   This is the same problem as 37.

36. $(-2.3, -2.3, 10.6)$ is the point farthest from the origin and $(1.3, 1.3, 3.4)$ is the point closest to the origin.

Chapter 16

16.1

1. 12.   3. $\frac{2}{3}$.   5. $\frac{-23}{3}$.   7. $\frac{\pi^2}{8}(e^{-1} - e)$.   9. 1.   11. $\frac{99}{2}$.

13. $\frac{\pi}{3}(e - e^{-1})$.   15. $(b-a)(d-c)(q-p)$, which is the volume of the cube.

16.2

1. $\frac{7895}{84}$.   3. $-\frac{1}{6}$.   5. $\frac{14}{3}$.   7. $\frac{4}{21}$.   9. $\frac{1}{3}[\frac{2000}{3}(5)^{\frac{3}{4}} - \frac{600}{7}(5)^{\frac{7}{4}} +$

$\frac{60}{11}(5)^{\frac{11}{4}} - \frac{4}{15(5)^{\frac{15}{4}}}] \approx 380.21$.   11. $\int_0^1 \int_{e^y}^e xe^y \, dx \, dy$.   13. 0.   15. 2π.

17. $\frac{2}{7}[-\frac{1}{3}(\frac{1}{\sqrt{2}})^3 + \frac{1}{5}(\frac{1}{\sqrt{2}})^5] \approx -.024$.   19. $\frac{3\pi}{2}$.   21. $\pi - 3\frac{\sqrt{3}}{2}$.   23. $\frac{\pi}{12}$.

16.3

1. $\frac{67}{630}$.   3. $\frac{1}{6}$.   5. 8π.   7. $\int_0^{\frac{\pi}{2}} \int_0^{\frac{\pi}{2}} \int_0^1 \rho^2 \sin\phi \, d\rho \, d\theta \, d\varphi = \frac{\pi}{6}$.   9. $\bar{x} = \frac{16}{35}$; $\bar{y} = \frac{16}{35}$; $\bar{z} = \frac{16}{35}$.   11. $M = \frac{1}{20}$.   13. $I_{xy} = I_{xz} = I_{yz} = \frac{2}{315}$.

15. 32π.

16.4

Exercises 1 and 3 use the following tabulated data:

(x, y)	$e^{x^2+y^2}$	(x, y)	$e^{x^2+y^2}$	(x, y)	$e^{x^2+y^2}$
$(0, 0)$	1	$(.2, .2)$	1.08	$(.1, .1)$	1.02
$(.2, 0)$	1.04	$(.4, .2)$	1.22	$(.1, .3)$	1.11
$(.4, 0)$	1.17	$(.6, .2)$	1.49	$(.1, .5)$	1.30
$(.6, 0)$	1.43	$(.8, .2)$	1.97	$(.1, .7)$	1.65
$(.8, 0)$	1.90	$(1, .2)$	2.83	$(.1, .9)$	2.27
$(0, .2)$	1.04	$(.2, .4)$	1.22	$(.3, .1)$	1.11
$(.2, .2)$	1.08	$(.4, .4)$	1.38	$(.3, .3)$	1.20
$(.4, .2)$	1.22	$(.6, .4)$	1.68	$(.3, .5)$	1.40
$(.6, .2)$	1.49	$(.8, .4)$	2.23	$(.3, .7)$	1.79
$(.8, .2)$	1.97	$(1, .4)$	3.19	$(.3, .9)$	2.46
$(0, .4)$	1.17	$(.2, .6)$	1.49	$(.5, .1)$	1.30
$(.2, .4)$	1.22	$(.4, .6)$	1.68	$(.5, .3)$	1.40
$(.4, .4)$	1.38	$(.6, .6)$	2.05	$(.5, .5)$	1.65
$(.6, .4)$	1.68	$(.8, .6)$	2.72	$(.5, .7)$	2.10
$(.8, .4)$	2.23	$(1, .6)$	3.90	$(.5, .9)$	2.89
$(0, .6)$	1.43	$(.2, .8)$	1.97	$(.7, .1)$	1.65
$(.2, .6)$	1.49	$(.4, .8)$	2.23	$(.7, .3)$	1.79
$(.4, .6)$	1.68	$(.6, .8)$	2.72	$(.7, .5)$	2.10
$(.6, .6)$	2.05	$(.8, .8)$	3.60	$(.7, .7)$	2.66
$(.8, .6)$	2.72	$(1, .8)$	5.16	$(.7, .9)$	3.67
$(0, .8)$	1.90	$(.2, 1)$	2.83	$(.9, .1)$	2.27
$(.2, .8)$	1.97	$(.4, 1)$	3.19	$(.9, .3)$	2.46
$(.4, .8)$	2.23	$(.6, 1)$	3.90	$(.9, .5)$	2.89
$(.6, .8)$	2.72	$(.8, 1)$	5.16	$(.9, .7)$	3.67
$(.8, .8)$	3.60	$(1, 1)$	7.40	$(.9, .9)$	5.05

1. ≈ 1.71 3. ≈ 2.11 5. $\approx .63$.

7.

(x, y)	$x^2 + y^2$
$(.5, .3)$	.34
$(.5, .5)$	.50
$(.7, .3)$	.58
$(.7, .5)$	.74
$(.9, .3)$	.90
$(.9, .5)$	1.06

$$\int_{.2}^{.6} \int_{.4}^{1} x^2 + y^2 \, dA \approx 4.12(.04) \approx .17.$$

Chapter 17

17.1

1. $\nabla f = \langle 2xy^3z^4, 3x^2y^2z^4, 4x^2y^3z^3 \rangle$.   3. $\nabla h = \langle yze^{xyz}, xze^{xyz}, xye^{xyz} \rangle$.
5. $\nabla \cdot \mathbf{F} = 0$.   7. $\nabla \cdot \mathbf{H} = yz + 2x^2yz^2 + 2z$.   9. $\nabla \times \mathbf{F} = \langle 2y - 2z, 2z - 2x, 2x - 2y \rangle$.   11. $\nabla \times \mathbf{H} = \langle 2y - 2x^2y^2z, xy - 2x, 2xy^2z^2 - xz \rangle$.   13.
$\nabla^2 f = 2y^3z^4 + 6x^2yz^4 + 12x^2y^3z^2$.   15. $\nabla^2 y = y^2z^2e^{xyz} + x^2z^2e^{xyz} + x^2y^2e^{xyz}$.
17. $\nabla^2 F = \langle 4, 4, 4 \rangle$.   19. $\nabla^2 H = \langle 0, 2y^2z^2 + 2x^2z^2 + 2x^2y^2, 6 \rangle$.   27. p
29. none   31. $\nabla \times \mathbf{F} = \langle 0, 0, -.25 \rangle$. The forces acting on different blades of the paddle wheel are not equal. Hence, the paddle wheel will tend to rotate.

17.2

1. $D\mathbf{F}(u, v) = \begin{bmatrix} 2 & -3 \\ 3 & 1 \end{bmatrix}$

3. $D\mathbf{H}(x, y, z) = \begin{bmatrix} yz & xz & xy \\ y^2z^3 & 2xyz^3 & 3xy^2z^2 \end{bmatrix}$

5. does not make sense

9. does not make sense.

11. $\mathbf{H} \circ \mathbf{G}(u, v) = \mathbf{H}(uv, u/v, v/u) = \langle uv, v^2 \rangle$;
$D(\mathbf{H} \circ \mathbf{G})(u, v) = \begin{bmatrix} v & u \\ 0 & 2v \end{bmatrix}$;

$D\mathbf{H}(x, y, z) = \begin{bmatrix} yz & xz & xy \\ y^2z^3 & 2xyz^3 & 3xy^2z^2 \end{bmatrix}$ and $D\mathbf{G}(u, v) = \begin{bmatrix} v & u \\ 1/v & -u/v^2 \\ -v/u^2 & 1/u \end{bmatrix}$;

$(D\mathbf{H} \circ \mathbf{G})D\mathbf{G} = \begin{bmatrix} 1 & v^2 & u^2 \\ v/u & 2v^3/u & 3uv \end{bmatrix}\begin{bmatrix} v & u \\ 1/v & -u/v^2 \\ -v/u^2 & 1/u \end{bmatrix} = \begin{bmatrix} v & u \\ 0 & 2v \end{bmatrix}$.

13. does not make sense

15. $\mathbf{P} \circ \mathbf{G}(u, v) = \mathbf{P}(uv, u/v, v/u) = \langle uv + u/v + v/u, uv + u^2/v^2 + v^3/u^3, u^3/v^3 - u^2/v^2 - v/u \rangle$;
$D(\mathbf{P} \circ \mathbf{G})(u, v) = \begin{bmatrix} v + 1/v - v/u^2 & u - u/v^2 + 1/u \\ v + 2u/v^2 - 3v^3/u^4 & u - 2u^2/v^3 + 3v^2/u^3 \\ 3u^2v^3 - 2u/v^2 + v/u^2 & 3u^3v^2 + 2u^2/v^3 - 1/u \end{bmatrix}$;

$$DP(x, y, z) = \begin{bmatrix} 1 & 1 & 1 \\ 1 & 2y & 3z^2 \\ 3x^2 & -2y & -1 \end{bmatrix} \text{ and } DG(u, v) = \begin{bmatrix} v & u \\ 1/v & -u/v^2 \\ -v/u^2 & 1/u \end{bmatrix};$$

$$(DP \circ G)(DG) = \begin{bmatrix} 1 & 1 & 1 \\ 1 & 2u/v & 3v^2/u^2 \\ 3u^2v^2 & -2u/v & -1 \end{bmatrix} \begin{bmatrix} v & u \\ 1/v & -u/v^2 \\ -v/u^2 & 1/u \end{bmatrix}$$

$$= \begin{bmatrix} v + 1/v - v/u^2 & u - u/v^2 + 1/u \\ v + 2u/v^2 - 3v^3/u^4 & u - 2u^2/v^3 + 3v^2/u^3 \\ 3u^2v^3 - 2u/v^2 + v/u^2 & 3u^3v^2 + 2u^2/v^3 - 1/u \end{bmatrix}.$$

17. $x = u/y$, $x = y/v \implies u/y = y/v \implies y^2 = uv \implies y = \sqrt{uv}$;

$y = u/x$, $y = vx \implies u/x = vx \implies x^2 = u/v \implies x = \sqrt{u/v}$;

$$J(u, v) = \begin{vmatrix} \frac{1}{2}(u/v)^{-1/2}(1/v) & \frac{1}{2}(u/v)^{-1/2}(-u/v^2) \\ \frac{1}{2}(uv)^{-1/2}(v) & \frac{1}{2}(uv)^{-1/2}(u) \end{vmatrix}$$

$$= \frac{1}{4}(u^2)^{-1/2}(u/v) - \frac{1}{4}(u^2)^{-1/2}(-u/v)$$

$$= \frac{1}{4}(u/v)(1/u + 1/u) = \frac{1}{2v}.$$

Borderlines of the new region lie on $v = 1$ (since $y/x = 1$) and $v = 2$ (since $y/x = 2$) and $u = 1$ and $u = 2$. We have

$$\int_1^2 \int_1^2 \frac{1}{2v} \, du \, dv = \int_1^2 \frac{u}{2v} \Big]_1^2 \, dv = \int_1^2 \left(\frac{1}{v} - \frac{1}{2v} \right) dv$$

$$= \ln v - \frac{1}{2} \ln v \Big]_1^2 = \ln 2 - \frac{1}{2} \ln 2 - \ln 1 + \frac{1}{2} \ln 1 = \frac{1}{2} \ln 2.$$

19. $J(r, \theta, z) = \begin{vmatrix} \cos \theta & -r \sin \theta & 0 \\ \sin \theta & r \cos \theta & 0 \\ 0 & 0 & 1 \end{vmatrix} = r \cos^2 \theta + r \sin^2 \theta = r.$

17.3

1. $\oint_C (xy + yz) \, ds = \dfrac{14\sqrt{3}}{3}$; $\oint_C \langle x, y^2, z^2 \rangle \cdot ds = \dfrac{37}{6}$.   **3.** $\oint_C (xy + z^2) ds = \dfrac{\sqrt{8}}{3}$;

$\oint_C \langle x^2, y^3, z \rangle \cdot ds = 0$.   **5.** $\oint_C (y^2 + z^2) \, ds = \dfrac{-50\sqrt{2}}{3}$; $\oint_C \langle x^2, xy, xz \rangle \cdot ds = \dfrac{4}{3}$.

7. ≈ 21.13.   **9.** $\approx$?   **11.** $r(t)$ and $p(t)$ have the same orientation. $q(t)$ has orientation opposite to that of $r(t)$ and $p(t)$.   **13.** $\oint_C f \, ds = \dfrac{2}{15}$.

15. $\oint_C f \, ds = \dfrac{2}{15}$.   **17.** $\oint_C \langle y, -x \rangle \cdot ds = \dfrac{-\pi}{2}$.   **19.** $\oint_C \langle y, -x \rangle \cdot ds = \dfrac{\pi}{2}$.

21. $\oint_C x^2 y^3 \, ds = \dfrac{\sqrt{2}}{60}$ **25.** $\oint_C \mathbf{F} \cdot \mathbf{T} \, ds = \int_a^b \mathbf{F}(\mathbf{r}(t)) \cdot \mathbf{r}'(t) \, dt = \int_a^b 0 \, dt = 0.$

17.4

1. True; a potential is $\varphi(x,y) = e^{xy} + \ln y$.

3. curl **F** is not defined, since $\mathbf{F} : \mathbb{R}^2 \to \mathbb{R}^2$.

5. True; $f(x,y) = e^{xy} + \ln y$ is an example.

7. does not make sense (see 3.)

9. True; $\varphi(x,y,z) = xyz + c$, where c is a constant.

11. True; a potential is $\varphi(x,y,z) = xyz$.

13. True; **F** is conservative.

15. $\mathbf{G}(x,y) = (\ln y \cos x)\mathbf{i} + (y \sin x)\mathbf{j}$.

17. **L** is conservative; a potential is $\varphi(x,y,z) = \dfrac{y^4}{4}\ln x + \sin z$.

19. $16/3$.

17.5

1. $\mathbf{r}(t) = \langle u, v, u^3 + v^2 \rangle$; $D = $ circle with radius 1 and center at $(0,0)$.

3. $\mathbf{r}(t) = \langle u, \sqrt{u+v}, v \rangle$; $D = $ triangle with vertices at the (u,v) coordinates $(0,0)$, $(1,1)$, and $(0,1)$.

5. $\mathbf{r}(t) = \langle 3 - 2u^2 + v^2, u, v \rangle$; $D = $ circle with radius 1 and center at $(0,0)$.

7. $\mathbf{T}_u = \langle -v \sin u, v \cos u, 0 \rangle$; $\mathbf{T}_u(\pi, -\pi) = \langle 0, \pi, 0 \rangle$;

$\mathbf{T}_v = \langle \cos u \sin u, 1 \rangle$; $\mathbf{T}_v(\pi, -\pi) = \langle -1, 0, 1 \rangle$;

$\mathbf{T}_u \times \mathbf{T}_v = \langle v \cos u, v \sin u, -v \rangle$; $\mathbf{T}_u \times \mathbf{T}_v(\pi, -\pi) = \langle \pi, 0, \pi \rangle$;

The equation of tangent plane is of the form $\pi(x - \pi) + \pi(z + \pi) = 0$.

9. $\mathbf{T}_u = \langle -\sin u, \cos u, 0 \rangle$; $\mathbf{T}_u(0,1) = \langle 0, 1, 0 \rangle$; $\mathbf{T}_v = \langle 0, 0, 1 \rangle$; $\mathbf{T}_v(0,1) = \langle 0, 0, 1 \rangle$;

$\mathbf{T}_u \times \mathbf{T}_v = \langle \cos u, \sin u, 0 \rangle$; $\mathbf{T}_u \times \mathbf{T}_v(0,1) = \langle 1, 0, 0 \rangle$;

The equation of the tangent plane is of the form $x - 1 = 0$.

11. $\mathbf{T}_u = \langle 1, 1, -1 \rangle$; $\mathbf{T}_v = \langle 1, -1, 1 \rangle$; $\mathbf{T}_u \times \mathbf{T}_v = \langle 0, -2, -2 \rangle$;

The equation of the tangent plane is of the form $-2(y + 3) - 2(z - 3) = 0$.

13. $\mathbf{T}_u = \langle 2u, 0, 2u \rangle$, $\mathbf{T}_v = \langle 0, 2v, 2v \rangle$.

$$r(1,1) = \langle 1,1,2 \rangle, \quad T_u(1,1) = \langle 2,0,2 \rangle, \quad T_v(1,1) = \langle 0,2,2 \rangle;$$

$$T_u \times T_v = \begin{vmatrix} i & j & k \\ 2 & 0 & 2 \\ 0 & 2 & 2 \end{vmatrix} = -4i - 4j + 4k = \langle -4, -4, 4 \rangle.$$

The equation of the tangent plane is of the form $-x - y + z = 0$.

17.6

1. $\|T_u \times T_v\| = v\sqrt{2}$; surface area $\iint\limits_S f \, dS = \dfrac{\pi}{3\sqrt{2}}$.

3. $\|T_u \times T_v\| = 1$; surface area $\iint\limits_S f \, dS = 2 + \dfrac{\pi}{2}$.

5. $\|T_u \times T_v\| = \sqrt{8}$; surface area $\iint\limits_S f \, dS = \dfrac{16\sqrt{8}}{3}$.

10. For each point on the sphere (x_0, y_0, z_0) there is another point directly opposite $(-x_0, -y_0, -z_0)$. Thus, $f(x_0, y_0, z_0) = x_0 + y_0 + z_0$ is "balanced" by $f(-x_0, -y_0, -z_0) = -(x_0 + y_0 + z_0)$.

11. $\dfrac{\sqrt{2}}{2}\pi$.

17.7

1. $\dfrac{\pi}{6}(5^{3/2} - 1)$.   3. -2π.   5. Stokes' Theorem holds since all hypotheses are satisfied.   7. $8/15$.   9. $1/6$.   11. $1/6$.   15. 27π.

CALCULUS, VOLUME 2

by Thomas Dick and Charles Patton
PRELIMINARY EDITION
Reviewer Survey

The Oregon State University Calculus Curriculum Project has developed this textbook with a great deal of input and feedback from manuscript reviewers and instructors at pilot test sites. We now seek student reaction. Your responses to the following questions will be of great help to both the authors and the publisher as a polished **First Edition** *is being assembled over the next few years.*

1. Please describe the ways in which you are expected to utilize technology in your calculus course. What materials and equipment are you required to purchase? What are supplied by your school or department?

_____
_____
_____

2. To what extent are technological tools incorporated in your class (i.e. daily, weekly, or monthly use; required for exercise sets or tests, etc.)?

_____
_____
_____

3. What are your impressions of the following elements in this textbook?

a. Writing style/explanations_____
_____
_____

b. Level_____
_____

c. Exercises_____
_____
_____

d. Illustrations_____
_____

4. What opinion do you have on the incorporation of technological methods in the text? Is it effective in teaching calculus concepts?

_____
_____
_____

5. Do you like the kinds of examples and applications in the text? Please comment briefly on their quantity, quality, and variety.

_____
_____
_____

6. Have you ever had previous training in calculus before? If so, when, and what are your impressions of this product juxtaposed to the one you used?

_____
_____

7. Are you glad that your instructor chose to use this text?

_____
_____
_____

8. How would you rate the overall quality of this product?

|⎯⎯⎯⎯⎯⎯⎯⎯|⎯⎯⎯⎯⎯⎯⎯⎯|⎯⎯⎯⎯⎯⎯⎯⎯|⎯⎯⎯⎯⎯⎯⎯⎯|

Low Opinion   High Opinion

Any additional comments?

_____
_____
_____

Optional:

Name_____

Address_____

School_____

City_____ State_____Zip_____

Please detach and return to:

Steve Quigley, Senior Editor
PWS-KENT Publishing Company
20 Park Plaza
Boston, Massachusetts 02116

Thank you for your time. We appreciate your assistance in the development of these materials.